AF355959

# Feeding Strategies and Nutritional Quality of Animal Products—Volume II

# Feeding Strategies and Nutritional Quality of Animal Products—Volume II

Editors

**Petru Alexandru Vlaicu**
**Arabela Elena Untea**
**Mihaela Saracila**

Basel • Beijing • Wuhan • Barcelona • Belgrade • Novi Sad • Cluj • Manchester

*Editors*

Petru Alexandru Vlaicu
National Research and
Development Institute for
Animal Biology and Nutrition
Balotesti
Romania

Arabela Elena Untea
National Research and
Development Institute for
Biology and Animal Nutrition
Balotesti
Romania

Mihaela Saracila
National Research and
Development Institute for
Biology and Animal Nutrition
Balotesti
Romania

*Editorial Office*
MDPI AG
Grosspeteranlage 5
4052 Basel, Switzerland

This is a reprint of articles from the Special Issue published online in the open access journal *Agriculture* (ISSN 2077-0472) (available at: https://www.mdpi.com/journal/agriculture/special_issues/QN2Z13IC36).

For citation purposes, cite each article independently as indicated on the article page online and as indicated below:

Lastname, A.A.; Lastname, B.B. Article Title. *Journal Name* **Year**, *Volume Number*, Page Range.

**ISBN 978-3-7258-2727-5 (Hbk)**
**ISBN 978-3-7258-2728-2 (PDF)**
**doi.org/10.3390/books978-3-7258-2728-2**

# Contents

# About the Editors

**Petru Alexandru Vlaicu**

Petru Alexandru Vlaicu is a Senior Researcher in the Feed and Food Quality Department at the National Research and Development Institute for Animal Biology and Nutrition, Romania. He holds a PhD in obtaining functional foods of animal origin, obtained from the University of Agronomic Sciences and Veterinary Medicine of Bucharest, Romania. His research interests include, but are not limited to, developing innovative and alternative feeding strategies for monogastric animals, especially poultry, by using different unconventional dietary feed ingredients; addressing animal-origin foods' nutritional quality; improving animal-origin food products' quality by dietary manipulation of animal feed. Recycling co- and by-products as well as vegetable waste (fruits and legumes) from the agro-food sector to recover valuable bioactive compounds and nutrients to reintegrate them into feed and food systems in view of a sustainable and circular economical approach is another of his research interests. He has published over 100 scientific papers in domestic and international peer-reviewed scientific journals, presented more than 40 communications at international conferences, and has authored 2 book chapters in relation to the results from the research projects in which he is involved. To date, he has obtained 4 national patent applications out of 13 deposited, which are under evaluation. His commitment to advancing knowledge in animal nutrition and food quality has been recognized through multiple awards, including gold, silver, and bronze medals, as well as section prizes for oral presentations, obtained at different national and international events. He was also awarded Best Reviewer (2023) from the Foods journal.

**Arabela Elena Untea**

Arabela Elena Untea (PhD in Analytical Chemistry Sciences, University of Bucharest, Romania, 2013) is currently a Senior Researcher and Head of the Feed and Food Quality Department at the National Research and Development Institute for Animal Biology and Nutrition, Romania. Her research activity is mainly focused on feeds and animal-origin food with functional characteristics designed through nutritional approaches; bioefficiency assessments of feed additives through laboratory experiments; studies of oxidative and reduction processes at the animal tissue level through nutrition; in vitro studies on the bioaccesibility of some nutrients in the gastrointestinal tract of monogastric animals. She is the author and co-author of more than 135 papers published in peer-viewed international journals, as well as producing 2 book chapters and 4 national patents. She has also participated in several EU-funded projects and has acted as scientific coordinator in 6 national projects and as a scientific team member in another 30 projects.

**Mihaela Saracila**

Mihaela Saracila is currently working as a postdoctoral researcher at the National Research and Development Institute for Animal Nutrition and Biology, within the Feed and Food Quality Department. She holds a PhD from the University of Agronomic Sciences and Veterinary Medicine of Bucharest in the development of innovative nutritional solutions to ensure the performance and the health of the digestive tract of broiler chickens reared under high heat stress conditions. In recent years, she has mainly focused on mitigating heat stress-related oxidative stress in monogastric animals and maintaining their gastrointestinal health; improving the nutritional composition and oxidative stability of animal products using natural dietary sources of antioxidants; the development and/or optimization of analytical methods for polyphenol extraction; and the determination of hydrophilic compounds and antioxidant capacity. She is currently the project leader of a national project and has participated in more than 28 scientific projects. She has published a laboratory book dedicated to students (collaborator), a book chapter, 1 patent, 8 national patent applications, over 60 peer-reviewed scientific papers, and has been awarded multiple awards at different scientific events.

# Preface

In the context of rapid global population growth, shifting dietary patterns, and the increasing demand for sustainable food systems, the production of high-quality animal-based foods has become more important and complex than ever. Ensuring that animal-derived foods, such as meat, milk, and eggs, are nutritionally rich and produced in a responsible manner is essential to meeting the dietary needs of all populations. Achieving this goal requires innovative, safe, and sustainable feeding strategies that not only optimize animal health and productivity, but also enhance the nutritional value of the foods produced, while minimizing environmental impacts.

This Special Issue is the second volume of "Feeding Strategies and Nutritional Quality of Animal Products," which addresses the critical intersections of animal nutrition, feeding practices, and food quality. It brings together cutting-edge research and insights from experts worldwide (Croatia, Bulgaria, Romania, Brazil, Mexico, and Poland), examining how various feeding strategies influence the nutritional profile, safety, and sensory qualities of animal-based foods. The studies presented in this Special Issue explore the impact of novel feed ingredients, supplementation techniques, and management practices on the nutritional composition of farm animal products. They also provide key insights into sustainable and ethical feeding solutions that could reduce dependency on traditional resources and promote a circular economy within the agricultural sector.

In this collection of papers, the authors aimed to provide a perspective on the current trends, challenges, and opportunities in the field of farm animals. The papers published examined different topics, including a revision of alternative feed ingredients for poultry for achieving zero hunger; the characterization and application of alternative feeding resource like oilseed cakes, yellow mealworm, and grape pomace silage; nutritional solutions for small ruminants (lambs and goats); and snails and chicken meat quality assessment under different storage conditions, showcasing their ability to improve both animal and human health outcomes.

This volume is intended for a broad audience of researchers, practitioners, policymakers, and students who are dedicated to advancing animal science, nutrition, and sustainable agriculture. By providing a deeper understanding of the ways that diet influences the quality of animal-derived foods, we hope to inspire new ideas and practices that will help to sustainably feed the world's growing population with nutritious, safe, and ethically produced animal products.

We extend our gratitude to the contributing authors, reviewers, and editorial team who made this Special Issue possible. Their expertise, dedication, and collaboration have created a valuable resource that we believe will contribute meaningfully to the fields of animal nutrition and food science.

**Petru Alexandru Vlaicu, Arabela Elena Untea, and Mihaela Saracila**
*Editors*

*Editorial*

# Feeding Strategies and Quality Assessments of Animal-Derived Products

**Petru Alexandru Vlaicu * and Arabela Elena Untea ***

Feed and Food Quality Department, National Research and Development Institute for Animal Biology and Nutrition, Calea Bucuresti 1, 077015 Balotesti, Romania
* Correspondence: alexandru.vlaicu@ibna.ro (P.A.V.); arabela.untea@ibna.ro (A.E.U.)

**Citation:** Vlaicu, P.A.; Untea, A.E. Feeding Strategies and Quality Assessments of Animal-Derived Products. *Agriculture* **2024**, *14*, 1949. https://doi.org/10.3390/agriculture14111949

Received: 29 October 2024
Accepted: 30 October 2024
Published: 31 October 2024

Feeding strategies play an important role in animal production systems by directly influencing animal health, productivity, and the quality of animal-derived products (meat, eggs, and milk). These techniques comprise a variety of practices designed to optimize the dietary feeds of animals while maintaining their well-being, production performance, and quality of end-products. Feeding practices can be executed using multiple techniques, including formulating balanced diets that meet the precise dietary requirements of various animal species by incorporating functional components such as pre/probiotics, food waste, by-products, recycled food waste, insects, or fermented ingredients [1] while adopting precision livestock feeding techniques that tailor diets to individual animal species [2]. Additionally, alternative feed sources are gaining attention as sustainable options to lessen dependence on conventional feed ingredients like soybean and corn.

Quality assessments of animal-derived products are important to assure food safety, nutritional value, and consumer acceptance while also meeting regulatory standards and market demands. In this context, feeding strategies have a high impact on animal end-product niches, such as those with the inclusion of specific nutrients (omega-3 fatty acids, antioxidants, carotenoids, minerals, and other nutrients), aligning with the consumers' request for safer products.

This Special Issue received a total of 30 papers, from which 10 have been published, comprising 8 research articles, 1 meta-analysis, and 1 review article. Based on the number of published works, the submitting authors are from six different countries: Croatia, Bulgaria, Romania, Brazil, Mexico, and Poland. The papers published in this Special Issue are the continuation of the first volume of the Feeding Strategies and Nutritional Quality of Animal Products [3]. The published papers in the second volume of the Special Issue are grouped into different categories: a review dealing with numerous feeding strategies for achieving zero hunger; alternative feeding resource (oilseed cakes; yellow mealworm, and grape pomace silage) characterization; nutritional solutions for small ruminants (lambs and goats); and snails and chicken meat quality assessment under different storage conditions.

The review paper by Vlaicu et al. [4] explores the potential of alternative sustainable poultry feeding strategies aimed at achieving SDG2 (Zero Hunger) while increasing production performance and food quality, focusing on the potential recycling of by-products, plants, and food waste derived from fruits, vegetables, and seeds, which, according to FAO and Eurostat reports [5], account for up to 35% annually. The paper provides a review analysis of the nutritional (protein, fat, fiber, and ash) and mineral (calcium, phosphorus, zinc, manganese, copper, and iron) contents as well as the bioactive compounds (polyphenols, antioxidants, carotenoids, fatty acids, and vitamins) of the reviewed alternative feed ingredients, which can contribute to resource efficiency, a reduction in dependency on conventional feeds, and lower production costs. The benefits of legumes, fruits, berries, seeds, and plant wastes have been reported to enhance the essential nutrients in poultry products and their performances [6–11]. Further, the circular economy approach in poultry farming, which emphasizes the efficient use of resources, waste minimization, and the recycling of discarded by-products, is also discussed. The paper also reports the challenges associated with the implementation of these alternative feeding strategies, including inconsistencies in quality and availability, the presence of anti-nutrients, and regulatory barriers. Despite these, future research directions in sustainable feeding strategies for animals with alternative feed ingredients should be continuously explored to better understand their usage.

The second category of papers explored the characterization of different alternative feeding sources and their potential usage in animal feeding. The study conducted by Rambu et al. [12] explored the solid-state fermentation efficiency of *Bacillus licheniformis* ATCC 21424 on various agro-industrial by-products (oilseed cakes of hemp, pumpkin, and flaxseed) by examining the nutritional composition, reduction in sugars, and in vitro protein digestibility for use in animal nutrition. The authors showed that this strategy produces a few significant changes in these by-products, especially by altering the crude protein, crude fat, crude fiber, carbohydrate content, neutral detergent fiber (NDF), and acid detergent fiber (ADF). Moreover, the authors demonstrate that solid-state fermentation with *Bacillus licheniformis*, under optimized conditions (72 h), increases the reduced sugar content for flax and pumpkin waste compared with hempseed waste. This study is important as it shows the advantages of oilseed waste bioprocessing by solid-state fermentation as an effective approach in yielding valuable products with probiotic and nutritional properties as suitable alternative ingredients for animal feeding. Although previous studies have reported the nutritional composition and the beneficial effects of feeding flax, pumpkin, and hemp waste in animals nutrition and product quality, the study conducted by Rambu et al. [12] shows additional potential usages of these wastes through fermentation.

Vlahova-Vangelova et al. [13] showed the impact of different drying treatments, such as freeze-drying, conventional drying, microwave drying, microwave drying without freezing prior to blanching, and microwave drying with the addition of 0.1% butylated hydroxytoluene during the blanching of the larvae, used in the production of flour from Tenebrio molitor larvae. Their results revealed that the freeze-drying and conventional drying of the larvae reduced the water activity of the derived flours but significantly reduced protein content and considerably increased the fat content compared to the flours after microwave treatments. Their research also revealed that the conventional drying and microwave treatment with butylated hydroxytoluene induced a significantly darker color in comparison to the other methods. The authors mentioned that despite the advantages of microwave drying as a fast and energy-efficient method, it displayed some negative effects associated with low lipid stability, such as higher acid value and increased secondary products of lipid oxidation (TBARS). By revealing both the positive and negative effects of the applied treatments, their study contributes to the development of optimal drying modes for the preparation of insect flours.

Further, Sokač Cvetnić et al. [14], in their study, monitored the chemical and fermentative characteristics during different treatments (raw, treated with an inoculum of lactic acid bacteria, and treated with zeolite) of grape pomace silage. Their results showed that after 30, 60, and 90 days of ensiling, the silages treated with inoculum and zeolite had better fermentation quality (lower pH and ammonia–nitrogen contents). Interestingly, the authors revealed that the tested additives (inoculum and zeolite) decreased the total polyphenols and tannins, confirming the degradation of polyphenols and tannins in grape pomace silage due to the lactic acid bacteria and zeolite presence. The conclusion of this study showed the importance of inoculant and zeolite supplementation impact on the quality of grape pomace silage as an alternative use in animal feed.

In the third category of papers, Antunović et al. [15] investigated how the feeding solutions that imply the inclusion of polyphenol-rich black oats in lamb diets affect their performance, metabolic health, antioxidant status, and carcass quality. Their results showed that the substitution of black oats did not influence production performances, carcass traits, or physical measures, except for reduced pH values in lamb meat. The metabolic analysis revealed that lambs fed black oats exhibited higher levels of red blood cells, hematocrit, and mean corpuscular volume, indicating improved overall metabolic health. The tested feeding strategy also improved the activities of the liver enzymes ALT and ALP, showing their potential to reduce hepatic stress in lambs consuming black oats. Additional benefit effects were reported for antioxidant status (higher glutathione peroxidase (GPx) and superoxide dismutase (SOD)) and DPPH radical scavenging activity in the liver of lambs. The potential of polyphenols to reduce oxidative stress in animals was also noted by the authors. The importance of this study stands in the potential of replacing 15% of conventional oats with black oats with beneficial effects on metabolic and antioxidant status without altering the production performances and carcass characteristics. As the authors concluded, the effects of polyphenols in weaning crises and older animals are a field of future studies.

The study of Dos Santos et al. [16] investigated how different feeding systems impact lamb growth, metabolic health, meat quality, and gene expression, subject to a distinct feeding strategy. They reported that lambs fed with forage and a diet of 72% concentrate and 28% Tifton-85 hay in a feedlot showed the highest daily weight gain and feed efficiency. On the other hand, the grazing system raised lambs fed 1.2% concentrate while grazing on Tifton-85 grass resulted in improved meat quality (higher concentration of omega-3 polyunsaturated fatty acids) and a more favorable n6/n3 ratio. Moreover, lambs in the feedlot systems fed a high-grain diet consisting of 85% whole-grain corn and 15% protein pellets showed higher insulin and glucose levels, reflecting the impact of the high energy content of grain-based diets. It was concluded that while feedlot systems, especially high-grain diets, improved growth performance and fat deposition, the grazing system produced lambs with healthier meat profiles, especially essential fatty acids composition and better sensory attributes. The results of this study suggest that producers can tailor feeding strategies based on their market goals.

Antunović et al. [17], in a different study, evaluated the effects of new nutritional solutions obtained by replacing yellow corn with red corn on dairy French Alpine goats' performance, metabolic health, and antioxidant status. Their results have indicated that the inclusion of red corn did not significantly impact dairy goats milk yield or nutritional composition, with only minor increases in milk lactose observed in the red corn-fed groups. However, the part and total (100%) red corn replacement for yellow corn improved the metabolic health of goats while also increasing superoxide dismutase (SOD) activity in their blood, pointing to enhanced antioxidant defenses. In terms of milk quality, total replacement with red corn resulted in a significant decrease in the thiobarbituric acid reactive substances (TBARS). These results are important, especially due to the richness of red corn in anthocyanins, which could improve dairy goat resilience to oxidative stress, but to test this hypothesis, long-term feeding trials are required.

A meta-analysis conducted by Orzuna-Orzuna et al. [18] examined the feeding solutions for dairy cows, including dietary supplementation with *Schizochytrium* sp., a microalgae rich in omega-3 fatty acids. The analysis conducted by the authors revealed that *Schizochytrium* sp. supplementation increased milk yield but simultaneously reduced dry matter intake, milk fat yield, and milk fat content. This analysis is of particular interest, especially understanding that although cows produce more milk, the concentration of fat in the milk is lowered, which could have implications for cheese production or milk products relying on higher fat content. Moreover, the fat content in milk gives the selling price, which represents an important economic aspect. The analyses also revealed that supplementation with *Schizochytrium* sp., a microalgae, decreased the levels of palmitic and stearic fatty acids, which are associated with important implications on cardiovascular risks in humans. Contrarily, the authors showed that the concentration of beneficial fatty acids like omega-3 polyunsaturated fatty acids and conjugated linoleic acid increased, making the milk more nutritious for consumers. This study shows that although beneficial effects are observed, the negative ones cannot be ignored, which suggests that further research is needed to optimize supplementation strategies that maintain or enhance milk fat content.

Also in this category, the research conducted by Rygało-Galewska et al. [19] aimed to develop new feeding solutions for snails by examining how varying concentrations of calcium (44.3, 66.1, 88.7, and 103.5 g/kg feed) and magnesium (3.3, 5.6, and 7.2 g/kg feed) in the diet of garden snails influenced their growth, shell quality, survival rate, and meat composition. The two research studies aimed to optimize the dietary levels of these minerals to improve the technological and production characteristics of snail farming. The results from the first experiment indicated that higher calcium levels led to significantly greater body and shell weights, with the snails receiving the highest calcium levels (103.5 g/kg) exhibiting the best growth and shell strength. However, an excess of calcium reduced overall body weight percentage relative to shell weight, and higher calcium levels were associated with a lower feed conversion ratio. The second experiment showed that moderate magnesium supplementation (5.6 g/kg) yielded better growth, while the highest magnesium level (7.2 g/kg) negatively impacted weight gain and feed efficiency. Shell strength improved with increasing magnesium levels, but too much magnesium was linked to a reduction in overall survival rates. From this study, it was concluded that although both calcium and magnesium are critical for the optimal growth and development of garden snails, excessive levels of either can lead to reduced performance. It is recommended to use a balanced approach with moderate levels of both minerals for optimal improvements. This study also shows the importance of snails for commercial production.

The fourth category of articles included a study conducted by Popova et al. [20], which aimed to evaluate the growth performance, economic aspects, and meat quality of Lohmann Brown Classic male layer-type chickens. Significant alterations were observed in the production performances, depending on the age of the birds. In terms of the economic aspect, they have reported that raising chickens until 9 weeks of age slightly reduced the costs per kg for live weight and per kg for the ready-to-cook carcass, mainly due to higher dressing percentages and lower labor and slaughtering costs. The authors mentioned that extending the rearing period could improve economic efficiency. The meat quality of broilers was also examined during frozen storage at 60 and 120 days. It was noted that lipid oxidation increased markers such as peroxide values over 60 days of frozen and TBARS after 120 days. However, the meat did not show significant deterioration, and the antioxidant parameters remained below thresholds for rancidity. The study recommends further research into optimizing the slaughter age to balance economic efficiency and meat quality.

**Author Contributions:** Conceptualization, investigation, writing—original draft preparation, writing—review and editing, and visualization: P.A.V. and A.E.U. All authors have read and agreed to the published version of the manuscript.

**Funding:** This research is supported by the Romanian Ministry of Research Innovation and Digitalization, project PN 2320-0301.

**Acknowledgments:** We would like to sincerely thank all authors who submitted papers to the Special Issue of *Agriculture* entitled *"Feeding Strategies and Nutritional Quality of Animal Products—Volume II"*, the reviewers of these papers for their constructive comments and thoughtful suggestions, and the editorial staff of *Agriculture*.

**Conflicts of Interest:** The authors declare no conflicts of interest.

## References

1. Vlaicu, P.A.; Untea, A.E.; Varzaru, I.; Saracila, M.; Oancea, A.G. Designing Nutrition for Health—Incorporating Dietary By-Products into Poultry Feeds to Create Functional Foods with Insights into Health Benefits, Risks, Bioactive Compounds, Food Component Functionality and Safety Regulations. *Foods* **2023**, *12*, 4001. [CrossRef] [PubMed]
2. Vlaicu, P.A.; Gras, M.A.; Untea, A.E.; Lefter, N.A.; Rotar, M.C. Advancing Livestock Technology: Intelligent Systemization for Enhanced Productivity, Welfare, and Sustainability. *AgriEngineering* **2024**, *6*, 1479–1496. [CrossRef]
3. Untea, A.E.; Saracila, M.; Vlaicu, P.A. Feeding Strategies and Nutritional Quality of Animal Products. *Agriculture* **2023**, *13*, 1788. [CrossRef]

4.  Vlaicu, P.A.; Untea, A.E.; Oancea, A.G. Sustainable Poultry Feeding Strategies for Achieving Zero Hunger and Enhancing Food Quality. *Agriculture* **2024**, *14*, 1811. [CrossRef]
5.  Food and Agriculture Organization of the United Nations (FAO). Development of a Code of Conduct on Food Loss and Food Waste Prevention. 2019. Available online: http://www.fao.org/fsnforum/es/node/4877 (accessed on 22 October 2024).
6.  Mustafa, A.F.; Baurhoo, B. Effects of feeding dried broccoli floret residues on performance, ileal and total digestive tract nutrient digestibility, and selected microbial populations in broiler chickens. *J. Appl. Poult. Res.* **2016**, *25*, 561–570. [CrossRef]
7.  Pliego, A.B.; Tavakoli, M.; Khusro, A.; Seidavi, A.; Elghandour, M.M.; Salem, A.Z.; Rene Rivas-Caceres, R. Beneficial and adverse effects of medicinal plants as feed supplements in poultry nutrition: A review. *Anim. Biotechnol.* **2022**, *33*, 369–391. [CrossRef] [PubMed]
8.  Vlaicu, P.A.; Panaite, T.D.; Turcu, R.P. Enriching laying hens eggs by feeding diets with different fatty acid composition and antioxidants. *Sci. Rep.* **2021**, *11*, 20707. [CrossRef] [PubMed]
9.  Panaite, T.D.; Nour, V.; Vlaicu, P.A.; Ropota, M.; Corbu, A.R.; Saracila, M. Flaxseed and dried tomato waste used together in laying hens diet. *Arch. Anim. Nutr.* **2019**, *73*, 222–238. [CrossRef] [PubMed]
10. Socas-Rodríguez, B.; Álvarez-Rivera, G.; Valdés, A.; Ibáñez, E.; Cifuentes, A. Food by-products and food wastes: Are they safe enough for their valorization? *Trends Food Sci. Technol.* **2021**, *114*, 133–147. [CrossRef]
11. Untea, A.E.; Oancea, A.-G.; Vlaicu, P.A.; Varzaru, I.; Saracila, M. Blackcurrant (Fruits, Pomace, and Leaves) Phenolic Characterization before and after In Vitro Digestion, Free Radical Scavenger Capacity, and Antioxidant Effects on Iron-Mediated Lipid Peroxidation. *Foods* **2024**, *13*, 1514. [CrossRef] [PubMed]
12. Rambu, D.; Dumitru, M.; Ciurescu, G.; Vamanu, E. Solid-State Fermentation Using *Bacillus licheniformis*-Driven Changes in Composition, Viability and In Vitro Protein Digestibility of Oilseed Cakes. *Agriculture* **2024**, *14*, 639. [CrossRef]
13. Vlahova-Vangelova, D.; Balev, D.; Kolev, N.; Dragoev, S.; Petkov, E.; Popova, T. Comparison of the Effect of Drying Treatments on the Physicochemical Parameters, Oxidative Stability, and Microbiological Status of Yellow Mealworm (*Tenebrio molitor* L.) Flours as an Alternative Protein Source. *Agriculture* **2024**, *14*, 436. [CrossRef]
14. Sokač Cvetnić, T.; Gunjević, V.; Damjanović, A.; Pušek, A.; Jurinjak Tušek, A.; Jakovljević, T.; Radojčić Redovniković, I.; Uher, D. Monitoring of Chemical and Fermentative Characteristics during Different Treatments of Grape Pomace Silage. *Agriculture* **2023**, *13*, 2264. [CrossRef]
15. Antunović, Z.; Klir Šalavardić, Ž.; Mioč, B.; Steiner, Z.; Đidara, M.; Sičaja, V.; Pavić, V.; Mihajlović, L.; Jakobek, L.; Novoselec, J. Dietary Effects of Black-Oat-Rich Polyphenols on Production Traits, Metabolic Profile, Antioxidative Status, and Carcass Quality of Fattening Lambs. *Agriculture* **2024**, *14*, 1550. [CrossRef]
16. dos Santos, I.J.; Dias Junior, P.C.G.; Alvarenga, T.I.R.C.; Pereira, I.G.; Gallo, S.B.; Alvarenga, F.A.P.; Furusho-Garcia, I.F. Impact of Feeding Systems on Performance, Blood Parameters, Carcass Traits, Meat Quality, and Gene Expressions of Lambs. *Agriculture* **2024**, *14*, 957. [CrossRef]
17. Antunović, Z.; Klir Šalavardić, Ž.; Novoselec, J.; Steiner, Z.; Đidara, M.; Pavić, V.; Jakobek Barron, L.; Ronta, M.; Mioč, B. Production Traits, Blood Metabolic Profile, and Antioxidative Status of Dairy Goats Fed a Red Corn Supplemented Feed Mixture. *Agriculture* **2024**, *14*, 82. [CrossRef]
18. Orzuna-Orzuna, J.F.; Godina-Rodríguez, J.E.; Garay-Martínez, J.R.; Reséndiz-González, G.; Joaquín-Cancino, S.; Lara-Bueno, A. Milk Yield, Composition, and Fatty Acid Profile in Milk of Dairy Cows Supplemented with Microalgae *Schizochytrium* sp.: A Meta-Analysis. *Agriculture* **2024**, *14*, 1119. [CrossRef]
19. Rygało-Galewska, A.; Zglińska, K.; Roguski, M.; Roman, K.; Bendowski, W.; Bień, D.; Niemiec, T. Effect of Different Levels of Calcium and Addition of Magnesium in the Diet on Garden Snails' (*Cornu aspersum*) Condition, Production, and Nutritional Parameters. *Agriculture* **2023**, *13*, 2055. [CrossRef]
20. Popova, T.; Petkov, E.; Dimov, K.; Vlahova-Vangelova, D.; Kolev, N.; Balev, D.; Dragoev, S.; Ignatova, M. Performance, Carcass Composition, and Meat Quality during Frozen Storage in Male Layer-Type Chickens. *Agriculture* **2024**, *14*, 185. [CrossRef]

 *agriculture*

*Review*

# Sustainable Poultry Feeding Strategies for Achieving Zero Hunger and Enhancing Food Quality

**Petru Alexandru Vlaicu *, Arabela Elena Untea and Alexandra Gabriela Oancea**

Feed and Food Quality Department, National Research and Development Institute for Animal Biology and Nutrition, 077015 Ilfov, Romania; arabela.untea@ibna.ro (A.E.U.); alexandra.oancea@ibna.ro (A.G.O.)
* Correspondence: alexandru.vlaicu@outlook.com

**Abstract:** As global demand increases for poultry products, innovative feeding strategies that reduce resource efficiency and improve food safety are urgently needed. This paper explores the potential of alternative sustainable poultry feeding strategies aimed at achieving SDG2 (Zero Hunger) while increasing production performance and food quality, focusing on the potential recycling of by-products, plants, and food waste derived from fruits, vegetables, and seeds, which account for up to 35% annually. The paper provides a review analysis of the nutritional (protein, fat, fiber, and ash) and minerals (i.e., calcium, phosphorus, zinc, manganese, copper, and iron) content as well as the bioactive compounds (polyphenols, antioxidants, carotenoids, fatty acids, and vitamins) of alternative feed ingredients, which can contribute to resource efficiency, reduce dependency on conventional feeds, and lower production costs by 25%. The nutritional benefits of these alternative feed ingredients, including their effects on poultry production and health, and their potential for improving poultry product quality, are presented. Carrot, paprika, rosehip, and some berry waste represent a great source of carotenoids, polyphenols, and vitamins, while the seed meals (flax, rapeseed, and sea buckthorn) have been reported to enhance the essential fatty acid composition in eggs and meat. Numerous plants (basil, sage, rosemary, and lettuce) are natural reservoirs of bioactive compounds with benefits for both animal and food products. Some challenges in implementing these alternative sustainable feeding strategies, including inconsistencies in quality and availability, the presence of anti-nutrients, and regulatory barriers, are also explored. In conclusion, future research directions in sustainable poultry feeding with alternative feed ingredients should be considered to achieve SDG2.

**Keywords:** food quality; poultry; eggs; meat; sustainability; zero hunger; SDG; feed ingredients

**Citation:** Vlaicu, P.A.; Untea, A.E.; Oancea, A.G. Sustainable Poultry Feeding Strategies for Achieving Zero Hunger and Enhancing Food Quality. *Agriculture* **2024**, *14*, 1811. https://doi.org/10.3390/agriculture14101811

Academic Editor: Zoltán Györi

Received: 21 August 2024
Revised: 8 October 2024
Accepted: 11 October 2024
Published: 14 October 2024

## 1. Introduction

The current global challenge of achieving food security and ensuring access to sufficient, nutritious, and safe food for all while maintaining or improving agricultural sustainability has been presented in the United Nations' Sustainable Development Goals (SDGs), particularly Goal 2, which refers to Zero Hunger. This goal emphasizes the necessity to end hunger while achieving food security through improved nutrition and promote sustainable agricultural practices by 2030.

In the agricultural sector, poultry production plays an important role in the global food system, providing consumers with a major source of high-quality animal protein through meat and eggs [1]. However, since conventional poultry feed raw materials, mainly corn and soybean meal, this poses challenges with respect to its sustainability, such as deforestation, biodiversity loss, and more greenhouse gases associated with their manufacture and transportation [2]. In this context, new strategies are needed to develop and implement alternative feeding strategies for poultry. In recent years, there has been a growing interest in the potential of alternative feed ingredients as viable and practical alternatives to conventional poultry diets. These alternatives not only aim to reduce the ecological footprint of poultry farming but also are great candidates to enhance the nutritional profile of poultry products and performances. The utilization of diverse novel and alternative feed ingredients, such as legume waste [3], oilseed meals [4,5], fruit waste [6,7], leaves or plants [8,9], and other agricultural by-/co-products [10], into poultry diets offers promising ways to mitigate the current challenges in poultry industries. Nevertheless, the transition to alternative feed ingredients for poultry is complex, and the nutritional adequacy, feed palatability, cost-effectiveness, and supply chain logistics should be considered. For example, understanding the nutritional requirements of poultry and the nutrient composition of various feed ingredients is crucial for formulating balanced diets that support optimal growth, health, productivity, and product quality. Moreover, the variability in nutrient content and the presence of anti-nutritional factors in some alternative ingredients require careful laboratory assessments and processing to ensure their effective utilization in poultry diets [11]. However, numerous research studies dealing with feed technology and animal nutrition science have facilitated the development of novel feed formulations that incorporate alternative ingredients without compromising poultry performance while improving poultry product quality, as further presented in the current paper.

In this context, this review paper aims to explore potential alternative and sustainable poultry feeding strategies that align with the Sustainable Development Goal of Zero Hunger (SDG 2). The paper examines the chemical composition, and bioactive compounds present in the alternative feed ingredient options and their potential to enhance poultry production performances, health, and food quality while contributing to a circular economy (Figure 1).

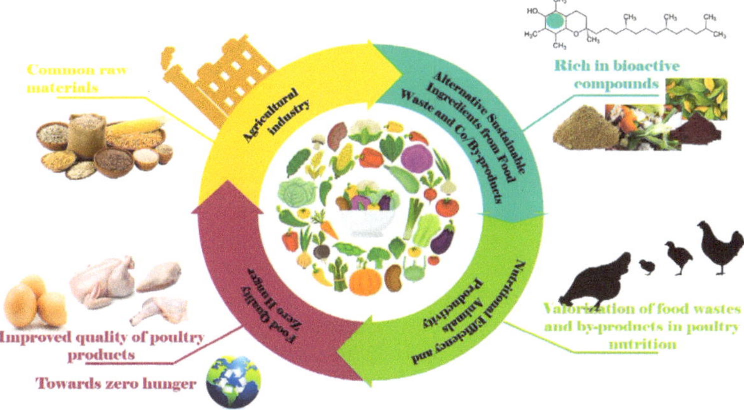


**Figure 1.** Overview of potential valorisation of co-/by-products and agro-food industry waste in poultry to achieve Zero Hunger.

## 2. SDG 2: Zero Hunger and the Role of Poultry Production in Global Food Security and the Circular Economy

### 2.1. SDG 2: Zero Hunger and Why It Is Relevant to Poultry Production

Zero Hunger is one of the 17 SDGs established by the United Nations in 2015 as part of the 2030 Agenda for Sustainable Development, officially known as SDG 2 [12]. This goal, as described in the agenda, addresses a range of interconnected issues related to hunger, food quality, and food security. The main goal is to ensure that all people, especially poor and vulnerable populations, have access to sufficient, nutritious, and safe food products all year round. This involves not only increasing animal-origin food production but also improving food system distribution and reducing food waste [13]. A potential strategy to achieve sustainable nutrition and health for the increasing population can be formulated by following the principles of bioeconomy, which involves the production of renewable biological resources and the conversion of these resources and waste streams into value-added products, such as food, feed, bio-based products, and bioenergy [14]. Furthermore, to increase the eco-sustainability of the food processing industry, food waste, by-products, co-products, and/or plants should be exploited before they become waste or neglected.

Dealing with food waste is of great importance, especially in terms of combating hunger, raising incomes, and improving food security in the poorest countries. Until a few decades ago, food waste was not considered either a cost nor a benefit. However, to maximize benefits for the environment, society, and economy, the FAO is currently developing a code of conduct (CoC) for the reduction of food loss and waste and has proposed an inverted pyramid, setting priorities on how to best reduce food waste and save natural resources [15]. To achieve this target, SDG 2 emphasizes the importance of sustainable agriculture, which involves promoting agricultural practices that increase productivity and production, among other objectives.

### 2.2. Poultry Production as a Pillar of Food Security

Poultry production plays a pivotal role in providing affordable, high-quality protein to millions of consumers globally, contributing directly to SDG 2 by enhancing food security and promoting sustainable agricultural practices [16].

From a nutritional point of view, poultry products (chicken meat and eggs) are among the most consumed animal-origin-derived products worldwide, being rich sources of protein, amino acids, fats, vitamins, and minerals, which are crucial for various bodily functions, including immune system support and cognitive development essential for human health, particularly in developing countries [17]. Furthermore, poultry meat generally contains lower levels of saturated fats compared to red meats [18], making it a healthier option for consumers and supporting efforts to reduce diet-related non-communicable diseases.

From an economic point of view, due to efficient feed conversion and short production cycles, poultry production is cost-effective, making poultry products more affordable compared to other animal protein sources [19], which increases their accessibility to a larger population, including low-income families. Eggs and chicken meat products are often included in school feeding programs, pregnant women, and emergency food aid due to their high nutritional value and ease of storage and preparation. Programs targeting such groups often use eggs and poultry meat as key components to improve nutritional outcomes, as recently reported in a study focused on programs from sub-Saharan Africa and South Asia regions [20]. Another important economic aspect is that poultry species are more resilient to climate variations compared to larger livestock, and they can be raised in diverse environmental conditions [21,22], from tropical to temperate climates, making poultry farming a reliable source of food even under changing climatic conditions. Lastly, poultry production generally emits fewer greenhouse gases per unit of meat compared to ruminant livestock [23], supporting sustainable food systems that balance food production with environmental conservation.

From a cultural point of view, poultry is widely accepted across different cultures and religions, making it a versatile and important component of global diets [24]. This cultural acceptance facilitates its inclusion in various food security initiatives without significant dietary restrictions. In this context, the importance of poultry products (eggs and meat) in global food security cannot be overlooked, as it supports the dietary needs of billions humans, particularly in regions where other forms of animal protein are scarce or too expensive for the average consumer [25]. Furthermore, poultry farming is often integrated into larger agricultural practices, providing a steady income stream for smallholder farmers, who represent a significant portion of the world's agricultural producers. These small-scale operations contribute not only to local food security but also to the livelihoods of millions of families, thereby reinforcing the socio-economic stability of rural areas, as shown in a study conducted in South Asia [26].

### 2.3. Poultry Production and the Circular Economy

The concept of a circular economy is gaining massive attention in discussions about sustainable agriculture, which includes poultry production. A circular economy approach in poultry farming emphasizes the efficient use of resources, waste minimization, and the recycling of discarded by-products [27]. This approach aligns closely with the principles of SDG by reducing environmental impacts, such as carbon footprint, and enhancing resource efficiency, which in the context of SDG 2 are considered critical aspects. Generally, traditional poultry feeding practices rely heavily on common grains such as corn, wheat, and soybeans, which are also staple foods for humans [28]. This competition for these common resources can amplify food insecurity, especially in regions where grain production is not sufficient to meet both human and animal needs. Therefore, the transition to more sustainable feeding practices is essential. Alternative feed ingredients such as insect protein, algae, and agricultural by-products and wastes [29–31] represent promising solutions for reducing the environmental footprint of poultry production. Insects, for example, can be produced using organic waste, which not only diverts waste from landfills but also creates a high-protein feed source for poultry [29]. Algae, another alternative, can be cultivated in environments that are unsuitable for traditional agriculture, making it a viable option in arid regions [30]. Agricultural by-products, such as oil seed meals [31], hulls [32], and other residues (fruits and legumes) from crop production [33–35], can also be repurposed as poultry feed ingredients, thereby closing the loop in agricultural systems and promoting a circular economy. The adoption of these alternative feeding strategies can lead to significant reductions in the environmental impacts of poultry farming. For instance, insect-based feeds have been shown to lower greenhouse gas emissions [36] and reduce the reliance on water-intensive crops like soy. Additionally, it was recently reported in a case study [37] that by utilizing waste materials as feed, the poultry industry can contribute to reducing overall food waste, which is a significant challenge in the global food system.

### 2.4. Challenges in Implementing Sustainable Feeding Practices

While the benefits of sustainable feeding strategies are clear, several challenges hinder their widespread adoption, particularly when focusing on by-products, food waste, and plant-based alternatives. As recently mentioned in other studies focused on the valorization of food waste, one of the primary economic barriers is the variability in the availability and quality of by-products and food waste [38,39]. Unlike conventional feed ingredients, which are produced in large quantities with a consistent quality, by-products and food waste can vary significantly depending on their source, processing methods, and seasonal availability [40]. This inconsistency poses a challenge for formulating balanced poultry diets that meet nutritional requirements consistently. Another challenge is the presence of anti-nutritional factors and contaminants in plant-based alternatives and by-products. Many of these alternative feeds contain compounds such as tannins, phytic acid, and mycotoxins, which can interfere with the bioavailability of various components, nutrient absorption, and poultry health [41]. Another challenge is given by the influence of soil, climatic conditions, vegetation, and seasons, which will result in wastes with variability in their chemical composition. However, as recently reported, these issues can be addressed by additional processing steps, such as fermentation, enzyme treatment, or heat treatment [42], which later can increase costs and complexity for farmers, particularly those with limited resources.

The logistical challenges associated with collecting, processing, and distributing food waste and by-products also present significant barriers. Farmers often struggle to access alternative feed ingredients reliably and affordably due to the absence of established supply chains. Unlike traditional feeds that are mass-produced and distributed through established networks, by-products and food waste may need new infrastructure for collection, transportation, and storage. The same applies to food waste, which must be treated to make sure it is safe and nutritionally appropriate for poultry to eat. As reported by Salvador et al. [43], solutions to this concerning behavior are crucial for an adequate circular bioeconomy strategy. In addition, regulatory barriers can impede the adoption of sustainable feeding practices. Moreover, consumer perceptions and market acceptance play a crucial role in the success of these alternative feeding strategies [44]. There may be resistance to poultry products raised on diets that include by-products or food waste, especially in markets where traditional feeding practices are well established and preferred by consumers [44,45]. Educating consumers about the sustainability and environmental advantages of these practices as well as ensuring that the final poultry products meet high standards of quality and safety will be essential to building trust and acceptance.

## 3. The Purpose of Developing Alternative Poultry Feeding Practices

Current poultry feeding includes a variety of traditional and innovative methods designed to meet the nutritional needs of poultry, while improving productivity. Traditional poultry feeding practices are primarily based on well-established ingredients that have been shown to support poultry health and yield. Poultry feeding protocols are largely a matter of tradition, relying on standard ingredients known to help promote poultry health and production. Ingredients that are used frequently among traditional feed ingredients include grains, like corn and wheat, as primary energy sources, while soybean meal is mainly utilized in order to provide protein [46]. Some of these end products are mixed with animal by-products, fats, vitamins, and minerals. This results in a well-balanced diet tailored to the specific needs and growth of different poultry [42]. Balanced macronutrients (proteins, carbohydrates, and fats), as well as perfect combination of essential micro-nutrients (vitamins and minerals), are key factors to be considered in diet formulation for poultry. Formulating complete feed is important for the best performance and so as to avoid nutrient deficiencies, which can help to produce better chickens', health-wise, regarding their production qualities. However, even the most effective traditional diets are not without their pitfalls. A major challenge is the high costs and price fluctuations of common feed ingredients, such as corn and soybean meal [47,48]. These costs can be limitative for resource-limited smallholder farmers, affecting their profitability and sustainability. Furthermore, the heavy reliance on particular crops raises concerns about the environmental impact of their agriculture on the environment, with issues such as deforestation, water consumption, including its use, and greenhouse gas emissions. To address these challenges, there is growing interest in exploring alternative feed ingredients that can supplement or replace entirely, if possible, the conventional ones [42,49]. These ingredients can provide valuable nutrients while potentially reducing feed costs and environmental impacts. However, their nutritional profiles as shown in Table 1 and effects on poultry health, performance, and product quality need to be thoroughly evaluated to ensure they meet the birds' needs.

**Table 1.** Chemical composition and bioactive compounds determined in different plants, waste, and their co-/by-products.

| Plant Family | Plant Common Name | Plant Part | Reported Proximal Composition | Reported Mineral Composition | Reported Lipid Composition | Reported Bioactive Compounds | Reference |
|---|---|---|---|---|---|---|---|
| *Apiaceae* | Carrot (*Daucus carota*) | waste | DM (88.29%), CP (6.11%), EE (0.92%), CF (7.03%), ash (6.27%) | NA | SFA (30.09%), MUFA (23.24%), PUFA 46.28%), n-3 (2.99%), n-6 (43.29%). | TPC (2.03 mg GAE/g dw), TFC 1.21 (mg QE/g dm), DPPH (4.28 mg TE/g dm), ABTS (14 mg TE/g dm), lycopene (3.19 µg/g dm), lutein (1.29 µg/g dm), β-carotene (59 µg/g dm), α-carotene (9.99 µg/g dm). | [33,50] |
| | | orange carrot | NA | NA | NA | Lutein (17.2 µg/g dw), α-carotene (255 µg/g dw), β-carotene (1016.35 µg/g dw), TEAC (63.82 µmol TE/100 g fw), ORAC (250.7 µmol TE/100 g fw). | [51] |
| | | black carrot | NA | NA | NA | Lutein (57.58 µg/g dw), α-carotene (21.87 µg/g dw), β-carotene (60.38 µg/g dw), anthocyanins (186.85 mg K Eq/100 g fw), TEAC (1026.43 µmol TE/100 g fw), ORAC (2159 µmol TE/100 g fw). | [51] |
| | | purple carrot | NA | NA | NA | Lutein (37.42 µg/g dw), α-carotene (47.26 µg/g dw), β-carotene (247.45 µg/g dw), anthocyanins (55.13 mg K Eq/100 g fw), TEAC (470.21 µmol TE/100 g fw), ORAC (866.3 µmol TE/100 g fw). | [51] |
| | | leaves | NA | NA | NA | TPC (3.26 mg GAE/g), TFC (1.80 mg RE/g), rutin (3738.50 µg/g dw), quercetin (236.53 µg/g dw). | [52] |
| | | flour | NA | Ca (34 to 80 mg/100 g), P (25 to 53 mg/100 g), Fe (0.4 to 2.2 mg/100 g), Mn (9 mg/100 g), | NA | TPC (13.16 to 18.57 mg GAE/g), thiamine (0.04 mg/100 g), riboflavin (0.02 mg/100 g). | [53] |
| | Parnship (*Pastinaca sativa*) | root | DM (16.8 to 18.8%), monosaccharides (6.5 to 8%), disaccharides (33.8 to 37.2%), | Ca (1716 to 2436 mg/kg dw), Fe (41.9 to 65.3 mg/kg dw), Zn (12.6 to 19.6 mg/kg dw), P (3518 to 4225 mg/kg dw), Mg (1437 to 1963 mg/kg dw). | NA | TPC (3.6 mg GAE/g dw), TAC (10.6 mg GAE/g dw), DPPH (0.080 µmol TE/g dw). | [54] |
| | Celery (*Apium graveolens*) | bulb | DM (91.72 to 90.38%), CP (9.4%), CF (2.18%). | Ca (0.34%), K (3.90%), Mg (0.21%), P (0.59%) Zn (31.2 mg/kg), Fe (21 mg/kg), | NA | TPC (265.44 to 368.51 (µM chlorogenic acid/g), EC50 (2.41 to 3.14 mg/mL). | [55,56] |
| | | leaves | NA | NA | NA | choline (1251 to 2224 µg/g), pantothenic acid (26 to 92 µg/g), riboflavin (37 to 79 µg/g), vitamin E (0.54 to 16.8 µg/g), rutin (143 to 267 µg/g), cyanidin (0.94 to 5.65 µg/g), elemicin (9.9 to 177 µg/g), xanthophyll (3 to 55 µg/g), ALA (3.26 to 21 µg/g), DPPH (84 to 90 inhibition ratio), O2 (24 to 97 inhibition ratio). | [57] |
| | | petioles | NA | NA | NA | choline (405 to 626 µg/g), pantothenic acid (16 to 245 µg/g), riboflavin (4.4 to 12.2 µg/g), vitamin E (0.25 to 0.46 µg/g), rutin (6.75 to 35 µg/g), cyanidin (0.16 to 0.23 µg/g), elemicin (0.55 to 5.8 µg/g), xanthophyll (1.3 to 12 µg/g), ALA (0.17 to 0.36 µg/g), DPPH (23 to 39 inhibition ratio), O2 (14 to 25 inhibition ratio). | [57] |
| | | entire plant | NA | NA | NA | Flavonoids (85.31 to 174.72 mg/100 g dw),apigenin (55.56 to 142.85 mg/100 g dw) luteolin (24.83 to 65.91 mg/100 g dw); phenolic acids (114.81 to 223.49 mg/100 g dw) caffeic (7.15 to 30.26 mg/100 g dw) ferulic (10.94 to 94.33 mg/100 g dw), p-coumaric (80.18 to 102.75 mg/100 g dw); TPC (3.48 to 5.02 mg GAE/100 g dw), DPPH (86.67 to 105.79 µmol TE/100 g dw), ABTS (81.90 to 114.38 µmol TE/100 g dw). | [58] |
| | Fennel (*Foeniculum vulgare*) | seed | CP (9.5 g/100 g), EE (10 g/100 g), carbohydrates (42.3 g/100 g), CF (18.5 g/100 g), | Ca (1.3 g/100 g) P (1.7 g/100 g) | NA | vitamin B1 and B2 (0.41 and 0.36 mg/100 g), niacin (6 mg/100 g), vitamin C (12 mg/100 g). | [59] |
| | | shoots | CP (1.33 g/100 g), EE (0.49 g/100 g), carbohydrates (21.49 g/100 g), sugars (6.57 g/100 g), | NA | SFA (19.95%), MUFA (2.72%), PUFA (77.33%), n-3 (36.96%), n-6 (39.99%). | NA | [60] |
| | | leaves | CP (1.16 g/100 g), EE (0.61 g/100 g), carbohydrates (18.44 g/100 g), sugars (1.29 g/100 g). | NA | SFA (27.99%), MUFA (4.96%), PUFA (67.05%), n-3 (43.72%), n-6 (23.25%). | NA | [60] |
| | | stems | CP (1.08 g/100 g), EE (0.45 g/100 g), carbohydrates (19.39 g/100 g), sugars (4.92 g/100 g). | NA | SFA (33.81%), MUFA (4.78%), PUFA (61.04%), n-3 (23.04%), n-6 (38.22%). | NA | [60] |
| | | inflorescences | CP (1.37 g/100 g), EE (1.28 g/100 g), carbohydrates (22.82 g/100 g), sugars (4.07 g/100 g), | NA | SFA (37.47%), MUFA (5.59%), PUFA (56.95%), n-3 (17.69%), n-6 (38.94%). | NA | [60] |

**Table 1.** *Cont.*

| Plant Family | Plant Common Name | Plant Part | Reported Proximal Composition | Reported Mineral Composition | Reported Lipid Composition | Reported Bioactive Compounds | Reference |
|---|---|---|---|---|---|---|---|
| Apiaceae | Anise (*Pimpinella anisum*) | seeds | CP (19.93 to 20.49 g/100 g DW), fiber (12.64 to 13.14 g/100 g), carbohydrates (49 g/100 g dw), | Fe (122.4 mg/kg), Ca (10.56 g/kg), K (33.16 g/kg). | NA | TPC (64.63 mg GAE/g), carotenoids (23.33 mg/100 g), tannins (83.31 mg), ALA (1.07%), catechin (3.31 ppm), chlorogenic (2.875 ppm), salicylic (1.704 ppm), coumarin (0.132 ppm), hypersoid (47.103 ppm), quercetin (17.239 ppm), luteolin (159.3 ppm), kaempferol (7.177 ppm), apigenin (6.88 ppm). | [61,62] |
| | Coriander (*Coriandrum sativum*) | leaves | CP 21.93 (g/100 g), EE 4.78 g/100 g, carbohydrate (52.10 g/100 g), CF (10.40 g/100 g), | Ca (1246 mg/100 g), Fe (42.46 mg/100 g), Mg (694 mg/100 g), P (481 mg/100 g), K (4466.mg/100 g), Na (211 mg/100 g), Zn (4.72 mg/100 g), | SFA (0.115 g/100 g), MUFA (2.232 g/100 g), PUFA (0.328 g/100 g) | vitamin C (566.7 mg/100 g), thiamine (1.252 mg/100 g), riboflavin (1.500 mg/100 g), niacin (10.707 mg/100 g), vitamin A, (293 µg/100 g). | [63] |
| | | seeds | CP (12.37 g/100 g), EE (17.77 g/100 g), carbohydrate (54.99 g/100 g), CF (41.9 g/100 g), | Ca (709 mg/100 g), Fe (16.32 mg/100 g), Mg (330 mg/100 g), P (409 mg/100 g), K (1267 mg/100 g), Na (35 mg/100 g), Zn (4.70 mg/100 g). | SFA (0.990 g/100 g), MUFA (13.58 g/100 g), PUFA (1.75 g/100 g), SFA (12.13 g/100 g), UFA (87.87 g/100 g), n-6/n-3 (0.009), | vitamin C (21.0 mg/100 g), thiamine (0.239 mg/100 g), riboflavin (0.290 mg/100 g), niacin (2.130 mg/100 g), TPC (14.81 to 89.81 mg GAE/g extract), TFC (4.89 to 19.11 mg QE/g extract). | [63,64] |
| | Parsley (*Petroselinum crispum*) | leaves | CP (23.49%), EE 1.40%), CF (8.73%), ash (20.17%). | Zn (56.16 mg/kg), Fe (3817.7 mg/kg), Cu (12.64 mg/kg), Mn (121.59 mg/kg), | NA | vitamin E (25.33 mg/kg), TPC (304.57 to 425.76 mg GAE/100 g fw), TPC (7.71 mg GAE/g dw), DPPH (13.05 mM Trolox), TFC (141.39 to 185.47 mg GAE/100 g fw), vitamin C (73.39 to 162.09 mg/100 g fw), TAC (2.19 to 2.29 mM TE/L), carotenoids (0.08 to 0.16 mg/g). | [65,66] |
| | | stem | NA | NA | NA | TPC (65.02 to 165.12 mg GAE/100 g fw), TFC (30.73 to 73.72 mg GAE/100 g fw), vitamin C (13.6 to 40.77 mg/100 g fw), TAC (1.25 to 2.24 mM TE/L), carotenoids (0.02 to 0.03 mg/g). | [66] |
| | | root | NA | NA | NA | TPC (55.21 to 75.01 mg GAE/100 g fw), TFC (25.83 to 35.67 mg GAE/100 g fw), vitamin C (9.37 to 26.93 mg/100 g fw), TAC (0.69 to 0.98 mM TE/L). | [66] |
| | Dill (*Anethum graveolens*) | leaf blade | NA | NA | NA | vitamin C (159 to 186 mg/100 g fw), carotenoids (27.8 to 34.9 mg/100 g fw), β-carotene (4.07 to 5.62 mg/100 g fw), TPC (173 to 331 mg/100 g fw). | [67] |
| | | petiole | NA | NA | NA | vitamin C (33 to 38 mg/100 g fw), carotenoids (6 to 6.8 mg/100 g fw), β-carotene (0.64 to 0.73 mg/100 g fw), TPC (54 to 92 mg/100 g fw). | [67] |
| | | whole leaf | NA | NA | NA | vitamin C (116 to 138 mg/100 g fw), carotenoids (20.5 to 25.3 mg/100 g fw), β-carotene (2.79 to 3.95 mg/100 g fw), TPC (129 to 248 mg/100 g fw). | [67] |
| | | stem | NA | NA | NA | vitamin C (29 to 39 mg/100 g fw), carotenoids (2 to 3.2 mg/100 g fw), β-carotene (0.2 to 0.25 mg/100 g fw), TPC (49 to 66 mg/100 g fw). | [67] |
| | | whole plant | NA | NA | NA | vitamin C (55 to 116 mg/100 g fw), carotenoids (8.8 to 16.9 mg/100 g fw), β-carotene (1.32 to 3.57 mg/100 g fw), TPC (100 to 129 mg/100 g fw). | [67] |
| | Lovage (*Levisticum officinale*) | leaves + stems | CP (3.01 g/100 g fw), EE (0.37 g/100 g fw), carbohydrates (5.7 g/100 g fw), organic acids (1.26 g/100 g fw), | NA | SFA (18%), MUFA (2.93%), PUFA (79%), n-3 (49.2%), n-6 (29.9%) | vitamin E (0.80 mg/100 g fw). | [68] |
| | | leaves | NA | NA | NA | protocatechuic (6.63 µg/g fw), hydroxybenzoic (3.56 µg/g fw), syringic (4.23 µg/g fw), vanillic (5.54 µg/g fw), synaptic (18.78 µg/g fw), salicylic (9.59 µg/g fw), caffeic (5.80 µg/g fw), ABTS (3.20 µM Trolox/g fw), RP (10.02 mg Trolox/g fw), iron chelation (711.71 mg EDTA/g fw). | [69] |

**Table 1.** *Cont.*

| Plant Family | Plant Common Name | Plant Part | Reported Proximal Composition | Reported Mineral Composition | Reported Lipid Composition | Reported Bioactive Compounds | Reference |
|---|---|---|---|---|---|---|---|
| *Adoxaceae* | American elderberry (*Sambucus canadensis* L.) | fruits | NA | NA | NA | TP (2898 to 4585 µg GAE/g fw), FRAP (13.4 to 31.7 µmol TE/g fw), DPPH (7 to 16.9 µmol TE/g fw), anthocyanins (1308 to 4004 ( µg Cy−3GE/g fw). | [70] |
| | European elderberry (*Sambucus nigra* L.) | fruits | CP (2.7 to 2.9%), glucose (33.33 to 50.23 g/kg fw), fructose (33.99 to 52.25 g/kg fw) | K (2953 to 5494 mg/kg fw), P (735 to 1337 mg/kg fw), Ca (574 to 1528 mg/kg fw), Na (13 to 146 mg/kg fw), Mg (396 to 739 mg/kg fw), Fe (12.4 to 84.7 mg/kg fw), Zn (1.9 to 11.3 mg/kg fw), Mn (3.6 to 9.5 mg/kg fw), Cu (1.7 to 2.9 mg/kg fw) | NA | TPC (364 to 582 mg GAE/100 g fw), TPC (4917 to 8974 mg GAE/100 g dw), sambunigrin (0.08 to 0.77 µg/g fw). | [71] |
| | | flowers | NA | Ca (2674 to 3334 µg/g), Mg (494 to 1556 µg/g), Fe (53 to 103 µg/g), Cu (6.5 to 13.8 µg/g), Zn (31.8 to 41.1 µg/g), Mn (19.7 to 49.7 µg/g), to 94.15%), | NA | TPC (1021.7 mg GAE/100 g fw), TPC (194 mg GAE/g dw), proteins (2.5%), sambunigrin (1.23 to 18.88 µg/g fw), DPPH (91.95 TFC (527 to 1319 mg RE/100 g dw). | [71,72] |
| | Jerusalem artichoke (*Helianthus tuberosus*) | tubers | CP (10.88 g/100 g dw), carbohydrate (81.67 g/100 g dw), inulin (78.22 g/100 g dw) | Ca (1.93 mg/g), K (10.56 mg/g), Zn (0.03 mg/g), P (4.17 mg/g), Na (0.15 mg/g), Fe (0.07 mg/g), | NA | TPC (4259.89 mg/kg). | [73] |
| | | leaves | CP (5.65 to 21.40%), CF (21 to 27.4%), EE (1.52 to 6.14%) | Ca (102 to 760 mg/100 g), K (3.600 to 4.500 mg/100 g), Mg (60 to 690 mg/100 g), P (7 to 105 mg/100 g), Na (4 to 7 mg/100 g), Zn (4.40 to 7.20 mg/10 g), Fe (0.10 to 8 mg/100 g), | NA | carotenoids (167.80 to 415.22 mg/100 g dw). | [74,75] |
| | Chicory (*Cichorium intybus*) | roots | DM (24.37%), CP (4.65%), ash (4.25%), EE (1.69%), carbohydrates (89.41%), inulin (44.69%) | Ca (181.26 mg/100 g), K (103.7 mg/100 g), Mg (20.14 mg/100 g), Na (67.42 mg/100 g), Fe (1.77 mg/100 g), Cu (0.36 mg/100 g), Mn (0.31 mg/100 g), Zn (0.39 mg/100 g), Pb (0.04 mg/100 g), | NA | TPC (20 mg GAE/g dw), protocatechuic acid (1.77%), chlorogenic acid (10.85%), caffeic acid (24.36%), m-coumaric acid (27.90%), p-coumaric acid (25.03%). | [76] |
| | | leaves | CP (86.03%), EE (3.68%), ash (10,91%), carbohydrates (70.71%), inulin (10.95%), | Ca (292.61 mg/100 g), K (166.57 mg/100 g), Mg (6.94 mg/100 g), Na (88.84 mg/100 g), Fe (9.18 mg/100 g), Cu (0.60 mg/100 g), Mn (0.90 mg/100 g), Zn (0.91 mg/100 g), Pb (0.03 mg/100 g), | NA | TPC (26.4 mg GAE/g dw), protocatechuic acid (2.50%), chlorogenic acid (17.84%), p-hydroxybenzoic acid (11.04%), caffeic acid (35.22%), isovanillic acid (1.97%), p-coumaric acid (9.65%). | [76] |
| | | seeds | CP (18.61%), EE (21.18%), ash (11.39%), CF (23.70%), carbohydrates (18.40%) | Cu (18.93 ppm), Zn (92.61 ppm), Mn (43.20 ppm), Fe (641 ppm) | NA | tannins (1.72 mg/g), vitamin C (23.85 mg/100 g). | [77] |
| *Asteraceae* | Calendula (*Calendula officinalis*) | flowers | CP (2.4 g/100 g), EE (5.6 g/100 g), ash (14 g/100 g), carbohydrates (78 g/100 g), sugars (11.7 g/100 g), organic acids (2830 mg/100 g), | NA | NA | tocopherols (23.33 mg/100 g), SFA (76.70%), MUFA (2.78%), PUFA (20.51%), TPC (4351 mg/100 g), TFC (4161 mg/100 g), TPA (190 mg/100 g), DPPH (4.6 EC50 mg/mL). | [78] |
| | | leaves | NA | NA | NA | Total phenylpropanoids (3.23 to 20.17 mg/g dw), total quercetin derivatives (4.58 to 12.16 mg/g dw), TFC (6.11 to 15.74 mg/g). | [79] |
| | Dandelion (*Taraxacum mongolicum*) | flower | NA | Ca (328.33 mg/kg), Mg (5.19 mg/kg). | NA | DPPH (118.24 to 130.49 mg Trolox/g dw), FRAP (88.16 to 103.14 mg Trolox/g dw), TPC (26.06 mg GAE/g dw), TFC (4.98 mg QE/g dw), vitamin C (58.54 mg/kg), total carotenoids (43.56 mg/kg). | [80] |
| | | leaves | NA | Ca (448.26 mg/kg), Mg (6.31 mg/kg). | NA | DPPH (121.18 to 135.14 mg Trolox/g dw), FRAP (98.23 to 119.27 mg Trolox/g dw), TPC (30.05 mg GAE/g dw), TFC (2.26 mg QE/g dw), vitamin C (106.49 mg/kg), total carotenoids (198.29 mg/kg), chlorophylls (427.18 mg/kg). | [80] |
| | | stem | NA | Ca (366.28 mg/kg), Mg (4.23 mg/kg). | NA | DPPH (98.27 to 109.28 mg Trolox/g dw), FRAP (90.28 to 108.14 mg Trolox/g dw), TPC (23.89 mg GAE/g dw), vitamin C (96.89 mg/kg), chlorophylls (47.21 mg/kg), total carotenoids (24.87 mg/kg). | [80] |
| | | roots | NA | Ca (408.21 mg/kg), Mg (2.17 mg/kg). | NA | DPPH (50.89 to 61.36 mg Trolox/g dw), FRAP (45.34 to 53.54 mg Trolox/g dw), TPC (4.23 mg GAE/g dw), vitamin C (18.02 mg/kg). | [80] |
| | Lettuce (*Lactuca sativa*) | red | NA | NA | NA | TPC (64.90 mg/100 g dw), TFC (291.6 mg/100 g dw), anthocyanins (23.7 mg/100 g dw), carotenoids (108.3 mg/g), lutein (51.5 mg/g), β-carotene (48.3 mg/g), DPPH (77.5 µL/mL). | [81] |
| | | green | NA | NA | NA | TPC (49.4 mg/100 g DW), TFC (223 mg/100 g dw), anthocyanins (7.4 mg/100 g dw), carotenoids (92.4 mg/g), lutein (39.4 mg/g), β-carotene (44.4 mg/g), DPPH (77.2 µL/mL). | [81] |
| | | red oak | NA | Fe (1.21 to 1.79 mg/kg), Zn (103.6 to 146.3 mg/kg). | NA | Chlorophyll (10.70 to 27.40 mg 100/g fw), β-carotene (11.98 to 16.45 mg 100/g fw), vitamin C (25 to 30.61 mg 100/g fw), TPC (1.61 to 2.81 mg GAE/100 g fw), DPPH (46.26 to 48.46 mg GAE/100 g fw), ABTS (2.64 to 6.05 mg TEAC/100 g fw), FRAP (126.75 to 127.46 mg TEAC/100 g fw). | [82] |

**Table 1.** *Cont.*

| Plant Family | Plant Common Name | Plant Part | Reported Proximal Composition | Reported Mineral Composition | Reported Bioactive Compounds | Reported Lipid Composition | Reference |
|---|---|---|---|---|---|---|---|
| *Asteraceae* | Lettuce (*Lactuca sativa*) | green oak | NA | Fe (1.17 to 1.82 mg/kg), Zn (99.1 to 122.7 mg/kg). | Chlorophyll (17.64 to 34.55 mg 100/g fw), β-carotene (9.22 to 13.12 mg 100/g fw), vitamin C (9.52 to 21.55 mg 100/g fw), TPC (0.71 to 0.86 mg GAE/100 g fw), DPPH (37.69 to 44.99 mg GAE/100 g fw), ABTS (0.88 to 3.02 mg TEAC/100 g fw), FRAP (15.59 to 87.74 mg TEAC/100 g fw). | NA | [82] |
| | | green curly | NA | Zn (87.4 to 118.4 mg/kg), Fe (1.09 to 1.73 mg/kg). | Chlorophyll (11.55 to 34.61 mg 100/g fw), β-carotene (6.59 to 13.90 mg 100/g fw), vitamin C (13.49 to 26.50 mg 100/g fw), TPC (0.39 to 1.14 mg GAE/100 g fw), DPPH (42.85 to 54.76 mg GAE/100 g fw), ABTS (1.31 to 2.53 mg TEAC/100 g fw), FRAP (16.85 to 110.24 mg TEAC/100 g fw). | NA | [82] |
| | | lollo rossa | NA | Fe (1.39 to 1.69 mg/kg), Zn (85.3 to 99.7 mg/kg). | Chlorophyll (16.40 to 25.73 mg 100/g fw), β-carotene (8.63 to 13.07 mg 100/g fw), vitamin C (27.67 to 29.97 mg 100/g fw), TPC (1.78 mg GAE/100 g fw), DPPH (45.43 to 50.89 mg GAE/100 g fw), ABTS (2.33 to 4.80 mg TEAC/100 g fw), FRAP (99.58 to 127.57 mg TEAC/100 g fw). | NA | [82] |
| | | bativa | NA | Fe (1.87 mg/kg), Zn (109 mg/kg). | Chlorophyll (23.84 mg 100/g fw), β-carotene (9.57 mg 100/g fw), vitamin C (25.11 mg 100/g fw), TPC (0.96 mg GAE/100 g fw), DPPH (44.76 mg GAE/100 g fw), ABTS (2.46 mg TEAC/100 g fw), FRAP (122.88 mg TEAC/100 g fw). | NA | [82] |
| | Artichoke (*Cynara cardunculus*) | receptacle | DM (9.9 to 16 g/100 g), CP (189 to 269 g/100 g) | Ca (3.4 to 7.3 g/kg), Mg (0.8 to 1.4 g/kg), Na (0.6 to 0.9 g/kg), K (16.4 to 18.6 g/kg), Fe (25.8 to 31.4 g/kg), Zn (19 to 28.8 g/kg), Mn (5.8 to 10.2 g/kg), Cu (4.5 to 7.5 g/kg) | TPC (2.5 to 6 g/kg fw), inulin (185 to 265 g/kg DM). | NA | [83] |
| | | heads | DM (16 to 25 g/100 g fw), EE (0.26 to 0.57 g/100 g fw), ash (1.01 to 1.67 g/100 g fw), CP (1.69 to 4.25 g/100 g fw), carbohydrates (13.49 to 19.09 g/100 g fw) | K (276 to 579 mg/100 g fw), Na (17 to 104 mg/100 g fw), Ca (158 to 861 mg/100 g fw), Mg (31 to 91 mg/100 g fw), Mn (0.68 to 1.17 mg/100 g fw), Fe (1.7 to 2.90 mg/100 g fw), Zn (0.48 to 1.17 mg/100 g fw) | NA | SFA (39.28 to 69.7%), MUFA (2.26 to 10.3%), PUFA (23.08 to 57.36%), n-6/n-3 (4.64 to 7.95) | [84] |
| | Yarrow (*Achillea millefolium*) | wild inflorescences and upper leaves | CP (12.53 g/100 g dw), EE (5.20 g/100 g dw), Ash (6.43 g/100 g dw), Carbohydrates (75.84 g/100 g dw), sugars (3.14 g/100 g dw), organic acids (4.55 g/100 g dw) | NA | Total tocopherols (16.62 mg/100 g dw). | SFA (22.09 g/100 g fat), MUFA (28.75 g/100 g fat), PUFA (49.16 g/100 g fat) | [85] |
| | | commercial inflorescences and upper leaves | CP (19.53 g/100 g dw), EE (8.03 g/100 g dw), Ash (8.54 g/100 g dw), Carbohydrates (63.90 g/100 g dw), sugars (4.86 g/100 g dw), organic acids (4.46 g/100 g dw) | NA | Vitamin E (15.16 mg/100 g dw). | SFA (44.06 g/100 g fat), MUFA (12.64 g/100 g fat), PUFA (43.30 g/100 g fat) | [85] |
| *Arecaceae* | Coconut (*Cocos nucifera*) | shell | DM (89.9%), CP 0.46%), ash (2.28%), CF 32.39%), EE (2.14%), carbohydrate 52.63%) | P (11.64 mg/100 g), Ca (16.02 mg/100 g), Mg (1.22 mg/100 g), Na (0.76 mg/100 g), K (3.30 mg/100 g), Fe (618 mg/100 g), Zn (1.20 mg/100 g), Mn (6 mg/100 g). | NA | NA | [86] |
| | | meal | DM (89.25%), CP (33.56%), EE (15.07%), CF (10.10%). | NA | TPC (7.95 mg GAE/g), TAC (24.57 mM Trolox/g), TFC (4.51 µg rutin/g). | SFA (16.83 mg/100 g), MUFA (42.90 mg/100 g), PUFA (40.26 mg/100 g), n-3 (4.42 mg/100 g), n-6 (35.85 mg/100 g). | [5] |
| *Brassicaceae* | Rapeseed (*Brassica napus*) | seeds | DM (92.99 to 93.98%), CP (17.42 to 21.01%), EE (40.58 to 54.20%), CF (7.16 to 23.20%), ash (3.60 to 9.02%). | NA | NA | n-3 (9.64 mg/100 g), n-6 (24.70 to 33.90 mg/100 g). | [87] |
| | | cake | DM (89.50 to 95.30%), CP (28 to 36.10%), EE (12.20 to 17.80%), CF (11.20 to 13.10%), ash (5.60 to 7.10%). | NA | NA | n-3 (13.05 mg/100 g), n-6 (21.96 mg/100 g). | [87] |
| | Kale (*Brassica oleracea*) | leaves | DM (17.08%), CP (4.06 g/100 g), EE (0.67 g/100 g), ash (2.11 g/100 g), carbohydrates (10.14 g/100 g), CF (8.39 g/100 g) | Na, (38.5 mg/100 g), K, (440.2 mg/100 g), Ca, (384.8 mg/100 g), Mg, (34.9 mg/100 g), Zn, (0.83 mg/100 g) | Vitamin C (62.27 mg/100 g), β-carotene, (6.40 mg/100 g), TPC (574.95 mg/100 g), ABTS (33.22 µm Trolox/g). | NA | [88] |
| | Mustard (*Brassica juncea*) | seeds | DM (92.85%), CP (33.93), EE (14.71%), CF (20.76%), ash (5.23%). | Cu (30.55 mg/kg), Fe (201.25 mg/kg), Mn (59.18 mg/kg), Zn (111.35 mg/kg). | Vitamin E (243.8 mg/kg), lutein 7.163 mg/kg), canthaxanthin (1.02 mg/kg), β-carotene (52.28 mg/kg), TPC (28.47 mg GAE/g), TAC (33.78 mM eq. Trolox). | NA | [89] |
| | Cabbage (*Brassica oleracea*) | waste | DM (89.65 g/100 g), CP (12.28 g/100 g), EE (0.80 g/100 g), ash (18.05 g/100 g), Carbohydrate (59.59 g/100 g). | Ca (1671.11 mg/100 g), Mg (292.68 mg/100 g), Fe (3.79 mg/100 g), Zn (3.64 mg/100 g), K (8582 mg/100 g), Cu (0.24 mg/100 g), Mn (6.04 mg/100 g), Na (75.41 mg/100 g). | NA | NA | [90] |

**Table 1.** *Cont.*

| Plant Family | Plant Common Name | Plant Part | Reported Proximal Composition | Reported Mineral Composition | Reported Lipid Composition | Reported Bioactive Compounds | Reference |
|---|---|---|---|---|---|---|---|
| *Cannabaceae* | Hemp (*Cannabis sativa*) | seeds | DM (94.47%), CP (23.93%), EE (31.86), NDF (54.08%), ADF (31.25%) | NA | SFA (10.70%), MUFA (10.92%), PUFA (77.72%), n-6 (57.35%), n-3 (20.37%). | α-tocopherol (20.05 μg/g), β-carotene (4.13 μg/g), TPC (6.41 mg GAE/g). | [91] |
| | Pumpkin (*Cucurbita pepo*) | flesh | DM (3.23 g/kg fw), CP (2.08 g/kg fw), EE (0.55 g/kg fw), CF (3.72 g/kg fw), ash (3.44 g/kg fw), | NA | NA | α-tocopherol (1.40 mg/kg fw), β-carotene (1.48 mg/kg fw). | [92] |
| | | peel | DM (6.40 g/kg fw), CP (9.25 g/kg fw), EE (4.71 g/kg fw), CF (12.28 g/kg fw), ash (6.30 g/kg fw), | NA | NA | α-tocopherol (4.49 mg/kg fw), β-carotene (39.48 mg/kg fw) | [92] |
| | | seeds | DM (25.94 g/kg fw), CP (308.83 g/kg fw), EE (439.88 g/kg fw), CF (148.42 g/kg fw), ash (55.02 g/kg fw), | NA | SFA (18.62%), MUFA (32.40%), PUFA (36.40%), | α-tocopherol (21.33 mg/kg fw), β-carotene (17.46 mg/kg fw). | [92] |
| *Cucurbitaceae* | Cucumber (*Cucumis sativus*) | fruits | DM (15.8%), CP (3.01%), CF (1.02%), ash (0.94%), EE (0.55%), carbohydrates (0.28%), | NA | NA | Tannins (1.26 mg/g), polyphenols (8.51 mg/g), phenols (7.72 mg/g), glycosides (32.23 mg/g), reducing sugars (574.36 mg/g), saponins (2.01 mg/g), alkaloids (2.22 mg/g), flavonoids (2.14 mg/g), terpenoids (26.27 mg/g), steroids (11.69 mg/g), resins (50.70 mg/g). | [93] |
| | Watermelon (*Citrullus lanatus*) | rind | DM (16.35%), ash (0.23%), CF (0.23%), CP (0.53%) | Ca (0.095 ppm), Fe (0.144 ppm), Mg (0.107 ppm), Zn (0.058 ppm), Na (0.085 ppm), K (0.114 ppm). | NA | NA | [94] |
| | | pulp | DM (5.53%), ash (0.31%), CF (0.45%), CP (0.34%), | Ca (0.136 ppm), Fe (0.242 ppm), Mg (0.167 ppm), Zn (0.086 ppm), Na (0.140 ppm), K (0.158 ppm). | NA | NA | [94] |
| | Pumpkin (*Cucurbita moschata*) | seed meal | DM (91.90%), CP (26.16%), EE (26.44%), CF (21.11%) | NA | PUFA (51.22 g/100 g), n-6 (48.75 g/100 g), n-3 (2.47 g/100 g), n-6/n-3 ratio (19.73) | TPC (25. 01 mg GAE/g), TAC (14. 80 mM Trolox), TFC (80 μg rutin/g). | [95] |
| | Sea buckthorn (*Hippophae rhamnoides*) | leaves | DM (91.63%), CP (14.48%), EE (5.12%), CF (13.68%), ash (6.37%) | NA | SFA (30.76%), MUFA (32.66%), PUFA (35.65%), UFA (68.31%), | TPC (58.61 mg/g GAE), TFC, 9.03 mg/g QE, lutein and zeaxanthin (583.4 μg/g), Vitamin E, 321.29 μg/g, TAC, 1147.91 μM Trolox, Co (3.05 mg/kg), Fe (334.79 mg/kg), Mg (159.59 mg/kg), Zn (126.78 mg/kg). | [96] |
| | | seed meal | DM (89.36%), CP (11.44%), EE (8.92%), CF (23.26%) | NA | SFA (23.69 g/100 g), MUFA (45.39 g/100 g), PUFA (30.44 g/100 g), n-3 (5.04 g/100 g), n-6 (25.40 g/100 g), | TPC (90.72 mg GAE/g), TAC (118.50 mM Trolox), TFC (120.01 μg rutin/g). | [31] |
| *Elaeagnaceae* | *Elaeagnus Angustifolia* | leaves | NA | NA | NA | TPC (7.78 to 10.91 mg GAE/100 g fw), TFC (4.81 to 5.20 mgQE/100 g fw). | [97] |
| | | flowers | NA | NA | NA | TPC (4.63 to 6. 24 mg GAE/100 g fw), TFC (1.43 to 2.35 mgQE/100 g fw). | [97] |
| | | fruits | DM (80 to 89.8%), EE (24.45 to 30.13%), | NA | SFA (8.82 to 12.16%), MUFA (28.71 to 34.18%), PUFA (54.58 to 59.08%) | TPC (46.1 to 138.7 mg GAE/100 g dw), TAC (1.4 to 50.3 μmol TE/g dw), TFC (8.01 to 135.2 mg CE/100 g dw). | [98] |
| *Ericaceae* | Cranberry (*Vaccinium macrocarpon*) | leaves | DM (93.11%), CP (6.63%), EE (2.52%), CF (20.15%), ash (3%), | Cu (2.18 mg/kg), Fe (114.6 mg/kg), Mn (448.9 mg/kg), Zn (33.88 mg/kg) | NA | NA | [99] |
| *Fabaceae* | Alfalfa (*Medicago sativa*) | plant pellets | CP (17.06%), EE (1.31%), CF (26.21%), ash (11.89%), | Cu (7.26 mg/kg), Fe (2318 mg/kg), Mn (57.12 mg/kg), Zn (31.24 mg/kg), | SFA (29.73 g/100 g), MUFA (9.87 g/100 g), PUFA (59.72 g/100 g), n-3 (41.97 g/100 g), n-6 (17.75 g/100 g). | TPC (31.24 mg GAE/g), TAC (17.78 mM Trolox), luetin and zeaxantin (28.53 mg/kg), vitamin E (53.17 mg/kg). | [100] |
| | Pea (*Pisum sativum*) | seeds | DM (93.1 to 90.82%), CP (20.51 to 23.80%), EE (2.19 to 2.63%), CF (9.14 to 11.24%), ash (3.16 to 3.72%), carbohydrates (50.86 to 56.54%), | NA | NA | α-tocopherol (10.9 to 13.3 mg/100 g). | [101] |
| | Lentil (*Lens culinaris*) | seeds | DM (94.4 to 92.12%), CP (20.5 to 25.5%), EE (0.78 to 1.25%), ash (2.59 to 3.40%), CF (20.90 to 29.11%), carbohydrates (64.3 to 69.8%), sugars (2.47 to 3.08 g/100 g FW), | NA | SFA (15.4 to 20.8%), MUFA (22.8 to 33.9%), PUFA (45.3 to 63.7%). | Tocopherols (6.46 to 10.1 mg/100 g fw). | [102] |

**Table 1.** *Cont.*

| Plant Family | Plant Common Name | Plant Part | Reported Proximal Composition | Reported Mineral Composition | Reported Lipid Composition | Reported Bioactive Compounds | Reference |
|---|---|---|---|---|---|---|---|
| *Fabaceae* | Mung Bean (*Vigna radiata*) | seeds | CP (17.36 to 24.89 g/100 g), EE (4.24 to 12.18 g/100 g), ash (2.78 to 3.53 g/100 g), | NA | NA | TPC (2.87 to 3.81 mg GAE/g), TFC (1.44 to 3.52 mg RE/g). | [103] |
| *Juglandaceae* | Walnuts (*Juglans regia*) | meal | DM (92.88%), CP (29.47%), EE (16.24%), CF (18.41%), ash (3.88%), | Cu (19.66 mg/kg), Fe (225.35 mg/kg), Mg (72.57 mg/kg), Zn (185.21 mg/kg), | NA | TAC (32.77 mM Trolox), vitamin E (69.32 mg/kg), lutein and zeaxanthin (2.90 mg/kg), phenolic acids (971.27 mg/100 g), Flavonoids (169.02 mg/100 g). | [99] |
| *Lamiaceae* | Spearmint (*Mentha spicata*) | leaves | DM (914 g/100 g dw), CP (2.3 g/100 g dw), EE (0.4 g/100 g dw), ash (1.7 g/100 g dw), | NA | NA | Carbohydrates (9.6 g/100 g DW), TPC (76.32 mg/g). | [104] |
| | Basil (*Ocimum basilicum*) | leaves | DM (91.35%), CP, 22.53%), EE, (1.51%), CF (12.22%), ash (14.12%), | Cu (27.69 mg/kg), Fe (624.51 mg/kg), Mn (78.46 mg/kg), Zn (54.63 mg/kg), | SFA (40.52%), MUFA (21.99%), PUFA (36.57%) | TPC (21.53 mg GAE/g), TAC (42.66 mM Trolox), vitamin E (291.71 mg/kg), lutein and zeaxanthin (267.91 mg/kg). | [24] |
| | Thyme (*Thymus vulgaris*) | leaves | DM (91.65%), CP (15.38%), EE (2.09%), CF (17.08%), ash (9.43%), | Cu (7.41 mg/kg), Fe (690.05 mg/kg), Mn (96.11 mg/kg), Zn (31.74 mg/kg), | SFA (43.86%), MUFA (11.98), PUFA (43.19%), | TPC (31.73 mg GAE/g), TAC (54.09 mM Trolox), vitamin E (379.37 mg/kg), lutein and zeaxanthin (535.79 mg/kg). | [24] |
| | Sage (*Savia officinalis*) | leaves | DM (90.64%), CP (9.56%), EE (3.15%), CF (27.92%), ash (10.36%), | Cu (7.89 mg/kg), Fe (732.72 mg/kg), Mn (68.92 mg/kg), Zn (38.87 mg/kg), | SFA (38.79%), MUFA (19.70%), PUFA (40.96%). | TPC (38.87 mg GAE/g), TAC (19.91 mM Trolox), vitamin E (148.07 mg/kg), lutein and zeaxanthin (99.89 mg/kg) | [24] |
| | Rosemary (*Rosmarinus officinalis*) | leaves | CP (5.35%), EE (3.62%), CF (22.25%), ash (6.61%), | Ca (0.14 mg/g), Fe (37.14 mg/g), Mn (2.20 mg/g), Zn (3.12 mg/g), | NA | TPC (53.42 mg GAE/g), vitamin E (15.61 mg/g), lutein and zeaxanthin (7.63 mg/g). | [105] |
| *Liliaceae* | Garlic (*Allium sativum*) | bulb | DM (41.9%), monosaccharides (2.6%), disaccharides (39.4%). | Ca (468 mg/kg dw), Fe (30 mg/kg dw), Zn (17.1 mg/kg dw), P (2825 mg/kg dw), Mg (540 mg/kg dw). | NA | TPC (3.7 mg GAE/g dw), TAC (8.3 mg GAE/g dw), DPPH (0.090 μMol TE/g dw). | [54] |
| | | leaves | DM (13.31%), CP (29.25 g/100 g), ash (10.36 g/100 g), CF (38.86 g/100 g), EE (3.18 g/100 g), carbohydrates (57.22 g/100 g), | NA | NA | Ascorbic acid (9.65 mg/100 g fw), TPC (241.73 mg CGA/g fw), TAC (29.51 μmol TEAC/g fw). | [106] |
| | Onion (*Allium cepa*) | waste | DM (6.3 to 51.9%), CP (2.3 to 15.6%), ash (4.4 to 10.6%), | Ca (1.8 to 30.7 mg/g), Mg (0.6 to 1.5 mg/g), Fe (19.6 to 888.9 μg/g), Zn (14.9 to 53.8 μg/g), Mn (6.5 to 28.8 μg/g), Se (0.03 to 0.93 μg/g), K (4.2 to 15.9 mg/g) | NA | TPC (9.4 to 52.7 mg GAE/g), TFC (7 to 43.1 mg QE/g). | [107] |
| | Asparagus (*Asparagus officinalis*) | whole plant | CP (2.2 g/100 g), EE (0.12 g/100 g), CF (2.1 g/100 g), sugars (1.9 g/100 g), | Ca (24 mg/100 g), Cu (0.19 mg/100 g), Fe (2.14 mg/100 g), Mg (14 mg/100 g), Mn (0.158 mg/100 g), K (202 mg/100 g), Se (2.3 μg/100 g), Na (2 mg/100 g), Zn (0.54 mg/100 g). | NA | Vitamin B1 (0.143 mg/100 g), vitamin B2 (0.141 mg/100 g), vitamin B3,(0.978 mg/100 g), vitamin B9, (52 μg/100 g), vitamin C (5.6 mg/100 g), vitamin E (1.13 mg/100 g), vitamin K (41.6 μg/100 g). | [108] |
| *Musaceae* | Plantain (*Musa paradisiaca*) | peel | DM (95.62%), ash (6.17%), CP (3.97%), CF (8.36%), EE (3.01%), carbohydrate (74.12%), | Cu (1.35 mg/100 g), Fe (5.06 mg/100 g), Mn (10.38 mg/100 g), Zn (11.60 mg/100 g), Ca (17.85 mg/100 g), Mg (49.32 mg/100 g), Na (58.16 mg/100 g), K (38.22 mg/100 g), P (22.64 mg/100 g). | NA | NA | [109] |
| | Banana (*Musa* spp.) | peel | DM (90.17%), ash (9.56%), CP (3.23%), CF (12.67%), EE (0.89%), carbohydrate (63.82%), | Cu (0.59 mg/100 g), Fe (7.89 mg/100 g), Mn (1.25 mg/100 g), Zn (13.30 mg/100 g), Ca (14.70 mg/100 g), Mg (45.21 mg/100 g), Na (76.88 mg/100 g), K (26.14 mg/100 g), P (28.95 mg/100 g). | NA | NA | [109] |
| *Oleaceae* | Olive (*Olea europaea*) | seeds | EE (30.4%), CP (17.2%), CF (47.6%), insoluble fibre (32.7%), ash (2.67%), carbohydrates (2.13%), | K (5579.0 mg/kg), Na (2758.2 mg/kg), Ca (2615.4 mg/kg), Mg (1878.5 mg/kg), P (745.5 mg/kg), Fe (12.8 mg/kg), Zn (45.6 mg/kg), Mn (31.5 mg/kg). | SFA (12.34%), MUFA (62.78%), PUFA (24.63%), | α-tocopherol (401 mg/kg), campesterol (72.7 mg/kg), stigmasterol (53.9 mg/kg), β-Sitosterol (1674.9 mg/kg), lanosterol (10.5 mg/kg), cycloartenol (109.7 mg/kg), citrostadienol (17.2 mg/kg), 24-methylenecycloartanol (365.3 mg/kg). | [110] |

**Table 1.** *Cont.*

| Plant Family | Plant Common Name | Plant Part | Reported Proximal Composition | Reported Mineral Composition | Reported Lipid Composition | Reported Bioactive Compounds | Reference |
|---|---|---|---|---|---|---|---|
| Rosaceae | Apple (*Malus domestica*) | peel | NA | K (695.3 to 980.9 mg/100 g dw), Ca (35.6 to 61.2 mg/100 g dw), Mg (18.5 to 65.9 mg/100 g dw), Na (2.9 to 7.3 mg/100 g dw), Zn (0.4 to 1.2 mg/100 g dw), Fe (1.1 to 2.4 mg/100 g dw). | NA | TPC (1907.5 to 2587.9 mg GA/100 g dw), TFC (1214.3 to 1816.4 mg catechin eq./100 g dw). | [111] |
| | Apple (*Malus domestica*) | pulp | NA | K (490.1 to 790.1 mg/100 g dw), Ca (19.8 to 36.7 mg/100 g dw), Mg (15.6 to 34.8 mg/100 g dw), Na (5.9 to 10.8 mg/100 g dw), Zn (0.2 to 0.9 mg/100 g dw), Fe (0.8 to 2.1 mg/100 g dw). | NA | TPC (1185.2 to 1475.5 mg GA/100 g dw), TFC (711.8 to 999.3 mg catechin eq/100 g dw). | [111] |
| | Rosehip (*Rosa canina*) | meal | DM (92.37%), CP (10.53%), EE (4.48%), CF (49.53%) | NA | PUFA (67.65 g/100 g), MUFA (22.80 g/100 g), SFA (9.55 g/100 g), n-3 (14.28 g/100 g), n-6 (53.07 g/100 g). | TPC (60.23 mg GAE/g), TAC (23.87 mM Trolox), TFC (12.18 mg eq rutin/g). | [6] |
| | Rosehip (*Rosa canina*) | seeds | DM (989.7 g/100 g fw), CP (2.99 g/100 g dw), EE (6.29 g/100 g dw), ash (1.64 g/100 g dw), carbohydrate (89.07 g/100 g dw) | NA | NA | TPC (2554 µg/g), carotenoids (2.92 µg/g), vitamin C (1798 µg/g). | [112] |
| | Raspberry (*Rubus* spp.) | leaves | DM (92.3 g/100 g dw), CP (19.54 g/100 g dw), EE (2.06 g/100 g dw), CF (18.18 g/100 g dw), ash (5.14 g/100 g dw), | Cu (4.09 mg/kg dw), Fe (172.5 mg/kg dry dw), Mn (75.23 mg/kg dry dw), Zn (46.14 mg/kg dry dw), Ca (0.66 g/100 g dw), P (0.28 g/100 g dw). | SFA (33.35%), MUFA (19.88%), PUFA (46.64%), n-3 (66.53%), n-6 (34.48%), | Lutein (261 mg/kg), zeaxanthin (1040 mg/kg), astaxanthin (38.52 mg/kg), canthaxanthin (1.12 mg/kg), vitamin E (149.7 mg/kg), TPC (26.19 mg GAE/g), TFC (10.6 mg/g). | [113] |
| | Blackberry (*Rubus* spp.) | leaves | DM (91.66 g/100 g dw), CP (18.37 g/100 g dw), EE (1.89 g/100 g dw), CF (20.48 g/100 g dw), ash (6.2 g/100 g dw), | Cu (6.51 mg/kg dw), Fe (115.6 mg/kg dry dw), Mn (80.63 mg/kg dry dw), Zn (23.81 mg/kg dry dw), Ca (1.01 g/100 g dw), P (0.29 g/100 g dw). | SFA (24.56%), MUFA (10.31%), PUFA (64.20%), n-3 (74.51%), n-6 (50.80%), | Lutein (547.1 mg/kg), zeaxanthin (3041 mg/kg), astaxanthin (38.52 mg/kg), canthaxanthin (3.04 mg/kg), vitamin E (179.9 mg/kg), TPC (14.57 mg GAE/g), TFC (5.961 mg/g). | [113] |
| | Strawberry | leaves | CP (80.63 mg/g) | NA | NA | TPC (108.83 mg GAE/g), TFC (10.25 mg QE/g), carotenoids (0.0074 mg/g dw). | [114] |
| | Aronia (*Aronia melanocarpa*) | fruits | DM (91.76%), CP (1.53%), EE (4.17%), CF (8.29%), ash (2.01%), carbohydrates (75.78%), | Fe (72.93 mg/kg), Mn (4.54 mg/kg), Zn (6.67 mg/kg). | SFA (10.78%), MUFA (23.08%), PUFA (66.13%), n-3 (2.04%), n-6 (64.09%) | NA | [115] |
| | Aronia (*Aronia melanocarpa*) | leaves | DM (90.92%), CP (10.11%), EE (6.75%), CF (13.33%), ash (7.82%), carbohydrates, (52.93%), | Fe (94.29 mg/kg), Mn (205.48%), Zn (20.13 mg/kg). | SFA (30.20%), MUFA (9.16%), PUFA (67.14%), n-3 (29.99%), n-6 (30.50%). | NA | [115] |
| | Aronia (*Aronia melanocarpa*) | pomace | DM (94.9%), CP (5.25%), EE (2.51%), CF (14.3%), ash (2.41%), carbohydrates (70.44%), | Fe (94.27 mg/kg), Mn (15.3 mg/kg), Zn (10.54 mg/kg). | SFA (12.09%), MUFA (20.21%), PUFA (67.14%), n-3 (3.45%), n-6 (63.69%). | NA | [115] |
| Rutaceae | Orange (*Citrus × aurantium*) | peel | CP (4.85%), EE (1.10%), CF (9.70%), ash (2.95%), | Zn (4.74 mg/kg). | NA | Vitamin E (100.5 mg/kg), lutein and zeaxanthin (81.52 mg/kg), TPC (8.035 mg GAE/g), TAC (238.51 mmoli/kg eq. vitamin C), TAC (231.33 mmoli/kg eq. vitamin E). | [116] |
| | Grapefruit (*Citrus × paradisi*) | peel | CP (5.39%), EE (0.95%), CF (11.82%), ash (3.51%), | NA | NA | Vitamin E (89.93 mg/kg), lutein and zeaxanthin (36.46 mg/kg), TPC (12.162 mg GAE/g), Zn (5.41 mg/kg), TAC (238.25 mmoli/kg eq. vitamin C), TAC (227.75 mmoli/kg eq. vitamin E). | [116] |
| | Lemons (*Citrus × limon*) | peel | CP (9.42%), CF (15.18%), EE (4.98%), ash (6.26%). | Na (755.5 mg/100 g), K (8600 mg/100 g), Ca (8452.5 mg/100 g), Cu (4.94 mg/100 g), Mg (1429.5 mg/100 g), Zn (13.94 mg/100 g), Fe (147.65 mg/100 g), P (6656.25 mg/100 g). | NA | NA | [117] |
| Solanaceae | Tomato (*Solanum lycopersicum*) | waste | DM (95.19%), CP (13.58%), EE (3.53), CF (43.6%), ash (3.59%), | NA | PUFA (57.31 g/100 g), n-6 (53.08 g/100 g), n-3 (4.23 g/100 g). | Astaxanthin (0.076 mg/kg), lutein (3.57 mg/kg), zeaxanthin (0.78 mg/kg), cantaxanthin (0.27 mg/kg), lycopene (105.38 mg/kg), β-carotene (9.50 mg/kg). | [50] |
| | Bell Pepper (*Capsicum annuum*) | fruit | DM (7.8 g), CP (0.99 g), EE (0.30 g), ash (0.47 g), CF (2.1 g), carbohydrates (6.03 g). | Na (4 mg), K (211 mg), Ca (7 mg), Mg (12 mg), P (26 mg), | NA | Niacin (0.979 mg), pyridoxine (0.291 mg), vitamin C (127.7 mg), vitamin E (1.58 mg), TPC (4.51 to 52.65 mg GAE/g), TFC (2.1 to 41 QE mg/g), carotenoids (1219 to 8800 µg/g). | [118] |
| | Eggplant (*Solanum melongena*) | fruit | CP (0.86 g/100 g fw), EE (0.05 g/100 g fw), Ash (0.56 g/100 g fw), carbohydrates (3 g/100 g fw), total sugars (3 g/100 g fw), | NA | SFA (83.8%), MUFA (4.5%), PUFA (11.8%). | NA | [119] |
| | Eggplant (*Solanum melongena*) | pulp | CP (0.78 g/100 g fw), EE (0.04 g/100 g fw), Ash (0.56 g/100 g fw), carbohydrates (2.89 g/100 g fw), total sugars (2.89 g/100 g fw). | NA | SFA (89.6%), MUFA (5.5%), PUFA (4.89%). | NA | [119] |

**Table 1.** *Cont.*

| Plant Family | Plant Common Name | Plant Part | Reported Proximal Composition | Reported Mineral Composition | Reported Lipid Composition | Reported Bioactive Compounds | Reference |
|---|---|---|---|---|---|---|---|
| *Vitaceae* | Grape (*Vitis vinifera*) | pomace | DM (89.92%), CP (12.33%), EE (5.95%), CF (35.17%), ash (2.83%), | NA | SFA (30.06 g/100 g), MUFA (42.63 g/100 g), PUFA (66.60 g/100 g), n-3 (1.12 g/100 g), n-6 (65.48 g/100 g), | TPC (26.65 mg GAE/g), TAC (148.35 mM TE/g 148.35). | [120] |
| | | seed meal | DM (91.85%), CP (12.9%), CF (7.22%), | NA | SFA (12.30 g/100 g), MUFA (20.39 g/100 g), PUFA (67.14 g/100 g), n-3 (0.68 g/100 g), n-6 (66.45 g/100 g). | TPC (90.42 mg GAE/g), TAC (496 mM Trolox), TFC (100.08 μg rutin/g). | [6,121] |
| *Zingiberaceae* | Ginger (*Zingiber officinale*) | rhizome | DM (10.86 to 15.84 g/100 g), CP (0.93 to 1.05 g/100 g), carbohydrate (97.19 to 97.26 g/100 g), EE (0.52 to 0.55 g/100 g), CF (1.01 to 1.05 g/100 g), ash (0.16 to 0.28 g/100 g) | NA | NA | Gingerol (5.54 to 6.11 mg/100 g). | [122] |

NA—not determined.

Another innovative approach in poultry feeding practices is the use of feed additives and supplements designed to enhance feed efficiency and animal health. Pliego et al. [123] reviewed the beneficial effects of medicinal and herbal plants, while other authors explored the uses of legumes [124,125], fruit pomaces and co-products [126,127], and other unexplored plants by/co-products [128,129] are nowadays used to improve digestion, nutrient absorption, and gut health. These additives can be particularly beneficial when incorporating alternative feed ingredients, as they help mitigate potential nutritional imbalances and enhance the bioavailability of nutrients, resulting in products with improved nutritional quality, which can further provide nutritious and healthier affordable food products. Sustainable feeding practices are also gaining attention as producers seek to reduce the environmental footprint of poultry production [130]. This involves optimizing feed formulations to minimize waste and enhance nutrient utilization, as well as adopting practices that promote resource efficiency. It was shown recently by Lefter et al. [131] that locally sourced ingredients that match the dietary supply of the birds' nutritional requirements can be a sustainable approach. Such a strategy will also contribute to reducing transportation emissions and environmental pollution.

These evolving practices are directly aligned with the SDG2 of Zero Hunger by adopting innovative feeding strategies, ensuring the availability of nutritious and affordable poultry products. These strategies not only enhance the animal's production performances and the quality of food but also support the sustainability of food systems, making a substantial impact on global food security.

## 4. Innovative Feeding Strategies for Sustainable Poultry Production

Sustainable feeding strategies in poultry production focus more on the use of alternative feedstocks to overcome the challenges of traditional feed practices, such as high cost, input restrictions, and environmental impact. These alternatives can replace conventional ingredients such as corn and soybean meals [132] and include plant substitutes such as wastes from legumes, oilseeds, plants, and various agricultural co-/by-products. Furthermore, some recent review reports showed that other unconventional sources like insects [133], algae [134], and food waste [135] are being investigated for their potential as sustainable alternative poultry feed ingredients. The benefits of using plant-based and by-product feeds are manifold. First, these innovations can significantly reduce feed costs, making chicken production economically viable, especially for smallholder farmers with limited resources. Secondly, many of these substitutes are by-products of other agricultural practices, which means they can be obtained cheaply and even contribute to reducing waste in the feed system. Moreover, additional feed supplementation to poultry feed can reduce competition between feed and food crops, resulting in more balanced agricultural sustainability [136].

Nutritionally, many alternative feed ingredients offer a diverse array of essential nutrients that can meet the dietary needs of poultry. For example, legumes and oilseeds are rich in proteins and amino acids, while by-products provide valuable energy, fibers, and numerous beneficial bioactive compounds [42]. Babatunde et al. [136] recently stated that the use of alternative feeds can enhance the resilience of poultry production systems. By diversifying their feed base, producers can reduce their vulnerability to market fluctuations and supply chain disruptions that commonly affect the availability and price of conventional feed ingredients. Other authors [137,138] showed that this resilience is particularly important in the context of global food security, as it helps ensure a steady supply of poultry products even in times of economic or environmental stress. All in all, such feeding strategies that incorporate alternative feed ingredients hold significant promise for advancing the goals of Zero Hunger and sustainable agriculture.

Furthermore, by reducing costs, minimizing environmental impacts, and improving the nutritional profile of poultry diets and final products, these feeding strategies can enhance the sustainability and productivity of poultry farming, aligning also to the principles of the 3 R's, which can now be considered as 5 R's (Figure 2).

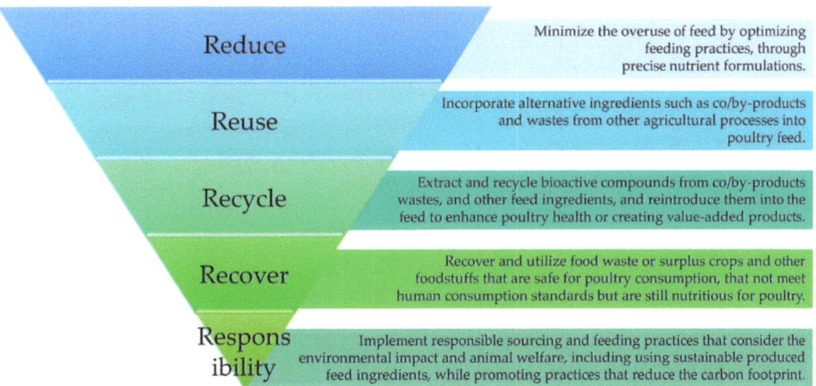


**Figure 2.** The principle of the 5 R's for alternative sustainable poultry production system.

### 4.1. Plant-Based as Alternative Feed Ingredients

Plant-based feeds have long been a cornerstone of poultry feed, primarily due to their availability, low cost, and high nutritional value. Traditional plant-based feeds, such as corn, barley, wheat, and soybean meals, are used as main sources of energy and protein, making them ideal for supporting fast growth and productivity in poultry [135]. But as demand for these crops increases worldwide for human consumption and animal feed, competition for agricultural products has intensified, as noted by Frona et al. [139]. This has raised concerns about sustainability as they rely heavily on these foods. In response to these challenges, researchers and industry stakeholders are looking for plant-based alternatives that are not only nutritious but also sustainable. These alternatives take less resources to manufacture, which are usually obtained as commodities resulting from other agricultural methods. For example, lupins, faba beans, chickpeas, and peas [140] are gaining attention as potential protein sources due to their high protein content and low environmental impact compared to soybeans. Similarly, oilseed meals like canola, flax, and rapeseed meal [87], which are by-products of oil extraction processes, offer substantial protein and fat content and can partially replace soybean meals in poultry diets.

Another promising approach is the use of agricultural residues and by-products as feed ingredients. These include rice bran, wheat bran, and other cereal by-products that are commonly discarded or used in low-value products [141]. These products can be recycled as poultry feed, contributing to a circular economy, which will result in reducing waste and improving food production. The challenge of using such by-products is in their variable nutrient content and potential presence of anti-nutritional compounds [142], which requires careful formulations and processing when it comes to poultry nutritional needs to ensure that they do not compromise health or performance.

In addition to diversifying the sources of plant-based feed ingredients, there is ongoing research into enhancing the nutritional quality of these alternatives. Some techniques, such as fermentation, enzyme supplementation, and genetic modification, are being explored to improve the digestibility and nutrient availability of plant-based feeds, as revived by Samtiya et al. [143]. The shift towards more sustainable plant-based feed alternatives also aligns with broader environmental goals, as they have a lower carbon footprint compared to animal-based feeds [144], and their production often requires less water and land. This makes them an attractive option for reducing the environmental impact of poultry farming, particularly in regions facing resource constraints. Nevertheless, the transition to these alternative plant-based feeds is not without challenges. Economic factors and challenges, like the cost and availability, the variability in nutrient content, and the presence of anti-nutritional factors of these ingredients, can be significant barriers to their worldwide implementation. Despite these challenges, the continuous research and development efforts in this area are paving the way towards more sustainable and resilient poultry production systems.

### 4.2. Insect-Based as Alternative Feed Ingredients

Over the past 10 years, insect-based feed has emerged as one of the most promising poultry alternative protein sources due to its high nutritional value, low environmental footprint, and potential to contribute to a circular economy. As the demand for sustainable livestock feed increases, insects such as black soldier fly larvae, mealworms, and crickets are increasingly recognized for their ability to efficiently convert organic wastes into high-quality protein sources [145,146]. Insects are highly nutritious, with protein content ranging from 30% to 80% depending on species and growth stage. These are rich in essential amino acids, fats, and minerals, making them a viable alternative to traditional protein sources such as fish meal and soybean meal [145]. For example, black soldier larvae have a balanced amino acids profile as reported by Khalifa et al. [133], while insect feed can provide essential fatty acids such as n-3 and n-6, which are essential for poultry health and productivity. Insects also have the advantage of being able to efficiently convert organic waste into biomass. They can be raised on a variety of natural materials, such as agricultural residues, food waste, and even garbage, making them an integral part of the waste management process. Not only does this diminish the environmental impact of waste disposal but also provides a valuable by-product of the process in the form of insect proteins [147]. By recycling waste into feed ingredients, insect farming supports the principles of a circular economy, reduces reliance on traditional feed crops, and reduces the overall carbon footprint of poultry production, as mentioned by Chavez et al. [148]. Despite the clear benefits, there are challenges associated with the use of insect-based feeds in poultry diets. One of the main barriers is the regulatory framework governing the use of insects in animal feed. In many regions, the approval process for insect-based feed ingredients is still in its early stages, and there are strict regulations regarding the substrates that can be used for insect rearing [149]. However, in a recent study conducted by Żuk-Gołaszewska et al. [150], it was reported that *H. illucens, M. domestica, T. molitor, A. diaperinus, G. sigillatus, A. domesticus,* and *G. assimilis* insect species fulfil safety conditions as insect production for feeding purposes, according to European Commission Regulation no. 893/2017, European Parliament Regulation no. 999/2001, and of the European Council and Commission Regulation no. 142/2011. This approval is important and ensures the safety and quality of insect meals in terms of contaminants that can potentially be transferred to the poultry. Another challenge is the scalability and economic viability of insect farming. While insect farming is less resource-intensive compared to conventional agriculture, Madau et al. [151] reported that it requires significant investment in infrastructure and technology to be scaled up to a level where it can meet the demands of the poultry industry. Other authors [152] mentioned that the cost of insect protein is currently higher than that of traditional feed ingredients, although this is expected to decrease as the industry matures and production processes become more efficient. All in all, ongoing research is exploring ways to optimize insect farming practices, improve the nutritional quality of insect meals, and assess the long-term effects of insect-based feeds on poultry health and productivity. Studies have shown that insect protein can successfully replace a significant portion of conventional protein sources in poultry diets without negatively affecting growth performance or feed efficiency [133,145]. In some cases, the inclusion of insect protein has been associated with improved gut health and immune function in poultry [153,154], suggesting potential additional benefits beyond basic nutrition.

### 4.3. Co-/By-Products and Wastes as Alternative Feed Ingredients—Nutritional Composition and Bioactive Compounds

The nutritional content of alternative poultry feed ingredients is an important factor in determining their suitability for poultry feed. Poultry species require a balanced diet of proteins (16 to 18% for laying hens; 20 to 23% for broilers), essential amino acids such as lysine (0.9 to 1.1% for laying hens; 1.1 to 1.3% for broilers), methionine (0.4 to 0.5% for laying hens; 0.5 to 0.6% for broilers), threonine (0.6 to 0.8% for laying hens, 0.8 to 1.0% for broilers), and tryptophan (0.2% for both hens and broilers), carbohydrates (50 to 60% of the total diets), fats (3 to 6% of the total diet), vitamins such as A (7500 to 12,000 IU/kg), D3 (2000 to 3000 IU/kg), and E (10 to 20 mg/kg)and minerals like Ca (3.5 to 4.5% for laying hens; 1% for broilers) and P (0.45 to 0.5% for both hens and broilers) to maintain optimal health, growth and productivity. These nutritional requirements are mentioned in each hybrid management breeding guide. The chemical composition of alternative feeds varies greatly depending on the source, but in general, these products provide essential nutrients that support poultry health and yield. A comprehensive understanding of the nutritional value of these new ingredients is essential for composing a balanced diet. These components include protein, fiber, fat, a mixture of vitamins and minerals, and several bioactive compounds, as shown in Table 1. After an extensive literature review, we classified several sources, compounds, and plant and vegetable food wastes based on their genus and identified them as potential sources of new alternative feed ingredients for poultry nutrition.

The literature revealed that the *Apiaceae* family includes plants, such as carrots, parsnips, celery, fennel, coriander, parsley, dill, anise, and lovage, which are known for their rich medicinal properties and potential bioactive compounds. Carrot (*Daucus carota*) waste exhibits a wide range of nutrients in different areas, including high levels of carotenoids such as β-carotene and α-carotene, especially in orange, black, and purple varieties, which also exhibit strong antioxidant activity [33,51]. Carrot flour and leaves contain significant amounts of TPC and TFC, contributing to their nutritional value [52,53]. Parsnip roots (*Pastinaca sativa*) are notable for their high dry matter content, monosaccharides, disaccharides, and minerals. They also contain substantial TPC and antioxidant capacities, enhancing their health benefits [54]. Celery (*Apium graveolens*) in its bulbs, leaves, and roots offers high levels of protein, fiber, and the essential minerals, with significant amounts of choline, pantothenic acid, and rutin in the leaves contributing to its resistance against infection [55–58]. Fennel (*Foeniculum vulgare*) seeds and shoots contain proteins, fats, carbohydrates, and essential fatty acids like PUFA and other essential fatty acids; the seeds also contain vitamins B1, B2, and niacin [59,60]. Anise (*Pimpinella anisum*) seeds contain protein and fiber content and are rich in TPC, carotenoids, tannins, and various phenolic compounds, making them more nutritious, as reported by Sun et al. [61] and Ghosh et al. [62]. In addition, the leaves and seeds of coriander (*Coriandrum sativum*) contain high levels of protein, fiber, essential minerals, and vitamins, with the leaves containing high levels of vitamin C and the seeds containing high levels of MUFA and PUFA [63,64]. Parsley (*Petroselinum crispum*) leaves are particularly rich in protein, fiber, vitamins, and phenolic compounds, which confer significant antioxidant properties, according to Cornescu et al. [65] and Dobricevich et al. [66]. In addition, the leaves and stems of dill (*Anethum graveolens*) contain large amounts of vitamin C, carotenoids, β-carotene, and TPC [67], while the leaves and stems of lovage (*Levisticum officinale*) provide essential proteins, carbohydrates, and fatty acids, with significant amounts of n-3 and n-6 fatty acids [68,69].

Another important alternative feed ingredient is represented by the fruits of American and European elderberries. These unexplored sources are rich in total phenolics, exhibit significant antioxidant capacity, and contain various essential nutrients such as glucose, fructose, proteins, and a variety of minerals, including K, P, Ca, and Mn [70]. They also contain notable amounts of anthocyanins, which contribute to their health benefits. However, they also have trace amounts of toxic compounds like sambunigrin [70,71]. The flowers of European elderberry are also rich in phenolics and proteins and exhibit strong antioxidant activity and minerals like Ca, Mn, Fe, Cu, and Zn [71,72].

Jerusalem artichoke (*Helianthus tuberosus*) are tubers from the *Asteraceae* family that are rich in carbohydrates and inulin, making them a significant source of dietary fiber. They are also a good source of Ca and K with high TPC content [73]. The Jerusalem artichoke leaves vary widely in protein content and are high in fiber. They are also rich in minerals (Ca, Mg, and K) and contain notable amounts of carotenoids [74,75]. Chicory (*Cichorium intybus*) roots have high inulin content (44.69%) and provide significant amounts of carbohydrates (89.41%). The roots are also rich in phenolic acids, such as chlorogenic and caffeic acids, which contribute to their antioxidant properties [76]. The leaves of chicory contain a notable amount of TPC and a variety of phenolic compounds, including protocatechuic and caffeic acids [76]. The chicory seeds are rich in protein (18.61%) and fat (21.18%) and contain high levels of Fe, Cu, and Zn, along with a modest amount of vitamin C [77]. Some authors [78,79] revealed that calendula (*Calendula officinalis*) flowers are high in TPC and TFC, with notable levels of tocopherols, containing significant amounts of carbohydrates and organic acids. Similarly, dandelion (*Taraxacum mongolicum*) has high antioxidant activity in its flower, leaves, and stem, with TPC values ranging from 23.89 to 30.05 mg GAE/g dry weight. The flowers and leaves are also rich in vitamin C and carotenoids, with notable levels of Ca and Mg [80]. Another important agricultural crop that produces waste is lettuce (*Lactuca sativa*). Although the results presented by Mampholo et al. [81] show variation in antioxidant content across different varieties, they present important bioactive compounds suitable as alternative feed ingredients in poultry [81,82]. Moreover, artichoke (*Cynara cardunculus*), particularly the receptacle and heads, were reported as rich sources of inulin and various minerals. The receptacle has high dry matter and protein content, while the heads contain a balance of fats, carbohydrates, and antioxidants [83,84]. Lastly, another important crop with high protein and fat content and a notable amount of PUFA in both wild and commercial varieties is the yarrow (*Achillea millefolium*). The commercial inflorescences have higher fat and protein levels, contributing to their nutritional profile [85].

In the *Arecaceae* family, coconut (*Cocos nucifera*) shell waste was identified as a source of high crude fiber (32.39%) and carbohydrate content (52.63%). Ewansiha et al. [86] showed that despite its low protein and fat levels, the shell is rich in Fe (618 mg/100 g), making it a significant source of this mineral.

The *Brassicaceae* family includes several nutritionally significant plants. Rapeseed (*Brassica napus*) is notable for its various components, such as meals, seeds, and cake. The meal is a rich source of protein and healthy fats, including significant amounts of MUFA and PUFA, and it contains valuable antioxidants and phenolic compounds [5]. Rapeseed seeds offer a substantial amount of crude protein and fat, with a noteworthy balance of n-3 and n-6 fatty acids. These seeds also provide essential minerals and fiber, making them a nutritious food source [87]. The cakes, which are by-products of oil extraction, are rich in protein and healthy fats, as well as essential fiber and minerals [87]. The kale (*Brassica oleracea*) leaves are another member of the Brassicaceae family, known well for having numerous essential nutrients (dietary fiber, vitamins, and minerals, including vitamin C and beta-carotene), and exhibit strong antioxidant properties due to their high phenolic content [88]. Similarly, mustard seeds (*Brassica juncea*) were reported by Oancea et al. [89] to contain high protein and fat content and are particularly rich in essential minerals such as Fe, Mg, and Zn. These seeds contain a variety of beneficial antioxidants and vitamins, contributing to their health-promoting properties. Commonly discarded cabbage (Brassica oleracea) waste is high in protein, fiber, and various minerals [90], making it a valuable source of nutrients.

Another important by-product belongs to the *Cannabaceae* family, which includes hemp (*Cannabis sativa*), whose seeds are highly nutritious, containing a balanced profile of proteins and fats, including a high proportion of PUFA. Mierlita et al. [91] reported recently that they offer significant amounts of dietary fiber and antioxidants, such as tocopherols and carotenoids, suitable for poultry nutrition.

In the *Cucurbitaceae* family, cucumbers (*Cucurbita pepo*) yield edible parts, such as flesh, peel and seeds, which are suitable for poultry feed, each with their own unique nutritional value. The flesh is low in calories but rich in vitamins and minerals. However, the peel contains high levels of protein, fiber, and antioxidants [92]. Cucumber (*Cucumis sativus*) and watermelon (*Citrullus lanatus*) also belong to this genus, where cucumber stands out for its high-water content and various mineral and vitamin properties in organic matter, extract, and juice, which are important [93,94]. Pumpkin seed meal (*Cucurbita moschata*) is also a rich source of protein, fat, and beneficial n-3 fatty acids [95].

Sea buckthorn (*Hippophae rhamnoides*) from the *Elaeagnaceae* family, whose leaves and fruits are rich in protein, fat, and antioxidants, contains important minerals and vitamins, making them highly nutritious and suitable for poultry nutrition [31,96]. Additionally, other important co-products are the leaves and flowers of *Elaeagnus angustifolia*, which were reported by Saboonchian et al. [97] as sources rich in phenolic and flavonoid compounds. The fruits of this plant provide a good source of fats, including a balanced ratio of SFA and UFA, along with significant amounts of antioxidants [98].

In a recent study, conducted by Untea et al. [99], it was revealed that the *Ericaceae* family cranberry (*Vaccinium macrocarpon*) offers significant nutritional benefits, particularly in its leaves. These coproducts contribute to a robust nutrient profile that includes protein, fat, fibers, and essential minerals such as Fe, Mn, and Zn.

The alfalfa plant, scientifically known as *Medicago sativa* and belonging to the *Fabaceae* family, is well known for its remarkable nutritional value. The alfalfa plant is abundant in protein and fiber, and it also has significant levels of healthy fats, such as a high proportion of PUFA. Alfalfa is additionally a beneficial supplier of antioxidants, such as tocopherols and carotenoids [100]. Pea (*Pisum sativum*) seeds are a beneficial legume, offering a good amount of protein and important nutrients such as tocopherols. Peas provide a beneficial combination of carbs, fiber, and fats [101]. Lentil seeds, containing high levels of protein and fiber, as well as a notable quantity of PUFA, are a healthy option for nutrition. Lentils have sugars and tocopherols that enhance their nutritional worth [102].

Mung bean (*Vigna radiata*) seeds are rich in protein and fat, and they offer antioxidant benefits through their TPC and TFC [103].

One important by-product from the *Juglandaceae* family includes walnuts (*Juglans regia*), whose meal is highly nutritious, contains protein and essential lipids, with a significant amount of antioxidants, including phenolic acids and flavonoids, as well as essential minerals, such as Mg and Zn, and has a high TAC [99].

The *Lamiaceae* family encompasses several aromatic herbs with notable nutritional profiles. Spearmint (*Mentha spicata*) leaves are characterized by their high dry matter content and significant levels of fiber and phenolic compounds [104]. Basil (*Ocimum basilicum*) leaves are rich sources of nutrients such as protein, fiber, and essential minerals and contain a balanced profile of fatty acids along with substantial antioxidant activity [24]. Thyme (*Thymus vulgaris*) leaves also have a high TPC and TAC, with a substantial amount of essential minerals and beneficial fatty acids [24]. Sage (*Salvia officinalis*) leaves are another member of this family with a high protein and fiber content and contain significant levels of antioxidants, including vitamin E and lutein [24]. Rosemary (*Rosmarinus officinalis*) leaves are noted for their high antioxidant content, including phenolic compounds and vitamin E, although they contain relatively lower levels of protein and fat compared to other herbs [105].

The *Liliaceae* family includes garlic (*Allium sativum*), which is valued for both its bulb and leaves. The bulb is noted for its high content of dry matter and minerals, alongside significant amounts of monosaccharides and disaccharides. The antioxidant properties of garlic are also noteworthy, with appreciable TPC and TAC [54]. In contrast, garlic leaves exhibit high protein content and crude fiber, along with notable antioxidant levels including TPC and TAC [106]. Onion (*Allium cepa*) waste provides a broad range of nutrients, with variability in dry matter, protein, and mineral content such as K, Ca, and Mg. The antioxidant properties are also significant, with variations in TPC and TFC as reported by Benítez et al. [107]. Further, another important crop, asparagus (*Asparagus officinalis*), offers a diverse nutritional profile, including fibers, sugars, proteins, and a variety of vitamins such as B and C vitamins and essential minerals like Ca and Mg [108].

In the *Musaceae* family, important food wastes are produced by plantain and banana peels. Plantain peel is rich in carbohydrates and contains essential minerals, including K and

Mg, with moderate levels of protein and fat [109]. Similarly, banana peel is characterized by its high carbohydrate content and significant levels of fiber and minerals, though with slightly lower protein and fat content compared to plantains [109].

The *Oleaceae* family includes olive (*Olea europaea*) seeds, which are rich in lipids, especially MUFA and PUFA. The seeds also provide various sterols and essential minerals, contributing to their nutritional richness as alternative feed ingredients [110].

Within the *Rosaceae* family, the nutritional profiles of apple (*Malus domestica*) and rosehip (*Rosa canina*) are distinct. Apple waste (peels, seeds, and pulp) is rich in TPC and TFC, with varying levels of K, Ca, and Mg [111]. Rosehip meal and seeds offer a high dry matter content and are rich in antioxidants, including high levels of PUFA, with substantial amounts of vitamin C and carotenoids [6,112]. The co-products of raspberry and blackberry (*Rubus* spp.) leaves have been recently shown to have a high dry matter content and are rich in proteins, fibers, and essential minerals. They also have significant levels of antioxidants and carotenoids [113]. Strawberry leaves are noted for their protein content and high TAC [114]. Another important crop that recently gained attention are the aronia (*Aronia melanocarpa*) fruits, leaves, and pomace, which were described by Saracila et al. [115] to be rich in dry matter and provide a balanced profile of nutrients including carbohydrates, proteins, and fats, along with high levels of Fe, Mn, and Zn.

Further, in the *Rutaceae* family, citrus fruits are prominent for their diverse nutritional benefits. *Citrus aurantium*, or orange, peels are known for having high levels of total polyphenol content (TPC) and large total antioxidant content (TAC), as expressed in terms of both vitamin C and E equivalents. The peel is also high in lutein and zeaxanthin, which are critical for eye health, and has a little amount of crude protein, fat, and fiber [116]. Similarly, the peel of grapefruits (*Citrus paradisi*) displays a strong TPC and TAC pattern, but it also has somewhat greater crude protein and fiber content than orange peel. Important antioxidants found in grapefruit peel include vitamin E and carotenoids like lutein and zeaxanthin [116]. Conversely, lemons (*Citrus × limon*) offer a more significant nutritional profile, including high levels of fiber, protein, and important minerals like Mn, Ca, and K. The high nutritional density of lemon peel, particularly with regard to minerals, underscores its potential to augment dietary consumption of these components [117].

As further highlighted by Panaite et al. [50], tomato (*Solanum lycopersicum*) waste is a member of the *Solanaceae* family and is rich in crude protein and fiber. It is also notable for its high amount of PUFA and carotenoids, such as lycopene and beta-carotene, which are known for their antioxidant effects. Waste from bell peppers (*Capsicum annuum*) is a good source of carotenoids and vitamins C and E. The varying TPC and TFC across different bell pepper samples underline its antioxidant potential [118]. Eggplant (*Solanum melongena*), both in its fruit and pulp forms, has relatively low levels of protein and fat but provides a balanced ratio of lipids [119].

Within the *Vitaceae* family, grape (*Vitis vinifera*) pomace and seed meal are the most important and studied sources rich in both PUFA and antioxidants. The pomace is high in crude protein and fiber, with a significant amount of fatty acids and a balanced profile of SFA, MUFA, and PUFA [120]. The seed meal, which contains high dry matter content, shows substantial antioxidant activity and a similar fatty acid profile, though with a higher proportion of unsaturated fats [121].

Lastly, in the *Zingiberaceae* family, ginger (*Zingiber officinale*) rhizome, known for its aromatic and medicinal properties, is primarily composed of carbohydrates and moisture, while its nutritional content is complemented by trace amounts of protein, fat, and fiber, and it contains gingerol, a compound with known health benefits [122].

All these reviewed plants, by-products, co-products, and plants, containing various amounts of nutrients and bioactive compounds, have been reported as safe and with multiple benefits when used in poultry diets, as presented further in Table 2.

## 5. Effects of Reviewed Alternative Feed Ingredients on Poultry Performance, Health Status, and Product Quality

In recent studies, various plants from different botanical families have been explored for their effects on poultry performance, health, and product quality. Each plant species tested by researchers showed unique impacts and effects depending on the poultry hybrid, dosage, and part or type of the plant used, as summarized in Table 2.

The reviewed alternative feed ingredients from the *Apiaceae* family demonstrate significant potential in improving the health and productivity of poultry through enhanced nutrient intake, better egg and meat quality, and improved immune responses. Carrot waste has shown positive effects on both laying hens and broilers. In laying hens, carrot waste was effective in improving internal and external egg quality [50]. Similarly, carrot leaf increased production performances and nutrient digestibility, as reported by Siti et al. [122], as well as egg quality. Anise seed, according to the results of Barakat et al. [155], was effective in improving the immune system of broilers and meat quality. For broiler chickens, coriander seed at a 1.5% dose was most effective in improving dressing percentage and the overall health status of the broilers [156]. Parsley leaves tested in laying hens raised under heat stress conditions improved production performances and antioxidant compounds in eggs [65], as well as the quality characteristics of eggs during 28 days of storage. In broiler chickens, parsley leaves in higher doses significantly increased feed intake and carcass quality and health parameters, according to Ali et al. [157]. The dill leaves in lower doses used in broiler chickens significantly improved production performance and improved lipid metabolism [158,159].

The reviewed alternative feed ingredients from the *Asteraceae* family exhibit various beneficial effects on poultry health and performance. The incorporation of these plants and their by-products into poultry diets can lead to improved growth rates, better immune responses, enhanced egg quality, and improved health status of poultry, aligning with sustainable feeding strategies and contributing to improved food quality and security. Jerusalem artichoke was evaluated for its effect on broiler chickens' diets, which significantly improved production performance, as reported by Al-Abboodi et al. [160]. Chicory has demonstrated several beneficial effects in broiler chickens, showing a significant reduction in abdominal fat pad and improved health status [161]. In laying hens, free access to chicory vegetation resulted in better production performance and improved lipid composition in the eggs [162]. Calendula flower supplement in broilers improved carcass yield; however, more than 1% could affect production performances [163]. In layers, petal and leaf supplements significantly increased carotenoid deposition in egg yolks, with no effect on egg production or quality characteristics [164]. Dandelion leaves and meal in different broiler hybrids and laying hens was demonstrated to be effective in improving product quality (meat and eggs) as well as production performance and health status [165–167]. However, in laying hens, a 4% dandelion meal had detrimental effects on feed consumption and egg weight, as shown by Saenz et al. [168]. Similar effects were reported for echinacea supplementation in broiler chickens, demonstrating its potential in improving health and performance [169].

Ginseng was not very effective when tested in laying except for a notable increase in egg production, according to Kang et al. [170], however, had a notable effect on the health status of laying hens.

Rapeseed has been extensively studied as an alternative feed ingredient for poultry. In broilers, rapeseed can be used up to 30% without detrimental effects on production performances, while in laying hens' lower levels < 20% are preferred [171]. The same effect was noted by others in laying hens [172]. In terms of egg quality, rapeseed meal had a significant impact on n-6 PUFA and health-related indices [5].

According to Mustafa et al. [8], broccoli waste up to 9% was suitable for broiler chickens' diets, with significant effects on production performance and nutrient digestibility. In a different study with broiler chickens, broccoli stem and leaf meal had no effect on growth performance but significantly increased antioxidant compounds in meat samples and enzyme activity [173]. For laying hens, broccoli stem and leaf meal did not affect production performance but significantly increased the xanthophyll content in egg yolk and decreased cholesterol content [174]. Further, cabbage waste in broiler chickens had no effect on production performance but improved nutrient digestibility [175]. In laying hens, cabbage waste up to 12% significantly improved egg quality, however, might negatively impact eggshell percentage [176]. These findings suggest that incorporating by-products from the *Araliaceae* and *Brassicaceae* families into poultry diets can enhance production performance, immune responses, and nutrient digestibility and health, contributing to sustainable feeding strategies and improved food quality.

Hemp seed by-products also have shown promising results when included in poultry diets. In broiler chickens, hemp seed cake significantly improved the fatty acid profile in the thigh and breast meat and health status but had no significant effects on performance, as shown by Tufarelli et al. [177]. In laying hens, significant effects were reported on both egg quality and performance [91].

Pumpkin (*Cucurbita pepo*) seed meal up to 20% inclusion levels in broiler chickens significantly increased production performance, with no effect on commercial parts [178]. The watermelon (*Citrullus lanatus*) rind in laying might have some detrimental effects on egg weight, but the overall health status was significantly improved [126]. However, further research is required. Pumpkin (*Cucurbita moschata*) seed meal significantly improved shelf-life and egg quality parameters in laying hens [95].

In broiler chickens, dietary supplements with cranberry leaves stimulated the deposition of bioactive compounds in meat samples and exhibited a strong effect in counteracting oxidative processes in broiler meat [99,128]. A 30% cranberry pomace in broilers diet significantly affected nutrient absorption but improved the plasma lipid profile [179]. These findings support the potential of these plant by-products from the *Cannabaceae*, *Cucurbitaceae*, and *Ericaceae* families as sustainable alternative feed ingredients in poultry nutrition, contributing to improved food quality and Zero Hunger initiatives.

Alfalfa (*Medicago sativa*) from the *Fabaceae* family has been extensively studied for its beneficial effects in poultry diets as a dietary fiber source. In broiler chickens, up to 5% was demonstrated to be effective in improving antioxidant compounds in meat except carotenoids content while maintaining the production performances [100,180]. In laying hens, 5 to 10% dietary alfalfa meal significantly decreased FCR, mortality, abdominal fat yield, and egg yolk cholesterol content while demonstrating potential health benefits [181,182]. Pea (*Pisum sativum*) has also been explored for its potential as an alternative protein and energy source. In broiler chickens, 4 to 48% raw pea was used as a replacement for soybean meal and corn, which had no detrimental or additional effects on meat quality, as mentioned by Dotas et al. [183]. Another group of authors reported significant effects on meat quality and significant improvements in health-related indices [184].

In a study by Untea et al. [128], walnut meal, when included in the diets of broiler chickens, improved meat lipid and nutritional composition; however, some minerals' ability to deposit in the tissue was affected [99]. These plant-based feed ingredients from the *Fabaceae*, *Cucurbitaceae*, and *Juglandaceae* families offer promising alternatives to traditional feed ingredients, improving meat and egg quality, enhancing health parameters, and contributing to sustainable poultry production practices.

Further, several plants from the *Lamiaceae* family, including peppermint, spearmint, basil, thyme, rosemary, and sage, have been investigated for their effects on broiler chicken diets. Each of these plants showed varying impacts on production performance, meat quality, and health parameters. Peppermint significantly improves oxidative stability and immune responses [185], while others showed its potential effects on production performance [186]. Spearmint also influenced the performance of broiler chickens while reducing cholesterol levels and increasing hemoglobin and superoxide dismutase activity, as showed by Abu Isha et al. [187]. Basil and thyme, when included in the diet of broilers, had similar effects. Both herbs decreased production performance; however, they significantly improved the deposition of bioactive compounds in breast and thigh meat samples [24,180]. Moreover, rosemary in broilers improves intestinal health, with potential negative effects above 1.5% inclusion on production performance [188,189]. Sage leaves had similar effects as basil and thyme [24,180]. Overall, these findings highlight the potential of *Lamiaceae* plants to modulate various aspects of broiler chicken health and production. While some herbs like peppermint and spearmint can improve production performance at optimal inclusion levels, others like basil, thyme, and sage offer more substantial benefits in enhancing meat quality, particularly in terms of antioxidant and lipid profiles. However, the inclusion rates are critical, as higher levels can sometimes lead to decreased performance metrics.

Avocado seed meal reduced BW without affecting FCR, as showed in the study by George et al. [190], while another group of authors [191] reported that using up to 8% seed meal found no significant effect on broilers health, indicating that avocado seed meal does not negatively affect broilers' health.

Garlic (*Allium sativum*) was found to have varying effects based on its preparation and dosage. In laying hens, garlic had no effect on productivity and egg characteristics but significantly improved the bird's health status [192]. Raw garlic powder slightly improved broiler performances but influenced meat aroma; however, boiled garlic powder did not show these beneficial effects, indicating that the method of preparation plays a critical role in the efficacy of garlic as a supplement [193]. Onion (*Allium cepa*) demonstrated similar effects to garlic in laying hens.

Plantain peel waste is a suitable maize substitute in broiler finisher diets, with a recommended inclusion rate of up to 10% for optimal blood characteristics [194].

Olive leaf powder in laying hens had no significant effect on production performances but improved egg yolk quality, making olive leaf powder a potential agent for reducing egg yolk cholesterol [195]. In broilers, olive pulp waste improved footpad dermatitis and feather cleanliness without affecting growth performance or health [196].

Rosehip (*Rosa canina*) showed promising effects in laying hens by improving TAC and TPC in PUFA-enriched eggs while extending the shelf life of stored eggs [6]. In broiler chickens, low levels of rosehip fruits improved BW, FI, protein intake, and dressing percentages, as reported by Monesa et al. [197]. Another recent study showed that the leaves of rosehip, which are co-products, can be used as feed additives in the first stage of laying hens and could potentially improve the production performance and some egg quality parameters [70].

Citrus waste, such as orange (*Citrus sinensis*) and grapefruit (*Citrus paradisi*) peels, was tested in broiler chickens, and showed promising effects on performances, meat quality, and health status [116]; however, grapefruit peel decreased final BW.

Tomato wastes up to 7.5% enriched egg yolk quality; however, more than 7.5% level depressed the absorption and deposition of n-3 fatty acids in the yolk, indicating that optimal dosing is critical [34]. The 5% inclusion rate showed the best effect in this study.

The grape seed meal from the *Vitaceae* family was beneficial for broilers, significantly improving performance, meat quality, and health [121]. In laying hens, improved production performance and antioxidant compounds in eggs, although n-6 fatty acids were more prevalent than n-3 [5]. Recently, Costa et al. [198] concluded that in poultry, the effect of grape by-products is more variable, and these sources should not be incorporated in broiler diets at more than 6–10% to prevent an impairment of animal growth.

Ginger (*Zingiber officinale*) powder, when tested in broilers, had no effect on production performance; however, it reduced gizzard weight, indicating potential benefits for gut health [199].

These findings highlight the diverse impacts of alternative feed ingredients on poultry, emphasizing the importance of selecting the appropriate dosage and plant part to achieve desired outcomes in production performance, health status, and product quality, as detailed in Table 2.

**Table 2.** Main effects of plants when administered in poultry diets.

| Plants Family | Plants Common Name | Poultry Species | Dose and Type | Main Effect on Laying Hens and/or Broilers | Reference |
|---|---|---|---|---|---|
| *Apiaceae* | Carrot (*Daucus carota*) | Lohmann Brown, laying hens | 2% carrot waste | Increased ADFI and had no significant effect on FCR. Increased the carotenoids content in egg yolk positively affected the physical properties (e.g., yolk pH, egg thickness). Decreased the cholesterol content in eggs and significantly improved oxidative stability. | [50] |
| | | Lohmann Brown laying hens | 2% carrot leaf | Increased ADFI, egg production, and feed efficiency. Better nutrients digestibility of DM, OM, and CP. The supplement improved egg yolk color, eggshell thickness, β-carotene and cholesterol contents of the eggs. | [122] |
| | Anise (*Pimpinella anisum*) | Cobb 500 broiler chickens | 0.5%, 1%, and 1.5% seed | Improved BW and FI of broilers, but higher FCR. The serum IgA, IgG, IL-2 and IL-10, IgM, and INF-γ of broilers were significantly increased. MDA levels decrease in breast and thigh samples with all three doses. | [200] |
| | Coriander (*Coriandrum sativum*) | Ross 308 broiler chickens | 1.5%, 2.5%, and 3.5 seed | No effect on growth performances. The 1.5% dose was the most effective in improving dressing percentage and health status of broilers. | [155] |
| | Parsley (*Petroselinum crispum*) | Tetra SL-LL laying hens | 2% leaves | Increased production performances. Egg yolk color and the antioxidant compounds in eggs were significantly improved (antioxidant capacity, vitamin E, lutein, and zeaxanthin) and quality characteristics of the eggs during storage. Affected Fe and Zn content in egg yolks. | [65] |
| | | Ross 308 broiler chickens | 3, 6, 9, and 12 g/kg leaves | Significant increase in BW at 9 and 12 g/kg of parsley supplement, improved FI, carcass weight and dressing percentage. However, the FCR was higher at 3 and 6 g/kg of supplements. | [201] |
| | | Ross 308 broilers | 0.5%, 1%, and 1.5% leaves | Significant improvement in broiler health parameters at 1% and 1.5% leaves were tested. The 0.5% was not very effective on any of the parameters. | [157] |
| | Dill (*Anethum graveolens*) | Ross 308 broiler chickens | 1%, 2%, and 3% leaves | The diet with 1% significantly improved the production performance. The diets with 2% and 3% were not very effective in any aspects. Using 1% significantly reduced cholesterol, triglyceride, and LDL levels in serum. | [158,159] |
| *Asteraceae* | Jerusalem artichoke (*Helianthus tuberosus*) | Ross 308 broiler chickens | 0.5%, 1%, 1.5% and 2% | Except for the 1% diet, all other three significantly improved production performances; however, the 2% was reported as the best option for Ross 308 broilers. | [160] |
| | Chicory (*Cichorium intybus*) | Ross 308 broiler chickens | 0.10%, 0.15%, and 0.20% | The BW of broilers fed the 0.10% chicory was significantly higher than those fed on the other treatments. The abdominal fat pad was significantly lower in all chicory groups. Blood triglycerides and LDL levels were reduced significantly while HDL increased significantly. Decreased the counts of *Escherichia coli* and increased the counts of *Lactobacillus*. | [161] |
| | | Lohmann Brown laying hens | Free-access chicory | Better production performances than control group. The fatty acids composition of the hens fed freely with chicory vegetation contributed to the production of eggs with higher PUFA and favorable n-6 to n-3 ratio. | [162] |
| | Calendula (*Calendula officinalis*) | Ross 308 broilers | 0.5% and 1.0% flower | The 0.5% flower improved BW and carcass yield, while 1% led to significantly lower BW. None of the tested doses exerted effect on immune parameters. | [163] |
| | | Hy-Line Brown layers | 0.5%, 1%, 2%, and 4% petal | Significant increase in egg yolk color; 4% supplement had the best egg yolk carotenoid deposition. No improvements in egg production or egg weight. The egg quality characteristics were not influenced. | [164] |
| | | | 1% leaves | Significant increase in egg yolk color, without any alterations on egg freshness or egg quality characteristics. | [164] |

**Table 2.** *Cont.*

| Plants Family | Plants Common Name | Poultry Species | Dose and Type | Main Effect on Laying Hens and/or Broilers | Reference |
|---|---|---|---|---|---|
| *Asteraceae* | Dandelion (*Taraxacum mongolicum*) | Arbor Acres broilers | 1% raw or enzymatically treated plant | Enzymatically treated significantly decreased FCR, increased apparent nutrient digestibility of nutrients. Same diet led to higher breast muscle rate, lower drip loss and water-holding capacity while lowered drip loss and water holding capacity. Improved serum IgA and IgG. Raw dandelion had no notable effect except for increased organic matter digestibility. | [165] |
| | | Arbor Acres broilers | 0.5% and 1% dandelion | The 0.5% and 1% increased eviscerated percentage, cooking loss and shear force. The 0.5% significantly increased the n-3 PUFA, while significantly lowered the TBARS values. | [166] |
| | | Cobb 273 broilers | 0.5% leaves | Significantly improved cecal health by reducing harmful bacteria. No effect on pH and moisture content of chicken litter. | [167] |
| | | Hisex laying hens | 1%, 2%, 3%, and 4% meal | The diet with 3% was the most effective in improving egg production and intestinal histomorphology parameters. The 4% showed detrimental effects on FI and egg weight. | [168] |
| | Echinacea (*Echinacea purpurea*) | Hubbard broiler chickens | 0.5% powder | Dietary supplementation with 0.5% echinacea improves the final BW and immune response of broiler chickens and had a significant effect on *Escherichia coli*, and hematological, serum biochemical adverse effects, and histopathological alterations that occur by E. coli infection. | [169] |
| *Araliaceae* | Ginseng (*Panax ginseng*) | Hy-Line Brown laying hens | 0.5% and 1% by-product | No effect on production performances, except for egg production which was significantly increased. No effect on triglyceride, aspartate aminotransferase, and alanine aminotransferase however, increased the serum IgG and IgM content. Improved the proliferation of intestinal *Lactobacillus* population but no effect on *Salmonella* and *Escherichia coli*. | [170] |
| *Brassicaceae* | Rapeseed (*Brassica napus*) | Cobb 500 broilers | 10% and 30% seed | A total of 30% rapeseed reduced BW and FI, cecal colonization and fecal shedding. Histomorphology showed that 30% had the highest duodenum and jejunum villus height and to crypt depth ratio. A total of 10% was more effective on production performance and laying hens' health. | [171] |
| | | Brown Nick laying hens | 20% and 30% cakes | Rapeseed cakes up to 20% have no adverse effects on productivity or egg quality. Both diets reduced digestibility of dry matter, gross energy, crude protein and the digestibility of indispensable amino acids except tryptophan, was reduced also reduced. | [172] |
| | | Tetra SL LL laying hens | 9% meal | The effect on production performance was not significant, only tendencies were noted. Also, no effect on egg weight and its components. The diet with 9% rapeseed meal had a significant effect on n-6 PUFA, as well as health related indices. | [5] |
| | Broccoli (*Brassica oleracea*) | Ross 508 broiler chickens | 3, 6, and 9% waste | All diets increased BW and FCR with no effect on FI. Apparent ileal crude protein and dry matter digestibility increased as the level of broccoli waste in the diet increased in the grower phase, however, in the finisher phase (>35 days) decreased. The 3 to 6% broccoli waste may improve the growth of broiler chickens with no detrimental effects on nutrient digestibility and retention while the 9% may affect ileal and total tract nutrient digestibility. | [8] |
| | | Ross 308 broiler chickens | 4, 8, and 12% stem and leaf meal | No effect on broilers growth performances. Significantly increased the yellowness in shank and breast skin, and the concentrations of xanthophylls in abdominal fat and breast skin. The TAC was improved significantly while lowering the MDA concentration. The activities of superoxide dismutase and catalase of breast muscle increased with 8% and 12% broccoli leaf and stem supplementation. | [173] |

**Table 2.** *Cont.*

| Plants Family | Plants Common Name | Poultry Species | Dose and Type | Main Effect on Laying Hens and/or Broilers | Reference |
|---|---|---|---|---|---|
| *Brassicaceae* | Broccoli (*Brassica oleracea*) | Roman brown shell laying hens | 3%, 6%, and 9% stem and leaf meal | No effect on production performance of laying hens. Significantly increased the xanthophyll content of egg yolk, while decreased the content of cholesterol in egg yolk. No effect on egg quality characteristics (albumen height, Haugh unit, shell thickness and shell strength of eggs). Hepatic hydroxymethylglutaryl-coenzyme A (HMG-CoA) reductase activity was decreased, and concentrations of cecal short chain fatty acids were increased with increasing broccoli stem and leaf meal supplementation. | [174] |
| | Cabbage (*Brassica oleracea var. capitata*) | Ross 508 broiler chickens | 3, 6, and 9% waste | Cabbage waste had no effects on BW, FI, or FCR. Inclusion of cabbage waste reduced apparent ileal dry matter, organic matter and crude protein in the grower phase, but no effect on finisher phase. Up to 9% had no negative impact on bird performance and apparent ileal digestibility and improved apparent total tract nutrient digestibility. | [175] |
| | | White Leghorn laying hens | 4, 8, and 12% waste | No effect on FI, egg production, FCA, egg yolk, and albumen percentage; however, the eggshell percentage decreased with increasing waste supplement. The α-tocopherol, PUFA and linolenic acid increased in egg yolks with increasing dietary waste, with no effect on egg yolk cholesterol concentration. It can be used up to 12% without adverse effects on production parameters and may improve total tract nutrient utilization and egg quality. | [176] |
| *Cannabaceae* | Hemp (*Cannabis sativa*) | Hubbard broiler chickens | 5% and 10% cake | Regardless of the level of hemp seed cake inclusion, no differences among groups were found for performance and meat quality traits. The thigh and breast fatty acid profile were significantly improved in both groups, with an increase of the long chain fatty acids of n-3 series and decrease in n-6/n-3 ratio. The MDA concentration and lipid hydroperoxides in breast meat decreased significantly. The tested diets improved intestinal health status in broilers. | [177] |
| | | Tetra SL laying hens | 8% hemp seed | The production performances in laying hens were improved using hemp seeds. The egg yolk had significantly lower cholesterol and SFA content, while the concentration of total and individual (ALA, EPA, and DHA) PUFA (n-6 and n-3 FAs) was significantly higher. | [91] |
| *Cucurbitaceae* | Pumpkin (*Cucurbita pepo*) | Anak 2000 broiler chickens | 5, 10, 15, and 20% seed meal | All four-level used led to significantly increased production performances in broilers, however significant differences among dietary treatments for live BW and dressed weight, were noted. Dressing percent, breast, thigh, abdominal fat, kidney, gizzard, liver and lungs weights did not differ significantly as the levels of pumpkin seed meal increased in the diets. | [178] |
| | Watermelon (*Citrullus lanatus*) | Tetra SL laying hens | 1% rind | Regarding the production performance, egg weight significantly decreased and improved FCR. The health status of animals was significantly improved by decreasing the concentration of cholesterol and triglycerides from blood samples. The intestinal histomorphology was improved, while alpha-amylase decreased in both duodenum and jejunum. The 1% supplement increased the *Firmicutes* and *Lactobacillus* spp. while reducing the counts of Bacteroidetes and Enterobacteriaceae. | [126] |
| | Pumpkin (*Cucurbita moschata*) | Tetra SL laying hens | 9% seed meal | Significantly improved ADFI but had no effect on egg quality characteristics. The egg yolk cholesterol decreased significantly while the total PUFA, especially the n-3 ALA and DHA, increased significantly. The shelf-life on eggs stored at room and refrigerator temperatures, was better than those from the control group. | [95] |
| *Ericaceae* | Cranberry (*Vaccinium oxycoccus*) | Cobb 500 broiler chickens | 1% and 2% leaves | The supplements led to higher concentrations of Cu and Fe deposition in breast meat samples compared with the control group. The lutein and zeaxanthin concentrations were also higher in the meat samples, while vitamin E concentrations decreased. The leaves stimulated the synthesis of n-3 PUFA and exhibited a powerful effect in counteracting the oxidative processes of broilers meat. | [99,128] |

**Table 2.** *Cont.*

| Plants Family | Plants Common Name | Poultry Species | Dose and Type | Main Effect on Laying Hens and/or Broilers | Reference |
|---|---|---|---|---|---|
| *Ericaceae* | Cranberry (*Vaccinium oxycoccus*) | Ross 708 broiler chickens | 30% pomace | The pomace had a significant effect for apparent retention of dry matter, nitrogen, neutral detergent fiber, gross energy, and apparent metabolizable energy. The plasma concentration of bile acid and cholesterol significantly decreased. | [179] |
| *Fabaceae* | Alfalfa (*Medicago sativa*) | Cobb 500 broiler chickens | 5% meal | Significantly increased FCR, thigh muscles and gizzard weights. Improves meat quality by enhancing antioxidant potential and n-3 PUFA, while significantly decreasing cholesterol content in breast and thigh meat samples. Significantly increased the TPC and vitamin E content in meat samples, however no effect on TAC and carotenoids content in meat. | [100,180] |
| | | Beijing-you laying hens | 5%, 8%, and 10% meal | Significantly decreased FCR, mortality, abdominal fat yield, and yolk cholesterol content. All diets improved meat and eggs protein quality. The diets with alfalfa stimulate the proliferation of beneficial bacteria in both duodenum and ileum, showing their potential in boosting health status. Up to 10% was recommended as the optimal inclusion level. | [181] |
| | | Zhuanghe Dagu chickens | 3%, 6%, and 9% meal | Dietary inclusion of alfalfa meal was beneficial to improve the laying performance, egg quality, small intestinal morphology, cecal microbiota diversity and cecal metabolic function, with the optimum dose being 6%. | [182] |
| | Pea (*Pisum sativum*) | Ross 308 broiler chickens | 4%, 8%, 12%, 16%, 18%, 24%, 36%, and 48% raw pea | The doses were used as a soybean meal and corn replacement. The production performances (BW, FI, and FCR) were not significantly altered. Also, carcass yield traits, skin color, and chemical composition of meat samples were not affected. Some significant differences were observed in fatty acid composition of breast and leg muscles. Up to 48% can be used as an alternative protein and energy source to replace soybean meals and corn in broiler chicken diets. | [183] |
| | | Hubbard broiler chickens | 19% and 40% dehulled peas | No significant effect on growth performance, dressing percentage, the percentage of breast or drumstick muscles, and abdominal fat. Significantly lower L* (lightness) and b* (yellowness, drumstick muscle) values and fat content. The PUFA concentration in breast and drumstick muscles was significantly increased, while lowering the n-6/n-3 ratio. Significant improvement in health-related indices. | [184] |
| *Juglandaceae* | Walnuts (*Juglans regia*) | Cobb 500 broiler chickens | 6% meal | Increased content of crude fat, Cu, vitamin E, and deoxymyoglobin in breast meat samples, while significantly decreased the Fe, Zn, and concentrations metmyoglobin. The content n-3 PUFA was double in samples of chickens fed walnut meal compared with the control samples. | [99,128] |
| *Lamiaceae* | Peppermint (*Mentha x piperita*) | Ross 308 broiler chickens | 1% and 2% plant | From the production performances, a significant effect was noted for final BW in broilers. Also, both experimental diets showed an antioxidative potential to improve oxidative stability and immune responses in broilers. | [185] |
| | | Ross 308 broiler chickens | 0.5%, 1%, and 1.5% leaves | The BW and FI increased corelated to the level of the leaves added, while FCR decreased in the same way. No effect on meat characteristics or organs development. A dose of 1.5% was recommended to improve production performances. | [186] |
| | Spearmint (*Mentha spicata*) | Arbor Acres broiler chickens | 0.25%, 0.5%, 1%, and 2% plant | The level of 2% spearmint significantly decreased BW, BWG, FCR and cholesterol levels while significantly increased the concentrations of hemoglobin and superoxide dismutase activity compared with the other groups. The 1% group had significantly higher concentration of total plasma lipid and TAC. | [187] |
| | Basil (*Ocimum basilicum*) | Cobb 500 broiler chickens | 1% plant | The production performances decreased, especially final BW and FI, while increasing the FCR. However, 1% basil led to deposition of significantly higher concentration of Zn, TPC, TAC, and vitamin E in thigh meat samples compared with the control group, while significantly improving the n-3 fatty acids deposition. The cholesterol concentration in breast samples was significantly lowered and the TPC and TAC were increased. | [24,180] |

**Table 2.** *Cont.*

| Plants Family | Plants Common Name | Poultry Species | Dose and Type | Main Effect on Laying Hens and/or Broilers | Reference |
|---|---|---|---|---|---|
| Lamiaceae | Thyme (*Thymus vulgaris*) | Cobb 500 broiler chickens | 1% plant | The production performance decreased, but not significantly. The 1% thyme significantly increased the concentration of Zn, TPC, TAC, and vitamin E in thigh meat samples compared with the control group, while significantly improving the n-3 fatty acids deposition. The cholesterol concentration in breast samples was significantly lowered and the antioxidant compounds were increased. | [24,180] |
| | Rosemary (*Rosmarinus officinalis*) | Ross 308 broiler chickens | 0.5% and 1% powder | No effect on antibody titters against viruses nor lymphoid tissues weight but has the potential to modulate the humoral immunity of broilers. | [188] |
| | | Ross 308 broiler chickens | 0.5%, 1%, and 1.5% plant | No significant effect on production performances. The gastrointestinal tract weight, relative to body weight, increases in all rosemary groups when compared with control. The intestinal health was significantly improved by increasing the Lactobacilli counts and decreased the *Escherichia coli*. A dose of more than 1.5% could have significant detrimental effects on production performance. | [189] |
| | Sage (*Savia officinalis*) | Cobb 500 broiler chickens | 1% plant | The final BW and FI were significantly decreased. The concentration of Zn, TPC, TAC, and vitamin E, and n-3 fatty acids deposition in thigh meat samples were significantly increased. The cholesterol concentration in breast samples was significantly lowered and the antioxidant compounds were increased. Significant alteration in breast meat color and texture quality were obtained. | [24,180] |
| Lauraceae | Avocado (*Persea americana*) | Cobb 500 broiler chickens | 0.5%, 1%, and 1.5% seed meal | The BW was decreased by 1.5% seed meal inclusion, with no effect on FCR. The Mg level in the serum of same group was significantly increased, while urea was significantly higher with 0.5% seed meal. The recommended level was 0.5%. | [190] |
| | | Cobb 500 broiler chickens | 2%, 4%, 6%, and 8% seed meal | The results showed no significant difference in the hematological parameters except lymphocytes and mean cell hemoglobin except at the normal range, which implied that the inclusion level of seed meal in broiler diets had no negative effect on the birds up to 8% inclusion level that was studied. It also has positive effect on the total protein, albumin, and globulin. | [191] |
| Liliaceae | Garlic (*Allium sativum*) | Bovan Brown layers | 0.5% and 1% | No significant effect on egg weight and FI but had a significant improvement in the number of eggs, egg production, egg mass, and FCR. The egg quality characteristics were not influenced, but the cholesterol content was significantly decreased in eggs. The general health status also improved significantly. | [192] |
| | | Shaver Starbo chicks | 0.5% and 5% raw powder | Raw garlic powder marginally improved BW, with the highest weight gain observed at the 5% level. While the carcass and organ characteristics were not significantly affected by the garlic supplementation, the abdominal fat content was significantly reduced. Additionally, the garlic aroma and palatability scores of the meat improved with higher levels of dietary garlic, with thigh muscle exhibiting the highest garlic aroma score. | [193] |
| | | Shaver Starbo chicks | 0.5% and 5% boiled powder | Boiled garlic powder did not produce any significant beneficial effects on broiler performance or meat quality. Specifically, there were no notable differences in weight gain, feed intake, or feed conversion ratio, and the carcass and organ characteristics, as well as the moisture content of the meat, were unaffected. | [193] |
| | Onion (*Allium cepa*) | Bovan Brown layers | 1% leaves | No significant effect on egg weight and FI but had a significant improvement in the number of eggs, egg production, egg mass, and FCR. The egg quality characteristics were not influenced, but the cholesterol content was significantly decreased in eggs. General health status was also improved significantly. | [192] |

**Table 2.** *Cont.*

| Plants Family | Plants Common Name | Poultry Species | Dose and Type | Main Effect on Laying Hens and/or Broilers | Reference |
|---|---|---|---|---|---|
| *Musaceae* | Plantain/Banana (*Musa* × *paradisiaca*) | Marshall broilers | 10% and 20% peel waste | No significant differences in the blood parameters except for erythrocyte sedimentation rate, white blood cell count and heterophil values of birds from control versus plantain supplements. Serum biochemical parameters analyzed except for serum globulin, albumin, urea, potassium, phosphorus, and aspartate aminotransferase were similar. It was concluded that plantain peel is a suitable substitute for maize in broiler finisher diet at inclusion rate not beyond 10% for optimal blood characteristics. | [194] |
| *Oleaceae* | Olive (*Olea europaea*) | Lohmann Brown laying hens | 1%, 2%, or 3% leaf powder | No effect on FI, egg weight, egg yield and FCR but increased BW. Yellowness in yolk color was increased without affecting other quality parameters. Yolk cholesterol content tended to decrease by about 10%. Suggested to be used for reducing egg yolk cholesterol content and egg yolk coloring agent in layer diets. | [195] |
| | | Ross 308 broiler chickens | 3% and 6% pulp waste | Improved foot pad dermatitis and feather cleanliness. No effects growth performance or fecal microbiota population. Changes of β-diversity in an age-dependent way were only observed. Both diets beneficially affected chickens' health and welfare. | [196] |
| *Rosaceae* | Rosehip (*Rosa canina*) | Tetra SL laying hens | 1.5% and 3% meal | Rosehip meal in eggs enriched with PUFA exhibited a positive effect on eggs protein and lipids quality. The TAC and TPC in eggs were significantly improved. The 3% dose was more effective on shelf life of eggs stored for 28 days at refrigerator and room temperature. | [6] |
| | | Cobb 500 broiler chickens | 0.10%, 0.20%, and 0.30% fruits | The 0.10% treatment improved BW, FI and protein intake. All treatments reduced percent GIT length of large intestine and ceca at higher supplement value. It was concluded that it can be safely used up to 0.30%, producing higher dressing percentage. | [197] |
| | | Lohmann Brown laying hens | 0.5% and 1% leaves | Significantly improved laying rate, FCR, egg mass, with no effect on egg weight and its components. The chromomeric parameters (L*, a* and b*) were significantly altered, as well as the n-3 and n-6 fatty acids. | [9] |
| *Rutaceae* | Orange (*Citrus sinensis*) | Cobb 500 broiler chickens | 2% peel | Significant increase in BW, health status and the total PUFA. Significantly reduced the oxidation process occurring during storage in thigh meat, and the growth of pathogenic *Escherichia coli* and *Staphylococcus* spp., proving their antimicrobial effect, while the beneficial bacteria, *Lactobacillus* spp. were increased. | [116] |
| | Grapefruit (*Citrus paradisi*) | Cobb 500 broiler chickens | 2% peel | Decreased BW, glucose, cholesterol, and triglyceride from blood serum and the total PUFA. Significantly reduced the oxidation process occurring during storage in thigh meat. Reduced the growth of pathogenic *Escherichia coli* and *Staphylococcus* spp., proving their antimicrobial effect, while the beneficial bacteria, *Lactobacillus* spp. were increased. | [116] |
| *Solanaceae* | Tomato (*Solanum lycopersicum*) | Tetra SL Laying hens | 2.5%, 5%, and 7.5% waste | Improved egg yolk color and carotenoids deposition. The 5% diet increased the oxidative stability of n-3 PUFA enriched eggs. The 7.5% tomato waste has depressed the absorption and deposition of n-3 fatty acids in egg yolk. | [34] |
| *Vitaceae* | Grape (*Vitis vinifera.*) | Hubbard broilers | 2% seed meal | BW, FI, and FCR improved significantly, as well as meat quality. The PUFA were significantly higher in breast samples, while cholesterol content was significantly lower in thigh samples. The plasma glucose, cholesterol and triglyceride levels were significantly lower. | [121] |
| | | Tetra SL laying hens | 3% seed meal | Improved production performances, TAC and TPC in eggs. The fatty acids profile was improved, however, the percentage on n-6 was higher than the n-3. | [5] |
| *Zingiberaceae* | Ginger (*Zingiber officinale*) | Ross 308 broiler chickens | 0.15%, 0.20%, and 0.25% powder | No effect on production performance, however, ginger at all levels resulted in a significant decrease in gizzard weight and abdominal fat. No effect on blood biochemistry and antibody production against sheep red blood cells. The *Lactobacillus* counts in the ileal content of birds fed 0.20 and 0.25% ginger were higher. | [199] |

## 6. Environmental and Economic Impacts and Limitations of the Presented Sustainable Poultry Feeding

### 6.1. Environmental and Economic Impacts of Sustainable Feeding Strategies

The shift towards alternative and sustainable feeding strategies in poultry production is driven not only by the need to improve nutrition and animal welfare but also by the imperative to reduce the environmental and economic impacts associated with conventional poultry farming. As the global demand for poultry products continues to rise, it is essential to adopt practices that minimize the carbon footprint, resource use, and economic costs of production [202]. The studies conducted by Campos et al. [203] and Osorio et al. [204] revealed that the use of agricultural by-products as feed ingredients contributes to a circular economy by recycling waste materials and reducing the environmental burden of waste disposal. This approach helps to lower the overall carbon footprint of poultry production. However sustainable feeding practices have obvious environmental benefits, chicken producers will only embrace them if they can be made economically viable [205]. However, the price of these substitute feeds should drop as the market for sustainable feed ingredients expands and manufacturing techniques advance. Economies of scale, advances in processing technologies, and increased competition in the market will likely drive down prices, making these feeds more accessible to a broader range of producers. Another economic consideration is the potential for alternative feeds to create new markets and revenue streams [206,207].

It is important to achieve a balance between environmental sustainability and economic viability for adopting alternative feeding strategies on a large scale. Moreover, there are surveys [208–210] suggesting that the rising awareness of environmental issues and preference for ethical food leads to increased customer demand for ecologically sustainable poultry products. As a result, such changes in consumer preferences can give rise to market opportunities for farmers who decide to go green in terms of their feed provisions, with possible high prices being charged on their goods.

However, moving away from conventional feeds poses challenges but also presents significant opportunities aimed at reducing the ecological burden from poultry farming as well as strengthening its economic resilience. Consequently, continued investments in research, technology, and policy support will be necessary towards unlocking the full potential of such approaches and making poultry production environmentally friendly in the long run.

### 6.2. Limitations of the Presented Alternative Sustainable Poultry Feeding

Feed control is a very important aspect in poultry diets, especially when incorporating fruit waste or other raw materials, due to their variability in nutritional composition. These materials offer potential nutritional benefits but also carry risks and limitations due to varying levels of anti-nutritive substances. In this regard, multiple levels of control are essential to ensure that the feed is both safe and nutritionally adequate for poultry. Poultry diets, as mentioned before, are composed of a balance of energy sources, proteins, vitamins, minerals, and additives, which all contribute to overall bird health and productivity. For these nutritional reasons, when including unconventional waste materials in poultry diets, several aspects must be carefully evaluated, as explained below.

The nutritional value of any ingredient, especially novel ones like fruit waste, needs to be thoroughly analyzed for proximate composition for basic nutrients (proteins, fats, and fibers), amino acids, fatty acids, and vitamins (like A, C, and E) content, knowing that they can vary in composition based on growing conditions and processing methods [211].

Anti-nutritional compounds present a significant challenge in alternative feed ingredients, especially tannins, saponins, phytic acid, and oxalates found in by-products, which can interfere with nutrient absorption and reduce feed efficiency, inhibit enzyme activity, reduce mineral bioavailability, and impair protein digestion [212].

Digestibility is a key factor when formulating poultry diets, particularly when using waste products. Ingredients with high fiber content, like many fruit peels, may lower feed digestibility if not processed correctly. Fiber-rich ingredients need to be included in controlled amounts to avoid increasing intestinal bulk, which can interfere with the digestion of essential nutrients like proteins and fats [212]. Palatability is another critical factor, as poultry may reject feed if it has an unpleasant taste, smell, or texture. Some by-products may contain bitter compounds or unfamiliar flavors, which could reduce feed intake as reported by others [211,213].

Considering the variability in the chemical composition of novel ingredients, they should be included in the compound feeds at levels that do not negatively affect the balance of the formulation. Although majority plant-derived waste can provide useful antioxidants or fiber, the excessive amounts may dilute energy-dense components or overwhelm the birds with non-digestible matter [213]. For this reason, formulation software and nutritional modeling are recommended to be used to adjust and balance the inclusion rates of novel ingredients tested based on their nutritional content.

### 7. Future Research Directions in Sustainable Poultry Feeding with Alternative Feed Ingredients

In order to fully exploit alternative feeding options, it is important that policy makers create an enabling environment that promotes research, innovation, and uptake of sustainable systems. Some aspects to consider for future research include the following:

Further investigations to improve upon alternative feed nutritional contents so that they can meet poultry nutrition requirements while still being cost-effective for farmers. This will require further studies on the digestibility, palatability, and nutritional advantages of new feed ingredients.

Comprehensive analyses should be performed to assess the environmental impacts of different poultry feeding strategies, including lifecycle analysis, potential carbon footprint savings, water usage, and other environmental indicators linked with alternative feeds.

The economic viability of alternative feeding strategies needs to be evaluated at various scales of production, ranging from small-holder farms to large commercial enterprises. It is important to understand such dynamics in terms of cost–benefit relationships when considering widespread adoption. Furthermore, consumer perception studies could identify barriers to market acceptance of poultry products raised on alternate feeds and suggest ways in which these could be overcome.

To maintain industry standards of production, there is a need for long-term health and productivity studies on the effect of alternate feeds on the health status, growth rate, reproduction indices, and the quality of products (meat and eggs) from poultry farms.

One example is that governments as well as international organizations must introduce economic incentives like grants or reduced taxes for chicken farmers who apply ecological principles when it comes to feeding their birds. For example, this could involve support for the establishment of infrastructure necessary for production or sourcing of alternative feeds, such as insect- and plant-based proteins.

Increased investment in research geared towards optimizing alternative feed formulations, their long-term effects on avian health status, and the existence of scalable production methods. Academia–industry–government collaborations can enable the development of affordable, sustainable feeding solutions.

Sustainable poultry feeding strategies should be explained to farmers, industry players, and consumers. Through awareness campaigns, resistance against new practices shall be overcome while building consumer demand for poultry products from birds fed using sustainable feeds.

It is vital to develop clear guidelines that ensure the safety, quality, and environmental benefits of alternative feed ingredients, in addition to setting standards for the use of insect-based feeds, which are still relatively new to the market.

## 8. Conclusions

In conclusion, it is important to embrace alternative sustainable feeding strategies in poultry production to achieve SDG 2, Zero Hunger. Through exploration of innovative alternatives and alignment of feeding practices with circular economic principles, the poultry industry can greatly contribute towards global food security as well as environmental sustainability. Further research supported by enabling policies will be critical in addressing challenges and ensuring successful implementation of these strategies.

**Author Contributions:** Conceptualization, P.A.V.; methodology, P.A.V., A.E.U. and A.G.O.; investigation, P.A.V.; resources, P.A.V.; graphics, P.A.V. and A.G.O.; writing—original draft preparation, P.A.V.; writing—review and editing, P.A.V., A.E.U. and A.G.O.; visualization, P.A.V. and A.E.U. All authors have read and agreed to the published version of the manuscript.

**Funding:** This research received no external funding.

**Institutional Review Board Statement:** Not applicable.

**Acknowledgments:** This research is supported by the Romanian Ministry of Research Innovation and Digitalization, project PN 2320-0301.

**Conflicts of Interest:** The authors declare no conflicts of interest.

## Abbreviations

| | |
|---|---|
| DM | dry matter |
| dw | dry weight |
| fw | fresh weight |
| CP | crude protein |
| EE | ether extract (crude fat) |
| CF | crude fiber |
| TPA | total phenolic acids |
| TPC | total phenolic compounds |
| TAC | total antioxidant capacity |
| TFC | total flavonoids content |
| SFA | saturated fatty acids |
| MUFA | monounsaturated fatty acids |
| PUFA | polyunsaturated fatty acids |
| n-3 | total omega 3 fatty acids |
| n-6 | total omega 6 fatty acids |
| GAE | gallic acids equivalent |
| QE | quercetin acids equivalent |
| DPPH | 2,2-diphenyl-1-picrylhydrazyl |
| ABTS | 2,2′-azino-bis (3-ethylbenzothiazoline-6-sulfonic acid) |
| FRAP | ferric reducing antioxidant power |
| TEAC | trolox equivalent antioxidant capacity |
| TBARS | thiobarbituric acid reactive substance |
| FCR | feed conversion ratio |
| BW | body weight |
| FI | feed intake |
| SDG | Sustainable Development Goal |

## References

1. Henchion, M.; Moloney, A.P.; Hyland, J.; Zimmermann, J.; McCarthy, S. Trends for meat, milk and egg consumption for the next decades and the role played by livestock systems in the global production of proteins. *Animal* **2021**, *15*, 100287. [CrossRef] [PubMed]
2. Siddiqui, S.A.; Elsheikh, W.; Ucak, İ.; Hasan, M.; Perlita, Z.C.; Yudhistira, B. Replacement of soy by mealworms for livestock feed-A comparative review between soy and mealworms considering environmental aspects. *Environ. Dev. Sustain.* **2024**, 1–44. [CrossRef]
3. Martens, S.D.; Tiemann, T.T.; Bindelle, J.; Peters, M.; Lascano, C.E. Alternative plant protein sources for pigs and chickens in the tropics–nutritional value and constraints: A review. *J. Agric. Rural Dev. Trop. Subtrop. (JARTS)* **2012**, *113*, 101–123.

4.  Juodka, R.; Nainienė, R.; Juškienė, V.; Juška, R.; Leikus, R.; Kadžienė, G.; Stankevičienė, D. Camelina (*Camelina sativa* (L.) crantz) as feedstuffs in meat type poultry diet: A source of protein and n-3 fatty acids. *Animals* **2022**, *12*, 295. [CrossRef] [PubMed]

5.  Vlaicu, P.A.; Panaite, T.D.; Turcu, R.P. Enriching laying hens eggs by feeding diets with different fatty acid composition and antioxidants. *Sci. Rep.* **2021**, *11*, 20707. [CrossRef]

6.  Vlaicu, P.A.; Untea, A.E.; Turcu, R.P.; Panaite, T.D.; Saracila, M. Rosehip (*Rosa canina* L.) meal as a natural antioxidant on lipid and protein quality and shelf-life of polyunsaturated fatty acids enriched eggs. *Antioxidants* **2022**, *11*, 1948. [CrossRef]

7.  Erinle, T.J.; Oladokun, S.; MacIsaac, J.; Rathgeber, B.; Adewole, D. Dietary grape pomace–effects on growth performance, intestinal health, blood parameters, and breast muscle myopathies of broiler chickens. *Poult. Sci.* **2022**, *101*, 101519. [CrossRef]

8.  Mustafa, A.F.; Baurhoo, B. Effects of feeding dried broccoli floret residues on performance, ileal and total digestive tract nutrient digestibility, and selected microbial populations in broiler chickens. *J. Appl. Poult. Res.* **2016**, *25*, 561–570. [CrossRef]

9.  Vlaicu, P.A.; Untea, A.E.; Lefter, N.A.; Oancea, A.G.; Saracila, M.; Varzaru, I. Influence of rosehip (*Rosa canina* L.) leaves as feed additive during first stage of laying hens on performances and egg quality characteristics. *Poult. Sci.* **2024**, *103*, 103990. [CrossRef]

10. Seidavi, A.; Azizi, M.; Swelum, A.A.; Abd El-Hack, M.E.; Naiel, M.A. Practical application of some common agro-processing wastes in poultry diets. *World's Poult. Sci. J.* **2021**, *77*, 913–927. [CrossRef]

11. Samtiya, M.; Aluko, R.E.; Dhewa, T. Plant food anti-nutritional factors and their reduction strategies: An overview. *Food Production. Process. Nutr.* **2020**, *2*, 6. [CrossRef]

12. Amoroso, L. Post-2015 Agenda and Sustainable Development Goals: Where are we now? Global opportunities to address malnutrition in all its forms, including hidden hunger. *Hidden Hunger Strateg. Improv. Nutr. Qual.* **2018**, *118*, 45–56.

13. Shahmohamadloo, R.S.; Febria, C.M.; Fraser, E.D.; Sibley, P.K. The sustainable agriculture imperative: A perspective on the need for an agrosystem approach to meet the United Nations Sustainable Development Goals by 2030. *Integr. Environ. Assess. Manag.* **2022**, *18*, 1199–1205. [CrossRef] [PubMed]

14. Lee, J.Y.; Lee, S.E.; Lee, D.W. Current status and future prospects of biological routes to bio-based products using raw materials, wastes, and residues as renewable resources. *Crit. Rev. Environ. Sci. Technol.* **2022**, *52*, 2453–2509. [CrossRef]

15. Food and Agriculture Organization of the United Nations (FAO). Development of a Code of Conduct on Food Loss and Food Waste Prevention. 2019. Available online: http://www.fao.org/fsnforum/es/node/4877 (accessed on 22 September 2024).

16. Sheffield, S.; Fiorotto, M.L.; Davis, T.A. Nutritional importance of animal-sourced foods in a healthy diet. *Front. Nutr.* **2024**, *11*, 1424912. [CrossRef]

17. Neumann, C.; Harris, D.M.; Rogers, L.M. Contribution of animal source foods in improving diet quality and function in children in the developing world. *Nutr. Res.* **2002**, *22*, 193–220. [CrossRef]

18. Bergeron, N.; Chiu, S.; Williams, P.T.; King, S.M.; Krauss, R.M. Effects of red meat, white meat, and nonmeat protein sources on atherogenic lipoprotein measures in the context of low compared with high saturated fat intake: A randomized controlled trial. *Am. J. Clin. Nutr.* **2019**, *110*, 24–33. [CrossRef]

19. Pesti, G.M.; Choct, M. The future of feed formulation for poultry: Toward more sustainable production of meat and eggs. *Anim. Nutr.* **2023**, *15*, 71–87. [CrossRef]

20. Masters, W.A.; Rosettie, K.L.; Kranz, S.; Danaei, G.; Webb, P.; Mozaffarian, D. Designing programs to improve diets for maternal and child health: Estimating costs and potential dietary impacts of nutrition-sensitive programs in Ethiopia, Nigeria, and India. *Health Policy Plan.* **2018**, *33*, 564–573. [CrossRef]

21. Renaudeau, D.; Collin, A.; Yahav, S.; De Basilio, V.; Gourdine, J.L.; Collier, R.J. Adaptation to hot climate and strategies to alleviate heat stress in livestock production. *Animal* **2012**, *6*, 707–728. [CrossRef]

22. Dumont, B.; Puillet, L.; Martin, G.; Savietto, D.; Aubin, J.; Ingrand, S.; Niderkorn, V.; Steinmetz, L.; Thomas, M. Incorporating diversity into animal production systems can increase their performance and strengthen their resilience. *Front. Sustain. Food Syst.* **2020**, *4*, 109. [CrossRef]

23. Scanes, C.; Pierzchała-Koziec, K. Poultry and Livestock Production: Environmental Impacts. In *Modern Technology and Traditional Husbandry of Broiler Farming*; IntechOpen: London, UK, 2024. [CrossRef]

24. Vlaicu, P.A.; Untea, A.E.; Turcu, R.P.; Saracila, M.; Panaite, T.D.; Cornescu, G.M. Nutritional Composition and Bioactive Compounds of Basil, Thyme and Sage Plant Additives and Their Functionality on Broiler Thigh Meat Quality. *Foods* **2022**, *11*, 1105. [CrossRef] [PubMed]

25. Leroy, F.; Abraini, F.; Beal, T.; Dominguez-Salas, P.; Gregorini, P.; Manzano, P.; Rowntree, J.; Van Vliet, S. Animal board invited review: Animal source foods in healthy, sustainable, and ethical diets—An argument against drastic limitation of livestock in the food system. *Animal* **2022**, *16*, 100457. [CrossRef] [PubMed]

26. Gauchan, D.; Shrestha, R.B. Improve Socio-Economic Inclusion, Resilience and Wellbeing of Family Farmers, Rural Households and Communities in South Asia. In *Regional Action Plan to Implement the UNDFF for Achieving the SDGs in South Asia*; Rudra, B.S., Pierre, F., Ma, E.P., Mohit, D., Younus, A., Eds.; SAARC Agriculture Center/FAO/AFA/ICA-AP: Dhaka, Bangladesh, 2021; pp. 161–173.

27. Doumkou, S. Circular Economy and Sustainable Agriculture: A Comparative Analysis of EU and MENA Regions. Master's Thesis, International Hellenic University, Thessaloniki, Greece, 2024.

28. Muleta, C.E. The Major Potential of Non-Conventional Feed Resources in Poultry Nutrition in Ethiopia: A Review. *Anim. Vet. Sci.* **2024**, *13*, 68–77. [CrossRef]

29. Murta, D. The future of animal feeding. In *Insects as Animal Feed: Novel Ingredients for Use in Pet, Aquaculture and Livestock Diets*; Publisher CABI, Brithish Library: London, UK, 2021; pp. 126–138.

30. Samani, S.A.; Jafari, M.; Sahafi, S.M.; Roohinejad, S. Applications of algae and algae extracts in human food and feed. *Recent. Adv. Micro Macroalgal Process. Food Health Perspect.* **2021**, 465–486. [CrossRef]

31. Vlaicu, P.A.; Panaite, T.D.; Dragotoiu, D.; Ropota, M.; Bobe, E.; Olteanu, M.; Criste, R.D. Feeding quality of the meat from broilers fed with dietary food industry by-products (flaxseed, rapeseeds and buckthorn meal, grape pomace). *Sci. Pap. Ser. D Anim. Sci.* **2017**, *60*, 123–130.

32. Wang, J.; Singh, A.K.; Kong, F.; Kim, W.K. Effect of almond hulls as an alternative ingredient on broiler performance, nutrient digestibility, and cecal microbiota diversity. *Poult. Sci.* **2021**, *100*, 100853. [CrossRef]

33. Bas-Bellver, C.; Barrera, C.; Betoret, N.; Seguí, L. Effect of Processing and In Vitro Digestion on Bioactive Constituents of Powdered IV Range Carrot (*Daucus carota* L.) Wastes. *Foods* **2023**, *12*, 731. [CrossRef]

34. Panaite, T.D.; Nour, V.; Vlaicu, P.A.; Ropota, M.; Corbu, A.R.; Saracila, M. Flaxseed and dried tomato waste used together in laying hens diet. *Arch. Anim. Nutr.* **2019**, *73*, 222–238. [CrossRef]

35. Malenica, D.; Kass, M.; Bhat, R. Sustainable Management and Valorization of Agri-Food Industrial Wastes and By-Products as Animal Feed: For Ruminants, Non-Ruminants and as Poultry Feed. *Sustainability* **2023**, *15*, 117. [CrossRef]

36. Vauterin, A.; Steiner, B.; Sillman, J.; Kahiluoto, H. The potential of insect protein to reduce food-based carbon footprints in Europe: The case of broiler meat production. *J. Clean. Prod.* **2021**, *320*, 128799. [CrossRef]

37. Kazancoglu, Y.; Ekinci, E.; Ozen, Y.D.O.; Pala, M.O. Reducing food waste through lean and sustainable operations: A case study from the poultry industry. *Rev. Adm. Empresas* **2021**, *61*, e2020-0226. [CrossRef]

38. Donner, M.; Verniquet, A.; Broeze, J.; Kayser, K.; De Vries, H. Critical success and risk factors for circular business models valorising agricultural waste and by-products. Resources. *Conserv. Recycl.* **2021**, *165*, 105236. [CrossRef]

39. Mahmudul, H.M.; Rasul, M.G.; Akbar, D.; Narayanan, R.; Mofijur, M. Food waste as a source of sustainable energy: Technical, economical, environmental and regulatory feasibility analysis. *Renew. Sustain. Energy Rev.* **2022**, *166*, 112577. [CrossRef]

40. Raak, N.; Symmank, C.; Zahn, S.; Aschemann-Witzel, J.; Rohm, H. Processing-and product-related causes for food waste and implications for the food supply chain. *Waste Manag.* **2017**, *61*, 461–472. [CrossRef] [PubMed]

41. Samtiya, M.; Aluko, R.E.; Dhewa, T.; Moreno-Rojas, J.M. Potential Health Benefits of Plant Food-Derived Bioactive Components: An Overview. *Foods* **2021**, *10*, 839. [CrossRef]

42. Vlaicu, P.A.; Untea, A.E.; Varzaru, I.; Saracila, M.; Oancea, A.G. Designing Nutrition for Health—Incorporating Dietary By-Products into Poultry Feeds to Create Functional Foods with Insights into Health Benefits, Risks, Bioactive Compounds, Food Component Functionality and Safety Regulations. *Foods* **2023**, *12*, 4001. [CrossRef]

43. Salvador, R.; Barros, M.V.; Donner, M.; Brito, P.; Halog, A.; Antonio, C. How to advance regional circular bioeconomy systems? Identifying barriers, challenges, drivers, and opportunities. *Sustain. Prod. Consum.* **2022**, *32*, 248–269. [CrossRef]

44. Socas-Rodríguez, B.; Álvarez-Rivera, G.; Valdés, A.; Ibáñez, E.; Cifuentes, A. Food by-products and food wastes: Are they safe enough for their valorization? *Trends Food Sci. Technol.* **2021**, *114*, 133–147. [CrossRef]

45. Focker, M.; Van Asselt, E.D.; Berendsen BJ, A.; Van De Schans, M.G.M.; Van Leeuwen, S.P.J.; Visser, S.M.; Van der Fels-Klerx, H.J. Review of food safety hazards in circular food systems in Europe. *Food Res. Int.* **2022**, *158*, 111505. [CrossRef]

46. Coon, C.N. Major feed ingredients: Feed management and analysis. In *Commercial Chicken Meat and Egg Production*; Springer: Boston, MA, USA, 2002; pp. 215–241.

47. Tallentire, C.W.; Mackenzie, S.G.; Kyriazakis, I. Can novel ingredients replace soybeans and reduce the environmental burdens of European livestock systems in the future? *J. Clean. Prod.* **2018**, *187*, 338–347. [CrossRef]

48. Van der Poel, A.F.B.; Abdollahi, M.R.; Cheng, H.; Colovic, R.; Den Hartog, L.A.; Miladinovic, D.; Page, G.; Sijssens, K.; Smillie, J.F.; Thomas, M.; et al. Future directions of animal feed technology research to meet the challenges of a changing world. *Anim. Feed. Sci. Technol.* **2020**, *270*, 114692. [CrossRef]

49. Piercy, E.; Verstraete, W.; Ellis, P.R.; Banks, M.; Rockström, J.; Smith, P.; OlWitard, C.; Hallett, J.; Hogstrand, C.; Knott, G.; et al. A sustainable waste-to-protein system to maximise waste resource utilisation for developing food-and feed-grade protein solutions. *Green. Chem.* **2023**, *25*, 808–832. [CrossRef]

50. Panaite, T.D.; Nour, V.; Saracila, M.; Turcu, R.P.; Untea, A.E.; Vlaicu, P.A. Effects of Linseed Meal and Carotenoids from Different Sources on Egg Characteristics, Yolk Fatty Acid and Carotenoid Profile and Lipid Peroxidation. *Foods* **2021**, *10*, 1246. [CrossRef] [PubMed]

51. Blando, F.; Marchello, S.; Maiorano, G.; Durante, M.; Signore, A.; Laus, M.N.; Soccio, M.; Mita, G. Bioactive Compounds and Antioxidant Capacity in Anthocyanin-Rich Carrots: A Comparison between the Black Carrot and the Apulian Landrace "Polignano" Carrot. *Plants* **2021**, *10*, 564. [CrossRef]

52. Kim, J.S.; Lim, J.H.; Cho, S.K. Effect of antioxidant and anti-inflammatory on bioactive components of carrot (*Daucus carota* L.) leaves from Jeju Island. *Appl. Biol. Chem.* **2023**, *66*, 34. [CrossRef]

53. Purewal, S.S.; Verma, P.; Kaur, P.; Sandhu, K.S.; Singh, R.S.; Kaur, A.; Salar, R.K. A comparative study on proximate composition, mineral profile, bioactive compounds and antioxidant properties in diverse carrot (*Daucus carota* L.) flour. *Biocatal. Agric. Biotechnol.* **2023**, *48*, 102640. [CrossRef]

54. Golubkina, N.; Zayachkovsky, V.; Stepanov, V.; Deryagina, V.; Rizhova, N.; Kirsanov, K.; Caruso, G. High temperature and humidity effect on biochemical characteristics of organically-grown parsnip roots compared to garlic bulbs. *Plant Foods Hum. Nutr.* **2020**, *75*, 292–297. [CrossRef]
55. Weselek, A.; Bauerle, A.; Zikeli, S.; Lewandowski, I.; Högy, P. Effects on Crop Development, Yields and Chemical Composition of Celeriac (*Apium graveolens* L. *var. rapaceum*) Cultivated Underneath an Agrivoltaic System. *Agronomy* **2021**, *11*, 733. [CrossRef]
56. Nikolić, N.; Cvetković, D.; Todorović, Z. A characterization of content, composition and antioxidant capacity of phenolic compounds in celery roots. *Ital. J. Food Sci.* **2011**, *23*, 214–219.
57. Liu, D.K.; Xu, C.C.; Zhang, L.; Ma, H.; Chen, X.J.; Sui, Y.C.; Zhang, H.Z. Evaluation of bioactive components and antioxidant capacity of four celery (*Apium graveolens* L.) leaves and petioles. *Int. J. Food Prop.* **2020**, *23*, 1097–1109. [CrossRef]
58. Yao, Y.; Sang, W.; Zhou, M.; Ren, G. Phenolic composition and antioxidant activities of 11 celery cultivars. *J. Food Sci.* **2010**, *75*, C9–C13. [CrossRef] [PubMed]
59. Grover, S.; Malik, C.P.; Hora, A.; Kushwaha, H.B. Botany, cultivation, chemical constituents and genetic diversity in fennel (*Foeniculum vulgare* Mill): A review. *LS Int. J. Life Sci.* **2013**, *2*, 128–139. [CrossRef]
60. Barros, L.; Carvalho, A.M.; Ferreira, I.C. The nutritional composition of fennel (*Foeniculum vulgare*): Shoots, leaves, stems and inflorescences. *LWT-Food Sci. Technol.* **2010**, *43*, 814–818. [CrossRef]
61. Sun, W.; Shahrajabian, M.H.; Cheng, Q. Anise (*Pimpinella anisum* L.), a dominant spice and traditional medicinal herb for both food and medicinal purposes. *Cogent Biol.* **2019**, *5*, 1673688. [CrossRef]
62. Ghosh, A.; Saleh-e-In, M.M.; Abukawsar, M.M.; Ahsan, M.A.; Rahim, M.M.; Bhuiyan, M.N.H.; Roy, S.K.; Naher, S. Characterization of quality and pharmacological assessment of *Pimpinella anisum* L. (Anise) seeds cultivars. *J. Food Meas. Charact.* **2019**, *13*, 2672–2685. [CrossRef]
63. Bhat, S.; Kaushal, P.; Kaur, M.; Sharma, H.K. Coriander (*Coriandrum sativum* L.): Processing, nutritional and functional aspects. *Afr. J. Plant Sci.* **2014**, *8*, 25–33.
64. Abbassi, A.; Mahmoudi, H.; Zaouali, W.; M'rabet, Y.; Casabianca, H.; Hosni, K. Enzyme-aided release of bioactive compounds from coriander (*Coriandrum sativum* L.) seeds and their residue by-products and evaluation of their antioxidant activity. *J. Food Sci. Technol.* **2018**, *55*, 3065–3076. [CrossRef]
65. Cornescu, G.M.; Panaite, T.D.; Untea, A.E.; Varzaru, I.; Saracila, M.; Dumitru, M.; Vlaicu, P.A.; Gavris, T. Mitigation of heat stress effects on laying hens' performances, egg quality, and some blood parameters by adding dietary zinc-enriched yeasts, parsley, and their combination. *Front. Vet. Sci.* **2023**, *10*, 1202058. [CrossRef]
66. Dobričević, N.; Šic Žlabur, J.; Voća, S.; Pliestić, S.; Galić, A.; Delić, A.; Fabek Uher, S. Bioactive compounds content and nutritional potential of different parsley parts (*Petroselinum crispum* Mill.). *J. Cent. Eur. Agric.* **2019**, *20*, 900–910. [CrossRef]
67. Lisiewska, Z.; Kmiecik, W.; Korus, A. Content of vitamin C, carotenoids, chlorophylls and polyphenols in green parts of dill (*Anethum graveolens* L.) depending on plant height. *J. Food Compos. Anal.* **2006**, *19*, 134–140. [CrossRef]
68. Spréa, R.M.; Fernandes, Â.; Calhelha, R.C.; Pereira, C.; Pires, T.C.; Alves, M.J.; Canan, C.; Barros, L.; Amaral, J.S.; Ferreira, I.C. Chemical and bioactive characterization of the aromatic plant *Levisticum officinale* WDJ Koch: A comprehensive study. *Food Funct.* **2020**, *11*, 1292–1303. [CrossRef] [PubMed]
69. Jakubczyk, A.; Złotek, U.; Szymanowska, U.; Rybczyńska-Tkaczyk, K.; Jęderka, K.; Lewicki, S. In vitro Antioxidant, Anti-inflammatory, Anti-metabolic Syndrome, Antimicrobial, and Anticancer Effect of Phenolic Acids Isolated from Fresh Lovage Leaves [*Levisticum officinale* Koch] Elicited with Jasmonic Acid and Yeast Extract. *Antioxidants* **2020**, *9*, 554. [CrossRef] [PubMed]
70. Özgen, M.; Scheerens, J.C.; Reese, R.N.; Miller, R.A. Total phenolic, anthocyanin contents and antioxidant capacity of selected elderberry (*Sambucus canadensis* L.) accessions. *Pharmacogn. Mag.* **2010**, *6*, 198. [CrossRef] [PubMed]
71. Młynarczyk, K.; Walkowiak-Tomczak, D.; Łysiak, G.P. Bioactive properties of *Sambucus nigra* L. as a functional ingredient for food and pharmaceutical industry. *J. Funct. Foods* **2018**, *40*, 377–390. [CrossRef]
72. Ferreira, S.S.; Silva, A.M.; Nunes, F.M. *Sambucus nigra* L. fruits and flowers: Chemical composition and related bioactivities. *Food Rev. Int.* **2022**, *38*, 1237–1265. [CrossRef]
73. Afoakwah, N.A.; Dong, Y.; Zhao, Y.; Xiong, Z.; Owusu, J.; Wang, Y.; Zhang, J. Characterization of Jerusalem artichoke (*Helianthus tuberosus* L.) powder and its application in emulsion-type sausage. *LWT-Food Sci. Technol.* **2015**, *64*, 74–81. [CrossRef]
74. Ersahince, A.; Kara, K. Nutrient composition and in vitro digestion parameters of Jerusalem artichoke (*Helianthus tuberosus* L.) herbage at different maturity stages in horse and ruminant. *J. Anim. Feed. Sci.* **2017**, *26*, 213–225. [CrossRef]
75. Wang, Y.; Zhao, Y.; Xue, F.; Nan, X.; Wang, H.; Hua, D.; Liu, J.; Yang, L.; Jiang, L.; Xiong, B. Nutritional value, bioactivity, and application potential of Jerusalem artichoke (*Helianthus tuberosus* L.) as a neotype feed resource. *Anim. Nutr.* **2020**, *6*, 429–437. [CrossRef]
76. Nwafor, I.C.; Shale, K.; Achilonu, M.C. Chemical composition and nutritive benefits of chicory (*Cichorium intybus*) as an ideal complementary and/or alternative livestock feed supplement. *Sci. World J.* **2017**, *2017*, 7343928. [CrossRef]
77. Jangra, S.S.; Madan, V.K. Proximate, mineral and chemical composition of different parts of chicory (*Cichorium intybus* L.). *J. Pharmacogn. Phytochem.* **2018**, *7*, 3311–3315.
78. Miguel, M.; Barros, L.; Pereira, C.; Calhelha, R.C.; Garcia, P.A.; Castro, M.Á.; Santos-Buelga, C.; Ferreira, I.C.F.R. Chemical characterization and bioactive properties of two aromatic plants: *Calendula officinalis* L.(flowers) and *Mentha cervina* L.(leaves). *Food Funct.* **2016**, *7*, 2223–2232. [CrossRef] [PubMed]

79. Olennikov, D.N.; Kashchenko, N.I. Componential profile and amylase inhibiting activity of phenolic compounds from *Calendula officinalis* L. leaves. *Sci. World J.* **2014**, *2014*, 654193. [CrossRef] [PubMed]
80. Dedić, S.; Džaferović, A.; Jukić, H. Chemical Composition and Antioxidant Activity of Water-Ethanol Extracts of Dandelion (*Taraxacum officinale*). *Hrana Zdr. Boles. Znan.-Stručni Časopis Nutr. Dijetetiku* **2022**, *11*, 8–14.
81. Mampholo, B.M.; Maboko, M.M.; Soundy, P.; Sivakumar, D. Phytochemicals and overall quality of leafy lettuce (*Lactuca sativa* L.) varieties grown in closed hydroponic system. *J. Food Qual.* **2016**, *39*, 805–815. [CrossRef]
82. Park, C.H.; Yeo, H.J.; Baskar, T.B.; Kim, J.K.; Park, S.U. Metabolic profiling and chemical-based antioxidant assays of green and red lettuce (*Lactuca sativa*). *Nat. Product. Commun.* **2018**, *13*, 1934578X1801300313. [CrossRef]
83. Pandino, G.; Lombardo, S.; Mauromicale, G. Chemical and Morphological Characteristics of New Clones and Commercial Varieties of Globe Artichoke (*Cynara cardunculus* var. scolymus). *Plant Foods Hum. Nutr.* **2011**, *66*, 291–297. [CrossRef]
84. Petropoulos, S.A.; Pereira, C.; Ntatsi, G.; Danalatos, N.; Barros, L.; Ferreira, I.C. Nutritional value and chemical composition of Greek artichoke genotypes. *Food Chem.* **2018**, *267*, 296–302. [CrossRef]
85. Dias, M.I.; Barros, L.; Dueñas, M.; Pereira, E.; Carvalho, A.M.; Alves, R.C.; Oliveira, M.B.P.; Santos-Buelga, C.; Ferreira, I.C. Chemical composition of wild and commercial *Achillea millefolium* L. and bioactivity of the methanolic extract, infusion and decoction. *Food Chem.* **2013**, *141*, 4152–4160. [CrossRef]
86. Ewansiha, C.J.; Ebhoaye, J.E.; Asia, I.O.; Ekebafe, L.O.; Ehigie, C. Proximate and mineral composition of coconut (*Cocos nucifera*) shell. *Int. J. Pure Appl. Sci. Technol.* **2012**, *13*, 57.
87. Gheorghe, A.; Vlaicu, P.A.; Olteanu, M.; Vișinescu, P.; Criste, R.D. Obtaining eggs enriched in polyunsaturated fatty acids (PUFA). 1. Use of vegetable sources rich in PUFA as functional ingredients in laying hens diets: A review. *Arch. Zootech.* **2019**, *22*, 54–85.
88. Sikora, E.; Bodziarczyk, I. Composition and antioxidant activity of kale (*Brassica oleracea* L. var. acephala) raw and cooked. *Acta Sci. Pol. Technol. Aliment.* **2012**, *11*, 239–248. [PubMed]
89. Oancea, A.-G.; Dragomir, C.; Untea, A.E.; Saracila, M.; Cismileanu, A.E.; Vlaicu, P.A.; Varzaru, I. The Effects of Flax and Mustard Seed Inclusion in Dairy Goats' Diet on Milk Nutritional Quality. *Agriculture* **2024**, *14*, 1009. [CrossRef]
90. Brito, T.B.N.; Pereira, A.P.A.; Pastore, G.M.; Moreira, R.F.A.; Ferreira, M.S.L.; Fai, A.E.C. Chemical composition and physicochemical characterization for cabbage and pineapple by-products flour valorization. *LWT* **2020**, *124*, 109028. [CrossRef]
91. Mierlita, D.; Teușdea, A.C.; Matei, M.; Pascal, C.; Simeanu, D.; Pop, I.M. Effect of Dietary Incorporation of Hemp Seeds Alone or with Dried Fruit Pomace on Laying Hens' Performance and on Lipid Composition and Oxidation Status of Egg Yolks. *Animals* **2024**, *14*, 750. [CrossRef]
92. Kim, M.Y.; Kim, E.J.; Kim, Y.N.; Choi, C.; Lee, B.H. Comparison of the chemical compositions and nutritive values of various pumpkin (*Cucurbitaceae*) species and parts. *Nutr. Res. Pract.* **2012**, *6*, 21–27. [CrossRef]
93. Agatemor, U.M.M.; Nwodo, O.F.C.; Anosike, C.A. Phytochemical and proximate composition of cucumber (*Cucumis sativus*) fruit from Nsukka, Nigeria. *Afr. J. Biotechnol.* **2018**, *17*, 1215–1219.
94. Olayinka, B.U.; Etejere, E.O. Proximate and Chemical Compositions of Watermelon (Citrullus lanatus (Thunb.) Matsum and Nakai cv Red and Cucumber (*Cucumis sativus* L. cv Pipino). *Int. Food Res. J.* **2018**, *25*, 1060–1066.
95. Vlaicu, P.A.; Panaite, T.D. Effect of dietary pumpkin (*Cucurbita moschata*) seed meal on layer performance and egg quality characteristics. *Anim. Biosci.* **2022**, *35*, 236. [CrossRef]
96. Saracila, M.; Untea, A.E.; Panaite, T.D.; Varzaru, I.; Oancea, A.-G.; Turcu, R.P.; Vlaicu, P.A. Effects of Supplementing Sea Buckthorn Leaves (*Hippophae rhamnoides* L.) and Chromium (III) in Broiler Diet on the Nutritional Quality and Lipid Oxidative Stability of Meat. *Antioxidants* **2022**, *11*, 2220. [CrossRef]
97. Saboonchian, F.; Jamei, R.; Sarghein, S.H. Phenolic and flavonoid content of *Elaeagnus angustifolia* L. (leaf and flower). *Avicenna J. Phytomed.* **2014**, *4*, 231. [PubMed]
98. Fakı, R.; Canbay, H.S.; Gürsoy, O.; Yılmaz, Y. Antioxidant activity, physico-chemical and fatty acid composition of oleaster (*Elaeagnus angustifolia* L.) varieties naturally grown in western mediterranean region of Turkey. *Akad. Gıda* **2022**, *20*, 329–335. [CrossRef]
99. Untea, A.E.; Turcu, R.P.; Saracila, M.; Vlaicu, P.A.; Panaite, T.D.; Oancea, A.G. Broiler meat fatty acids composition, lipid metabolism, and oxidative stability parameters as affected by cranberry leaves and walnut meal supplemented diets. *Sci. Rep.* **2022**, *12*, 21618. [CrossRef] [PubMed]
100. Vlaicu, P.A.; Untea, A.E.; Turcu, R.P.; Saracila, M.; Varzaru, I.; Oancea, A.G. Chemical composition of dietary alfalfa and its effectiveness on broiler chicken thigh meat quality. *Czech J. Food Sci.* **2023**, *41*, 279–286. [CrossRef]
101. Zia-Ul-Haq, M.; Ahmad, S.; Amarowicz, R.; Ercisli, S. Compositional studies of some pea (*Pisum sativum* L.) seed cultivars commonly consumed in Pakistan. *Ital. J. Food Sci. /Riv. Ital. Di Sci. Degli Aliment.* **2013**, *25*, 295–302.
102. Liberal, Â.; Almeida, D.; Fernandes, Â.; Pereira, C.; Ferreira, I.C.; Vivar-Quintana, A.M.; Barros, L. Nutritional, chemical and antioxidant evaluation of Armuña lentil (*Lens culinaris* spp.): Influence of season and soil. *Food Chem.* **2023**, *411*, 135491. [CrossRef]
103. Wang, F.; Huang, L.; Yuan, X.; Zhang, X.; Guo, L.; Xue, C.; Chen, X. Nutritional, phytochemical and antioxidant properties of 24 mung bean (*Vigna radiate* L.) genotypes. *Food Prod. Process. Nutr.* **2021**, *3*, 28. [CrossRef]
104. Scherer, R.; Lemos, M.F.; Lemos, M.F.; Martinelli, G.C.; Martins, J.D.L.; da Silva, A.G. Antioxidant and antibacterial activities and composition of Brazilian spearmint (*Mentha spicata* L.). *Ind. Crops Prod.* **2013**, *50*, 408–413. [CrossRef]
105. Paula, T.R.; Olteanu, M.; Untea, A.E.; Saracila, M.; Varzaru, I.; Vlaicu, P.A. Nutritional characterization of some natural plants used in poultry nutrition. *Arch. Zootech.* **2020**, *23*, 58–72. [CrossRef]

106. Jędrszczyk, E.; Kopeć, A.; Bucki, P.; Ambroszczyk, A.M.; Skowera, B. The enhancing effect of plants growth biostimulants in garlic cultivation on the chemical composition and level of bioactive compounds in the garlic leaves, stems and bulbs. *Not. Bot. Horti Agrobot. Cluj-Napoca* **2019**, *47*, 81–91. [CrossRef]

107. Benítez, V.; Mollá, E.; Martín-Cabrejas, M.A.; Aguilera, Y.; López-Andréu, F.J.; Cools, K.; Terry, L.A.; Esteban, R.M. Characterization of industrial onion wastes (*Allium cepa* L.): Dietary fibre and bioactive compounds. *Plant Foods Hum. Nutr.* **2011**, *66*, 48–57. [CrossRef] [PubMed]

108. Pegiou, E.; Mumm, R.; Acharya, P.; de Vos, R.C.H.; Hall, R.D. Green and White Asparagus (*Asparagus officinalis*): A Source of Developmental, Chemical and Urinary Intrigue. *Metabolites* **2020**, *10*, 17. [CrossRef] [PubMed]

109. Tsado, A.N.; Okoli, N.R.; Jiya, A.G.; Gana, D.; Saidu, B.; Zubairu, R.; Salihu, I.Z. Proximate, minerals, and amino acid compositions of banana and plantain peels. *BIOMED Nat. Appl. Sci.* **2021**, *1*, 032–042.

110. Maestri, D.; Barrionuevo, D.; Bodoira, R.; Zafra, A.; Jiménez-López, J.; Alché, J.D.D. Nutritional profile and nutraceutical components of olive (*Olea europaea* L.) seeds. *J. Food Sci. Technol.* **2019**, *56*, 4359–4370. [CrossRef]

111. Manzoor, M.; Anwar, F.; Saari, N.; Ashraf, M. Variations of Antioxidant Characteristics and Mineral Contents in Pulp and Peel of Different Apple (*Malus domestica* Borkh.) Cultivars from Pakistan. *Molecules* **2012**, *17*, 390–407. [CrossRef]

112. Ilyasoğlu, H. Characterization of rosehip (*Rosa canina* L.) seed and seed oil. *Int. J. Food Prop.* **2014**, *17*, 1591–1598. [CrossRef]

113. Varzaru, I.; Oancea, A.G.; Vlaicu, P.A.; Saracila, M.; Untea, A.E. Exploring the Antioxidant Potential of Blackberry and Raspberry Leaves: Phytochemical Analysis, Scavenging Activity, and In Vitro Polyphenol Bioaccessibility. *Antioxidants* **2023**, *12*, 2125. [CrossRef]

114. Salas-Arias, K.; Irías-Mata, A.; Sánchez-Kopper, A.; Hernández-Moncada, R.; Salas-Morgan, B.; Villalta-Romero, F.; Calvo-Castro, L.A. Strawberry *Fragaria x ananassa* cv. Festival: A Polyphenol-Based Phytochemical Characterization in Fruit and Leaf Extracts. *Molecules* **2023**, *28*, 1865. [CrossRef]

115. Saracila, M.; Untea, A.E.; Oancea, A.G.; Varzaru, I.; Vlaicu, P.A. Comparative Analysis of Black Chokeberry (*Aronia melanocarpa* L.) Fruit, Leaves, and Pomace for Their Phytochemical Composition, Antioxidant Potential, and Polyphenol Bioaccessibility. *Foods* **2024**, *13*, 1856. [CrossRef]

116. Vlaicu, P.A.; Untea, A.E.; Panaite, T.D.; Turcu, R.P. Effect of dietary orange and grapefruit peel on growth performance, health status, meat quality and intestinal microflora of broiler chickens. *Ital. J. Anim. Sci.* **2020**, *19*, 1394–1405. [CrossRef]

117. Janati, S.S.F.; Beheshti, H.R.; Feizy, J.; Fahim, N.K. Chemical composition of lemon (*Citrus limon*) and peels its considerations as animal food. *Gida* **2012**, *37*, 267–271.

118. Anaya-Esparza, L.M.; Mora, Z.V.-d.l.; Vázquez-Paulino, O.; Ascencio, F.; Villarruel-López, A. Bell Peppers (*Capsicum annum* L.) Losses and Wastes: Source for Food and Pharmaceutical Applications. *Molecules* **2021**, *26*, 5341. [CrossRef] [PubMed]

119. Silva, G.F.P.; Pereira, E.; Melgar, B.; Stojković, D.; Sokovic, M.; Calhelha, R.C.; Pereira, C.; Abreu, R.M.V.; Ferreira, I.C.F.R.; Barros, L. Eggplant Fruit (*Solanum melongena* L.) and Bio-Residues as a Source of Nutrients, Bioactive Compounds, and Food Colorants, Using Innovative Food Technologies. *Appl. Sci.* **2021**, *11*, 151. [CrossRef]

120. Olteanu, M.; Panaite, T.D.; Turcu, R.P.; Ropota, M.; Vlaicu, P.A.; Mitoi, M. Using grapeseed meal as natural antioxidant in slow-growing *Hubbard broiler* diets enriched in polyunsaturated fatty acids. *Rev. Mex. De Cienc. Pecu.* **2022**, *13*, 43–63. [CrossRef]

121. Turcu, R.P.; Olteanu, M.; Criste, R.D.; Panaite, T.D.; Ropotă, M.; Vlaicu, P.A.; Drăgotoiu, D. Grapeseed meal used as natural antioxidant in high fatty acid diets for Hubbard broilers. *Braz. J. Poult. Sci.* **2019**, *21*, eRBCA-2018. [CrossRef]

122. Yeh, H.Y.; Chuang, C.H.; Chen, H.C.; Wan, C.J.; Chen, T.L.; Lin, L.Y. Bioactive components analysis of two various gingers (*Zingiber officinale* Roscoe) and antioxidant effect of ginger extracts. *LWT-Food Sci. Technol.* **2014**, *55*, 329–334. [CrossRef]

123. Pliego, A.B.; Tavakoli, M.; Khusro, A.; Seidavi, A.; Elghandour, M.M.; Salem, A.Z.; Márquez-Molina, O.; Rene Rivas-Caceres, R. Beneficial and adverse effects of medicinal plants as feed supplements in poultry nutrition: A review. *Anim. Biotechnol.* **2022**, *33*, 369–391. [CrossRef]

124. Harouna, D.V.; Kawe, P.C.; Mohammed, E.M.I. Under-utilized legumes as potential poultry feed ingredients: A mini-review. *Arch. Anim. Poult. Sci.* **2018**, *1*, 1–3.

125. Nalluri, N.; Karri, V.R. Grain legumes and their by-products: As a nutrient rich feed supplement in the sustainable intensification of commercial poultry industry. In *Sustainable Agriculture Reviews 51: Legume Agriculture and Biotechnology*; Springer Nature: Cham, Switzerland, 2021; Volume 2, pp. 51–96. [CrossRef]

126. Panaite, T.D.; Vlaicu, P.A.; Saracila, M.; Cismileanu, A.; Varzaru, I.; Voicu, S.N.; Hermenean, A. Impact of Watermelon Rind and Sea Buckthorn Meal on Performance, Blood Parameters, and Gut Microbiota and Morphology in Laying Hens. *Agriculture* **2022**, *12*, 177. [CrossRef]

127. Erinle, T.J.; Adewole, D.I. Fruit pomaces—Their nutrient and bioactive components, effects on growth and health of poultry species, and possible optimization techniques. *Anim. Nutr.* **2022**, *9*, 357–377. [CrossRef]

128. Untea, A.E.; Varzaru, I.; Saracila, M.; Panaite, T.D.; Oancea, A.G.; Vlaicu, P.A.; Grosu, I.A. Antioxidant Properties of Cranberry Leaves and Walnut Meal and Their Effect on Nutritional Quality and Oxidative Stability of Broiler Breast Meat. *Antioxidants* **2023**, *12*, 1084. [CrossRef] [PubMed]

129. Saracila, M.; Panaite, T.D.; Predescu, N.C.; Untea, A.E.; Vlaicu, P.A. Effect of Dietary Salicin Standardized Extract from Salix alba Bark on Oxidative Stress Biomarkers and Intestinal Microflora of Broiler Chickens Exposed to Heat Stress. *Agriculture* **2023**, *13*, 698. [CrossRef]

130. Ponnampalam, E.N.; Holman, B.W. Sustainability II: Sustainable animal production and meat processing. In *Lawrie's Meat Science*; Woodhead Publishing Series in Food Science, Technology and Nutrition; Woodhead Publishing: Cambridge, MA, USA; Kidlington, UK, 2023; pp. 727–798. [CrossRef]
131. Lefter, N.A.; Gheorghe, A.; Habeanu, M.; Ciurescu, G.; Dumitru, M.; Untea, A.E.; Vlaicu, P.A. Assessing the effects of microencapsulated Lactobacillus salivarius and cowpea seed supplementation on broiler chicken growth and health status. *Front. Vet. Sci.* **2023**, *10*, 1279819. [CrossRef] [PubMed]
132. Alshelmani, M.I.; Abdalla, E.A.; Kaka, U.; Basit, M.A. Nontraditional feedstuffs as an alternative in poultry feed. In *Advances in Poultry Nutrition Research*; IntechOpen: London, UK, 2021. [CrossRef]
133. Khalifah, A.; Abdalla, S.; Rageb, M.; Maruccio, L.; Ciani, F.; El-Sabrout, K. Could Insect Products Provide a Safe and Sustainable Feed Alternative for the Poultry Industry? A Comprehensive Review. *Animals* **2023**, *13*, 1534. [CrossRef]
134. Kusmayadi, A.; Leong, Y.K.; Yen, H.W.; Huang, C.Y.; Chang, J.S. Microalgae as sustainable food and feed sources for animals and humans–biotechnological and environmental aspects. *Chemosphere* **2021**, *271*, 129800. [CrossRef]
135. Brunetti, L.; Leuci, R.; Colonna, M.A.; Carrieri, R.; Celentano, F.E.; Bozzo, G.; Loiodice, F.; Selvaggi, M.; Tufarelli, V.; Piemontese, L. Food Industry Byproducts as Starting Material for Innovative, Green Feed Formulation: A Sustainable Alternative for Poultry Feeding. *Molecules* **2022**, *27*, 4735. [CrossRef]
136. Babatunde, O.O.; Park, C.S.; Adeola, O. Nutritional Potentials of Atypical Feed Ingredients for Broiler Chickens and Pigs. *Animals* **2021**, *11*, 1196. [CrossRef]
137. Davis, K.F.; Downs, S.; Gephart, J.A. Towards food supply chain resilience to environmental shocks. *Nat. Food* **2021**, *2*, 54–65. [CrossRef]
138. Barrett, C.B. Overcoming global food security challenges through science and solidarity. *Am. J. Agric. Econ.* **2021**, *103*, 422–447. [CrossRef]
139. Fróna, D.; Szenderák, J.; Harangi-Rákos, M. The Challenge of Feeding the World. *Sustainability* **2019**, *11*, 5816. [CrossRef]
140. David, L.S.; Nalle, C.L.; Abdollahi, M.R.; Ravindran, V. Feeding Value of Lupins, Field Peas, Faba Beans and Chickpeas for Poultry: An Overview. *Animals* **2024**, *14*, 619. [CrossRef] [PubMed]
141. Rehal, J.; Kaur, K.; Kaur, P. Cereals and Their By-Products. In *Cereal Grains*; CRC Press: Boca Raton, FL, USA, 2023; pp. 147–176.
142. Purohit, P.; Rawat, H.; Verma, N.; Mishra, S.; Nautiyal, A.; Anshul; Bhatt, S.; Bisht, N.; Aggarwal, K.; Bora, A.; et al. Analytical approach to assess anti-nutritional factors of grains and oilseeds: A comprehensive review. *J. Agric. Food Res.* **2023**, *14*, 100877. [CrossRef]
143. Samtiya, M.; Aluko, R.E.; Puniya, A.K.; Dhewa, T. Enhancing Micronutrients Bioavailability through Fermentation of Plant-Based Foods: A Concise Review. *Fermentation* **2021**, *7*, 63. [CrossRef]
144. Kustar, A.; Patino-Echeverri, D. A Review of Environmental Life Cycle Assessments of Diets: Plant-Based Solutions Are Truly Sustainable, even in the Form of Fast Foods. *Sustainability* **2021**, *13*, 9926. [CrossRef]
145. Sajid, Q.U.A.; Asghar, M.U.; Tariq, H.; Wilk, M.; Płatek, A. Insect Meal as an Alternative to Protein Concentrates in Poultry Nutrition with Future Perspectives (An Updated Review). *Agriculture* **2023**, *13*, 1239. [CrossRef]
146. Kolev, N.; Vlahova-Vangelova, D.; Balev, D.; Dragoev, S.; Dimov, K.; Petkov, E.; Popova, T. Effect of the Addition of Soybean Protein and Insect Flours on the Quality of Cooked Sausages. *Foods* **2024**, *13*, 2194. [CrossRef]
147. Jagtap, S.; Garcia-Garcia, G.; Duong, L.; Swainson, M.; Martindale, W. Codesign of Food System and Circular Economy Approaches for the Development of Livestock Feeds from Insect Larvae. *Foods* **2021**, *10*, 1701. [CrossRef]
148. Chavez, M. The sustainability of industrial insect mass rearing for food and feed production: Zero waste goals through by-product utilization. *Curr. Opin. Insect Sci.* **2021**, *48*, 44–49. [CrossRef]
149. Cadinu, L.A.; Barra, P.; Torre, F.; Delogu, F.; Madau, F.A. Insect Rearing: Potential, Challenges, and Circularity. *Sustainability* **2020**, *12*, 4567. [CrossRef]
150. Żuk-Gołaszewska, K.; Gałęcki, R.; Obremski, K.; Smetana, S.; Figiel, S.; Gołaszewski, J. Edible Insect Farming in the Context of the EU Regulations and Marketing—An Overview. *Insects* **2022**, *13*, 446. [CrossRef]
151. Madau, F.A.; Arru, B.; Furesi, R.; Pulina, P. Insect Farming for Feed and Food Production from a Circular Business Model Perspective. *Sustainability* **2020**, *12*, 5418. [CrossRef]
152. Veldkamp, T.; Meijer, N.; Alleweldt, F.; Deruytter, D.; Van Campenhout, L.; Gasco, L.; Roos, N.; Smetana, S.; Fernandes, A.; van der Fels-Klerx, H.J. Overcoming Technical and Market Barriers to Enable Sustainable Large-Scale Production and Consumption of Insect Proteins in Europe: A SUSINCHAIN Perspective. *Insects* **2022**, *13*, 281. [CrossRef] [PubMed]
153. Colombino, E.; Biasato, I.; Ferrocino, I.; Bellezza Oddon, S.; Caimi, C.; Gariglio, M.; Dabbou, S.; Caramori, M.; Battisti, E.; Zanet, S.; et al. Effect of Insect Live Larvae as Environmental Enrichment on Poultry Gut Health: Gut Mucin Composition, Microbiota and Local Immune Response Evaluation. *Animals* **2021**, *11*, 2819. [CrossRef] [PubMed]
154. Elahi, U.; Xu, C.C.; Wang, J.; Lin, J.; Wu, S.G.; Zhang, H.J.; Qi, G.H. Insect meal as a feed ingredient for poultry. *Anim. Biosci.* **2022**, *35*, 332. [CrossRef] [PubMed]
155. Barakat, D.; El-Far, A.; Sadek, K.; Mahrous, U.; Ellakany, H.; Abdel-Latif, M. Anise (*Pimpinella anisum*) enhances the growth performance, immunity and antioxidant activities in broilers. *Int. J. Pharm. Sci. Rev. Res.* **2016**, *37*, 134–140.
156. Khubeiz, M.M.; Shirif, A.M. Effect of coriander (*Coriandrum sativum* L.) seed powder as feed additives on performance and some blood parameters of broiler chickens. *Open Vet. J.* **2020**, *10*, 198–205. [CrossRef]

157. Ali, N.A.L.; Gaakd, M.; AL-Nasrawi, A.M. Effect of addition different levels of Parsley leaves powder (*Petroselinum sativum*) to the ration on some blood serum biochemical traits of broiler Ross 308. *J. Nat. Sci. Res.* **2016**, *6*, 18–21.
158. Hammod, A.J.; Abd El-Aziz, A.H.; Areaaer, A.H.; Alfertosi, K.A. Effect of Dill Powder (*Anethum graveolens*) as a Dietary Supplement on Productive Performance, Mortality and Economic Figure in Broiler. *IOP Conf. Ser. Earth Environ. Sci.* **2020**, *553*, 012018. [CrossRef]
159. Abadi, K.M.A.; Andi, M.A. Effects of using coriander (*Coriandrum sativum* L.), savory (*Satureja hortensis* L.) and dill (*Anethum graveolens* L.) herb powder in diet on performance and some blood parameters of broilers. *Open Vet J.* **2014**, *5*, 95–103. [CrossRef]
160. Al-Abboodi, A.A.; Jawad, H.S. Effect of supplementing different levels of Jerusalem artichoke (*Helianthus tuberosus* L.) on broiler production performance. *Plant Arch.* **2018**, *18*, 1570–1574.
161. Khoobani, M.; Hasheminezhad, S.-H.; Javandel, F.; Nosrati, M.; Seidavi, A.; Kadim, I.T.; Laudadio, V.; Tufarelli, V. Effects of Dietary Chicory (*Chicorium intybus* L.) and Probiotic Blend as Natural Feed Additives on Performance Traits, Blood Biochemistry, and Gut Microbiota of Broiler Chickens. *Antibiotics* **2020**, *9*, 5. [CrossRef] [PubMed]
162. Kop-Bozbay, C.; Akdag, A.; Bozkurt-Kiraz, A.; Gore, M.; Kurt, O.; Ocak, N. Laying Performance, Egg Quality Characteristics, and Egg Yolk Fatty Acids Profile in Layer Hens Housed with Free Access to Chicory- and/or White Clover-Vegetated or Non-Vegetated Areas. *Animals* **2021**, *11*, 1708. [CrossRef] [PubMed]
163. Foroutankhah, M.; Toghyani, M.; Landy, N. Evaluation of *Calendula officinalis* L.(marigold) flower as a natural growth promoter in comparison with an antibiotic growth promoter on growth performance, carcass traits and humoral immune responses of broilers. *Anim. Nutr.* **2019**, *5*, 314–318. [CrossRef] [PubMed]
164. Mim, Y.Z.; Sultana, F.; Dey, B.; Ray, B.C.; Nishat, N.J. Effects of Marigold Petal and Leaves on Yolk Pigmentation in Laying Hens. *J. Bangladesh Agric. Univ.* **2021**, *19*, 325–331. [CrossRef]
165. Du, J.; Zhao, Y.; Wang, Y.; Xie, M.; Wang, R.; Liu, N.; An, X.; Qi, J. Growth, carcase characteristics, meat quality, nutrient digestibility and immune function of broilers fed with enzymatically treated or raw dandelion (*Taraxacum mongolicum* hand.-mazz.). *Ital. J. Anim. Sci.* **2022**, *21*, 1117–1125. [CrossRef]
166. Wang, Y.; Duan, T.; Wang, W.; Mao, J.; Yin, N.; Guo, T.; Liu, N.; An, X.; Qi, J. Impact of dietary dandelion (*Taraxacum mongolicum* Hand.-Mazz.) supplementation on carcase traits, breast meat quality, muscle fatty and amino acid composition and antioxidant capacity in broiler chickens. *Ital. J. Anim. Sci.* **2023**, *22*, 441–451. [CrossRef]
167. Qureshi, S.; Banday, M.T.; Shakeel, I.; Adil, S.; Khan, A.A. Effect of raw and enzyme-treated dandelion leaves and fenugreek seed supplemented diet on gut microflora of broiler chicken. *Appl. Biol. Res.* **2016**, *18*, 76–79. [CrossRef]
168. Saenz, F.M.C.; Saucedo-Uriarte, J.A.; Sotelo-Mendez, A.; Zamora-Huamán, S.J. A prebiotic diet based on dandelion (*Taraxacum officinale*) improves the productive performance and intestinal morphology of laying hens. *Sci. Agropecu.* **2021**, *12*, 403–410. [CrossRef]
169. Hashem, M.A.; Neamat-Allah, A.N.; Hammza, H.E.; Abou-Elnaga, H.M. Impact of dietary supplementation with Echinacea purpurea on growth performance, immunological, biochemical, and pathological findings in broiler chickens infected by pathogenic *E. coli*. *Trop. Anim. Health Prod.* **2020**, *52*, 1599–1607. [CrossRef]
170. Kang, H.K.; Park, S.B.; Kim, C.H. Effect of dietary supplementation of red ginseng by-product on laying performance, blood biochemistry, serum immunoglobulin and microbial population in laying hens. *Asian-Australas. J. Anim. Sci.* **2016**, *29*, 1464. [CrossRef]
171. Yadav, S.; Teng, P.Y.; Choi, J.; Singh, A.K.; Vaddu, S.; Thippareddi, H.; Kim, W.K. Influence of rapeseed, canola meal and glucosinolate metabolite (AITC) as potential antimicrobials: Effects on growth performance, and gut health in *Salmonella* Typhimurium challenged broiler chickens. *Poult. Sci.* **2022**, *101*, 101551. [CrossRef] [PubMed]
172. Oryschak, M.A.; Smit, M.N.; Beltranena, E. *Brassica napus* and *Brassica juncea* extruded-expelled cake and solvent-extracted meal as feedstuffs for laying hens: Lay performance, egg quality, and nutrient digestibility. *Poult. Sci.* **2020**, *99*, 350–363. [CrossRef] [PubMed]
173. Hu, C.H.; Wang, D.G.; Pan, H.Y.; Zheng, W.B.; Zuo, A.Y.; Liu, J.X. Effects of broccoli stem and leaf meal on broiler performance, skin pigmentation, antioxidant function, and meat quality. *Poult. Sci.* **2012**, *91*, 2229–2234. [CrossRef] [PubMed]
174. Hu, C.H.; Zuo, A.Y.; Wang, D.G.; Pan, H.Y.; Zheng, W.B.; Qian, Z.C.; Zou, X.T. Effects of broccoli stems and leaves meal on production performance and egg quality of laying hens. *Anim. Feed. Sci. Technol.* **2011**, *170*, 117–121. [CrossRef]
175. Mustafa, A.F.; Baurhoo, B. Evaluation of dried vegetables residues for poultry: II. Effects of feeding cabbage leaf residues on broiler performance, ileal digestibility and total tract nutrient digestibility. *Poult. Sci.* **2017**, *96*, 681–686. [CrossRef]
176. Mustafa, A.F.; Baurhoo, B. Evaluation of dried vegetable residues for poultry: III Effects of feeding cabbage leaf residues on laying performance, egg quality, and apparent total tract digestibility. *J. Appl. Poult. Res.* **2018**, *27*, 145–151. [CrossRef]
177. Tufarelli, V.; Losacco, C.; Tedone, L.; Passantino, L.; Tarricone, S.; Laudadio, V.; Colonna, M.A. Hemp seed (*Cannabis sativa* L.) cake as sustainable dietary additive in slow-growing broilers: Effects on performance, meat quality, oxidative stability and gut health. *Vet. Q.* **2023**, *43*, 1–12. [CrossRef]
178. Wafar, R.; Hannison, M.; Abdullahi, U.; Makinta, A. Effect of Pumpkin (*Cucurbita pepo* L.) seed meal on the performance and carcass characteristics of broiler chickens. *Asian J. Adv. Agric. Res.* **2017**, *2*, 1–7. [CrossRef]
179. Kithama, M.; Ross, K.; Diarra, M.S.; Kiarie, E.G. Utilization of grape (*Vitis vinifera*), cranberry (*Vaccinium macrocarpon*), wild blueberry (*Vaccinium angustifolium*), and apple (*Malus pumila*/domestica) pomaces in broiler chickens when fed without or with multi-enzyme supplement. *Can. J. Anim. Sci.* **2022**, *103*, 15–25. [CrossRef]

180. Vlaicu, P.A.; Panaite, T.D.; Untea, A.E.; Idriceanu, L.; Cornescu, G.M. Herbal plants as feed additives in broiler chicken diets. *Arch. Zootech.* **2021**, *24*, 76–95. [CrossRef]

181. Zheng, M.; Mao, P.; Tian, X.; Guo, Q.; Meng, L. Effects of dietary supplementation of alfalfa meal on growth performance, carcass characteristics, meat and egg quality, and intestinal microbiota in Beijing-you chicken. *Poult. Sci.* **2019**, *98*, 2250–2259. [CrossRef] [PubMed]

182. Cui, Y.; Diao, Z.; Fan, W.; Wei, J.; Zhou, J.; Zhu, H.; Li, D.; Guo, L.; Tian, Y.; Song, H.; et al. Effects of dietary inclusion of alfalfa meal on laying performance, egg quality, intestinal morphology, caecal microbiota and metabolites in Zhuanghe Dagu chickens. *Ital. J. Anim. Sci.* **2022**, *21*, 831–846. [CrossRef]

183. Dotas, V.; Bampidis, V.A.; Sinapis, E.; Hatzipanagiotou, A.; Papanikolaou, K. Effect of dietary field pea (*Pisum sativum* L.) supplementation on growth performance, and carcass and meat quality of broiler chickens. *Livest. Sci.* **2014**, *164*, 135–143. [CrossRef]

184. Laudadio, V.; Tufarelli, V. Growth performance and carcass and meat quality of broiler chickens fed diets containing micronized-dehulled peas (*Pisum sativum* cv. Spirale) as a substitute of soybean meal. *Poult. Sci.* **2010**, *89*, 1537–1543. [CrossRef] [PubMed]

185. Arab Ameri, S.; Samadi, F.; Dastar, B.; Zerehdaran, S. Effect of peppermint (*Mentha piperita*) powder on immune response of broiler chickens in heat stress. *Iran. J. Appl. Anim. Sci.* **2016**, *6*, 435–445.

186. Abdel-Wareth, A.A.; Kehraus, S.; Südekum, K.H. Peppermint and its respective active component in diets of broiler chickens: Growth performance, viability, economics, meat physicochemical properties, and carcass characteristics. *Poult. Sci.* **2019**, *98*, 3850–3859. [CrossRef]

187. Abu Isha, A.A.; Abd El-Hamid, A.E.; Ziena, H.M.; Ahmed, H.A. Effect of spearmint (*Mentha spicata*) on productive and physiological parameters of broiler chicks. *Egypt. Poult. Sci. J.* **2018**, *38*, 815–829. [CrossRef]

188. Rostami, H.; Seidavi, A.; Dadashbeiki, M.; Asadpour, Y.; Simões, J.; Shah, A.A.; Laudadio, V.; Losacco, C.; Perillo, A.; Tufarelli, V. Supplementing dietary rosemary (*Rosmarinus officinalis* L.) powder and vitamin E in broiler chickens: Evaluation of humoral immune response, lymphoid organs, and blood proteins. *Environ. Sci. Pollut. Res.* **2018**, *25*, 8836–8842. [CrossRef]

189. Norouzi, B.; Qotbi AA, A.; Seidavi, A.; Schiavone, A.; Marín AL, M. Effect of different dietary levels of rosemary (*Rosmarinus officinalis*) and yarrow (*Achillea millefolium*) on the growth performance, carcass traits and ileal micro-biota of broilers. *Ital. J. Anim. Sci.* **2015**, *14*, 3930. [CrossRef]

190. George, O.S.; Allison, G.H.; Ekine, O.A. Performance and biochemical parameters of broiler chickens fed avocado (*Persea americana*) seed meal based diet. *Niger. J. Anim. Prod.* **2020**, *47*, 188–193. [CrossRef]

191. Adelowo, O.V.; Shon, E.M.; Kwaghaondo, B.A.; Gambo, C.D. Haematology and serum biochemistry of broiler chickens fed diets with avocado (*Persea americana*) seed meal. In Proceedings of the 49th Conference, Nigeria Society for Animal Production, Ibadan, Nigeria, 24–27 March 2024; pp. 484–487.

192. Omer, H.A.; Ahmed, S.M.; Abdel-Magid, S.S.; El-Mallah, G.M.; Bakr, A.A.; Abdel Fattah, M.M. Nutritional impact of inclusion of garlic (*Allium sativum*) and/or onion (*Allium cepa* L.) powder in laying hens' diets on their performance, egg quality, and some blood constituents. *Bull. Natl. Res. Cent.* **2019**, *43*, 23. [CrossRef]

193. Onibi, G.E.; Adebisi, O.E.; Fajemisin, A.N.; Adetunji, A.V. Response of broiler chickens in terms of performance and meat quality to garlic (*Allium sativum*) supplementation. *Afr. J. Agric. Res.* **2009**, *4*, 511–517.

194. Uchegbu, M.C.; Ogbuewu, I.P.; Ezebuiro, L.E. Blood chemistry and haematology of finisher broilers fed with plantain (*Musa paradisiaca* L.) peel in their diets. *Comp. Clin. Pathol.* **2017**, *26*, 605–609. [CrossRef]

195. Cayan, H.; Erener, G. Effect of olive leaf (*Olea europaea*) powder on laying hens performance, egg quality and egg yolk cholesterol levels. *Asian-Australas. J. Anim. Sci.* **2015**, *28*, 538. [CrossRef]

196. Dedousi, A.; Kotzamanidis, C.; Kritsa, M.-Z.; Tsoureki, A.; Andreadelli, A.; Patsios, S.I.; Sossidou, E. Growth Performance, Gut Health, Welfare and Qualitative Behavior Characteristics of Broilers Fed Diets Supplemented with Dried Common (*Olea europaea*) Olive Pulp. *Sustainability* **2023**, *15*, 501. [CrossRef]

197. Monesa, S.B.; Oluremi OI, A. Effect of Adding Different Levels of Undecorticated Rosehip (*Rosa canina* L.) Fruit in the Diets on Productive Performance of Broiler Chickens. *Asian J. Res. Anim. Vet. Sci.* **2024**, *7*, 122–133. [CrossRef]

198. Costa, M.M.; Alfaia, C.M.; Lopes, P.A.; Pestana, J.M.; Prates, J.A.M. Grape By-Products as Feedstuff for Pig and Poultry Production. *Animals* **2022**, *12*, 2239. [CrossRef]

199. Qorbanpour, M.; Fahim, T.; Javandel, F.; Nosrati, M.; Paz, E.; Seidavi, A.; Ragni, M.; Laudadio, V.; Tufarelli, V. Effect of Dietary Ginger (*Zingiber officinale* Roscoe) and Multi-Strain Probiotic on Growth and Carcass Traits, Blood Biochemistry, Immune Responses and Intestinal Microflora in Broiler Chickens. *Animals* **2018**, *8*, 117. [CrossRef]

200. Siti, N.W.; Bidura, I.G.N.G. Effects of carrot leaves on digestibility of feed, and cholesterol and β-carotene content of egg yolks. *S. Afr. J. Anim. Sci.* **2021**, *51*, 786–792. [CrossRef]

201. Majeed, R.H.; Aziz, A.A.; Aziz KO, H.; Faraj, H.A. Utilization of Parsley (*Petroselinum crispum*) as feed additive for broiler chickens performance. *J. Anim. Poult. Prod.* **2021**, *12*, 363–366. [CrossRef]

202. Marmelstein, S.; Costa, I.P.d.A.; Terra, A.V.; Silva, R.F.d.; Capela, G.P.d.O.; Moreira, M.Â.L.; Junior, C.d.S.R.; Gomes, C.F.S.; Santos, M.d. Advancing Efficiency Sustainability in Poultry Farms through Data Envelopment Analysis in a Brazilian Production System. *Animals* **2024**, *14*, 726. [CrossRef]

203. Campos, D.A.; Gómez-García, R.; Vilas-Boas, A.A.; Madureira, A.R.; Pintado, M.M. Management of Fruit Industrial By-Products—A Case Study on Circular Economy Approach. *Molecules* **2020**, *25*, 320. [CrossRef] [PubMed]

204. Osorio, L.L.D.R.; Flórez-López, E.; Grande-Tovar, C.D. The Potential of Selected Agri-Food Loss and Waste to Contribute to a Circular Economy: Applications in the Food, Cosmetic and Pharmaceutical Industries. *Molecules* **2021**, *26*, 515. [CrossRef] [PubMed]
205. Khoshnevisan, B.; Duan, N.; Tsapekos, P.; Awasthi, M.K.; Liu, Z.; Mohammadi, A.; Angelidaki, I.; Tsang, D.C.W.; Zhang, Z.; Pan, J. A critical review on livestock manure biorefinery technologies: Sustainability, challenges, and future perspectives. *Renew. Sustain. Energy Rev.* **2021**, *135*, 110033. [CrossRef]
206. De Corato, U.; De Bari, I.; Viola, E.; Pugliese, M. Assessing the main opportunities of integrated biorefining from agro-bioenergy co/by-products and agroindustrial residues into high-value added products associated to some emerging markets: A review. *Renew. Sustain. Energy Rev.* **2018**, *88*, 326–346. [CrossRef]
207. Gómez-García, R.; Campos, D.A.; Aguilar, C.N.; Madureira, A.R.; Pintado, M. Valorisation of food agro-industrial by-products: From the past to the present and perspectives. *J. Environ. Manag.* **2021**, *299*, 113571. [CrossRef]
208. Busse, M.; Kernecker, M.L.; Zscheischler, J.; Zoll, F.; Siebert, R. Ethical concerns in poultry production: A German consumer survey about dual purpose chickens. *J. Agric. Environ. Ethics* **2019**, *32*, 905–925. [CrossRef]
209. Kleyn, F.J.; Ciacciariello, M. Future demands of the poultry industry: Will we meet our commitments sustainably in developed and developing economies? *World's Poult. Sci. J.* **2021**, *77*, 267–278. [CrossRef]
210. Henchion, M.M.; De Backer, C.J.; Hudders, L.; O'Reilly, S. Ethical and sustainable aspects of meat production; consumer perceptions and system credibility. In *New Aspects of Meat Quality; Woodhead Publishing Series in Food Science*; Technology and Nutrition; Woodhead Publishing: Cambridge, MA, USA; Kidlington, UK, 2022; pp. 829–851. [CrossRef]
211. Patra, J.K.; Shin, H.-S.; Yang, I.-J.; Nguyen, L.T.H.; Das, G. Sustainable Utilization of Food Biowaste (Papaya Peel) Extract for Gold Nanoparticle Biosynthesis and Investigation of Its Multi-Functional Potentials. *Antioxidants* **2024**, *13*, 581. [CrossRef]
212. Pathak, P.D.; Mandavgane, S.A.; Kulkarni, B.D. Waste to Wealth: A Case Study of Papaya Peel. *Waste Biomass Valor.* **2019**, *10*, 1755–1766. [CrossRef]
213. Han, Z.; Park, A.; Su, W.W. Valorization of papaya fruit waste through low-cost fractionation and microbial conversion of both juice and seed lipids. *RSC Adv.* **2018**, *8*, 27963–27972. [CrossRef] [PubMed]

 *agriculture*

*Article*

# Solid-State Fermentation Using *Bacillus licheniformis*-Driven Changes in Composition, Viability and In Vitro Protein Digestibility of Oilseed Cakes

Dan Rambu [1,2,*], Mihaela Dumitru [2], Georgeta Ciurescu [2] and Emanuel Vamanu [1]

[1]  Faculty of Biotechnology, University of Agricultural Sciences and Veterinary Medicine, 011464 Bucharest, Romania; emanuel.vamanu@bth.usamv.ro
[2]  Animal Nutrition and Biotechnology Department, National Research Development Institute for Biology and Animal Nutrition, 077015 Balotesti, Romania; mihaela.dumitru@ibna.ro (M.D.); ciurescu@ibna.ro (G.C.)
*  Correspondence: dan.rambu@ibna.ro

**Abstract:** The solid-state fermentation (SSF) efficiency of *Bacillus licheniformis* ATCC 21424 (BL) on various agro-industrial by-products such as oilseed cakes [hemp (HSC), pumpkin (PSC), and flaxseed (FSC)] was evaluated by examining the nutritional composition, reducing sugars, and in vitro protein digestibility (IVPD) for use in animal nutrition. SSF significantly decreased crude protein, along with changes in the total carbohydrates ($p < 0.05$) for all substrates fermented. An increase in crude fat for HSC (1.04%) and FSC (1.73%) was noted, vs. PSC, where the crude fat level was reduced ($-3.53$%). Crude fiber does not differ significantly between fermented and nonfermented oilseed cakes ($p > 0.05$). After fermentation, neutral detergent fiber (NDF) and acid detergent fiber (ADF) significantly increased for HSC and FSC ($p < 0.05$), as well as for PSC despite the small increase in ADF (4.46%), with a notable decrease in NDF ($-10.25$%). During fermentation, pH shifted toward alkalinity, and after drying, returned to its initial levels for all oilseed cakes with the exception of PSC, which maintained a slight elevation. Further, SSF with BL under optimized conditions (72 h) increases the reducing sugar content for FSC (to 1.46%) and PSC (to 0.89%), compared with HSC, where a reduction in sugar consumption was noted (from 1.09% to 0.55%). The viable cell number reached maximum in the first 24 h, followed by a slowly declining phase until the end of fermentation (72 h), accompanied by an increase in sporulation and spore production. After 72 h, a significant improvement in water protein solubility for HSC and FSC was observed ($p < 0.05$). The peptide content (mg/g) for oilseed cakes fermented was improved ($p < 0.05$). Through gastro-intestinal simulation, the bacterial survivability rate accounted for 90.2%, 101.5%, and 85.72% for HSC, PSC, and FSC. Additionally, IVPD showed significant improvements compared to untreated samples, reaching levels of up to 65.67%, 58.94%, and 80.16% for HSC, PSC, and FSC, respectively. This research demonstrates the advantages of oilseed cake bioprocessing by SSF as an effective approach in yielding valuable products with probiotic and nutritional properties suitable for incorporation into animal feed.

**Keywords:** solid-state fermentation; Bacillus; proximate composition; digestibility; IVPD; fatty acid profile; soluble proteins; peptides; spores; hemp; pumpkin; linseed; flaxseed

**Citation:** Rambu, D.; Dumitru, M.; Ciurescu, G.; Vamanu, E. Solid-State Fermentation Using *Bacillus licheniformis*-Driven Changes in Composition, Viability and In Vitro Protein Digestibility of Oilseed Cakes. *Agriculture* **2024**, *14*, 639. https://doi.org/10.3390/agriculture14040639

Academic Editor: Gang Shu

Received: 19 March 2024
Revised: 16 April 2024
Accepted: 18 April 2024
Published: 22 April 2024



## 1. Introduction

Considering the increasing demand of alternative high-protein sources of foods and feeds, oilseed cakes (OSCs), derived from the industry of edible oil extraction as an under-rated by-product, could serve as a drive for sustainability through enhancing the circular economy and food security [1]. Replacing soybean meal, which has long been considered as the golden standard in animal nutrition [2], with locally sourced oilseed cakes could contribute to the amelioration of the negative impact related with its cultivation such as poor socioeconomics, loss of biodiversity, fertilizers, long transport routes, etc. [3]. With a high protein content, hemp, pumpkin, and flaxseed cakes represent a tempting alternative to traditional feeding systems. However, the presence of anti-nutritional factors such as trypsin inhibitors, phytic acid, and tannins can reduce protein bioavailability and digestibility, impairing OSC efficient utilization [4]. In this context, an integration of the knowledge acquired from traditional fermentation and modern innovation can drive meticulous changes in nutritional and bioactive composition, shaping the path towards precision.

Solid-state fermentation (SSF), a process caried out by one or a diverse array of microorganisms on a solid substrate with a moisture typically between 30 and 80% [5], has gained more attention in recent years due to a variety of advantages that this technology poses in biotransformation. By using solid matrixes, microbial fermentation can create a unique microenvironment that promotes a microorganism's resistance to catabolic repression (i.e., inhibition of enzyme synthesis) [6], the biosynthesis of high yields of enzymes (amylases, proteases, xylanases, phytases, etc.) [7–9], and bioactive compounds' [10] improvement of their nutrient profile and absorption, followed by a reduction in anti-nutritional factors [11]. Despite the numerous advantages of SSF, the process of scaling-up encounters several constraints, including temperature build-up, pH regulation, oxygen transfer, moisture regulation, and an uneven distribution of cell mass and nutrients. These challenges underscore the critical significance of rational design and meticulous process control [12].

*Bacillus licheniformis* is a Gram-positive, endospore-forming bacterium which is generally regarded as safe (GRAS) and that has a high extracellular protein secretory capacity [13]. It is known for a diverse range of proteases, including alkaline proteases exhibiting optimal activity within the pH range of 8–12, and neutral proteases showing efficacy between pH 5.0 and 8.0 [14]. The enzymatic action of proteases facilitates the hydrolysis of protein substrates in fermentation media into smaller peptides with improved functionality, such as solubility, digestibility, and biologic activity [15].

Furthermore, SSF is considered one of the most efficient techniques used in altering the functional properties and/or bioactive compounds of a target substrate [16]. Moreover, fermentation has the potential to improve the nutritional composition of various substrates or to obtain value-added products. Several studies found that *B. licheniformis* increased peptide content [17] and bioactivity, and significantly improved the nutritional profile of co-substrates of brewer's spent grain and soybean meal (SBM) [18]. In addition, data from the literature present successful results of *B. licheniformis* in SSF, for example, the SSF of SBM and rapeseed meal to increase peptide, soluble protein content, and also to eliminate anti-nutritional factors [19,20]. Through this enzymatic activity, they contribute to the degradation of substrates, the conversion of ingredients, and the synthesis of new compounds [20,21].

Some reports indicate that fermented products derived from *Bacillus licheniformis* exhibit promising effects on gut health. These products have been found to positively influence gut conditions by modulating the microbiota composition. They demonstrate the ability to inhibit the proliferation of harmful enteric pathogens while promoting the growth of beneficial bacteria such as lactobacilli. Additionally, these fermented products have been associated with improvements in nutrient digestibility and have shown potential in alleviating various gut-related diseases [22]. The proposed mechanisms are related to the enhancement of gut barrier functions, enhanced immunity (lymphocyte activation, increased levels of immune globulins, improved cellular or humoral immunity), as well as the production of antibiotic proteins/peptides [23]. Animal studies have shown a positive impact of solid-state fermentation (SSF) using *Bacillus licheniformis* in alleviating necrotic enteritis and improving intestinal morphology in broilers challenged with Clostridium perfringens [24,25] as well as in reducing diarrhea incidence in weaning piglets [26].

The next step toward precision agriculture involves targeted probiotic interventions, as described here, owing to the ability of the *Bacillus licheniformis* to form spores under environmental stress; the bacterium demonstrates a unique resilience that enables it to withstand harsh conditions encountered within the gastrointestinal tract of animals [27,28].

Considering these aspects, this study aims to bridge the gap between solid-state fermentation, probiotics, and sustainable protein sources by exploring the dynamics of *Bacillus licheniformis* growth and sporulation patterns throughout the fermentation process. Additionally, we studied the driven changes in proximate composition and fatty acid profiles to gain a deeper understanding of how this process influences these nutritional aspects. By assessing alterations in protein solubility and reducing sugars, we seek to provide an overview of the hydrolysis process. Finally, implementing gastrointestinal simulation allows us to study protein bioavailability and bacterial survival which are crucial factors in developing probiotic-enhanced feed formulations.

## 2. Materials and Methods

### 2.1. Chemicals

Enzymes used in this study were $\alpha$-amylase (10080; 52.2 U/mg), pepsin (P7000, 605 U/mg), and pancreatin (P3292, 4 × USP) bought from Sigma-Aldrich (Saint-Louis, MO, USA). Culture media and bile salts were bought from Oxoid™ (Basingstoke, UK). All other reagents used in the study were of analytical grade, purchased from Sigma-Aldrich (Saint-Louis, MO, USA) and Carl Roth (Karlsruhe, Germany).

### 2.2. Plant Material and Microorganism

Hempseed cake (*Cannabis sativa* L.) was kindly provided by Canah International S.R.L. (Bihor, Romania), while pumpkin (*Cucurbita* spp.) and flaxseed (*Linum usitatissimum* L.) cakes were kindly supplied by Dachim S.R.L. (Cluj, Romania), both of which resulted as byproducts from oil cold-pressing production processes. The microbial strain used in fermentation, *Bacillus licheniformis* ATCC 21424, was acquired from The American Type Culture Collection (Manassas, VA, USA).

### 2.3. Solid-State Fermentation (SSF)

SSF was carried out for 3 different timepoints (24, 48, and 72 h) using separate 500 mL Erlenmeyer flasks for each timepoint, where 100 g from each oilseed cake was added individually. Flasks were plugged with cotton wool covered by aluminum foil and autoclaved at 121 °C for 15 min. Frozen bacterial culture was revitalized, and after 2 passages on Nutrient Agar, a working stock culture of *Bacillus licheniformis* was prepared in LB (Luria–Bertani) broth and grown overnight for approx. 20 h at 37 °C, 150 rpm to an $OD_{600}$ of 4.5 (this value was extrapolated from a 10-fold dilution). SSF flasks were inoculated with 18 mL of working stock culture and 82 mL of autoclaved distilled water was added to achieve a final $OD_{600}$ of 0.8/g equivalent to a viable cell number of $7.75 \times 10^8$ cfu/g and 50% moisture content. After a thorough mix using a sterile spatula, inoculated flasks were incubated at 37 °C. After the designated incubation period elapsed, samples were collected for viability assessment and spore number quantitation according to a pH of 2.4. Subsequently, the remaining quantity of samples was stabilized for future analysis by subjecting it to freeze-drying; afterwards, the samples were blended for 30 s in order to achieve a powdery consistency and stored at 4 °C.

### 2.4. Microbial Viability, Spore Number, Colony Morphology, and pH Value

2.4.1. Determination of Microbial Viability

Approximately 1 g of sample was prelevated at 24, 48, and 72 h, mixed with autoclaved distilled water (1:9 $w/v$), and shaken at 150 rpm for 30 min at room temperature. For the determination of viable cell number, 1 mL of supernatant was sampled and serially diluted in 0.85% saline solution, and 0.1 mL was plated on Nutrient Agar and expressed as log cfu/g after 24 h incubation at 37 °C.

2.4.2. Determination of Spore Number

Spore number was assayed at 24, 48, and 72 h, with 1 mL of the supernatant consisting of vegetative cells, and the spores were diluted 10-fold in 0.85% saline solution and incubated at 80 °C for 15 min [28] in order to eliminate vegetative forms. Afterwards, it was serially diluted and plated on Nutrient Agar, and spore number was determined in the same manner as for microbial viability assay.

### 2.4.3. Determination of pH Value

The pH of the fermented samples at 0, 24, 48, and 72 h was measured for the supernatant using a calibrated pH meter (pH 7.0 + DHS, XS Instruments, Carpi, Italy).

### 2.5. Proximate Composition

The proximate composition of oilseed cakes at 0 and 72 h was assayed based on the recommended methods of the AOAC (Association of Official Analytical Chemists) [29]. In brief, dry matter was determined gravimetrically by the drying oven method at 105 °C (Method 925.09); crude protein was determined by the Kjeldahl method (Method 979.09) on a Kjeltec auto 1030 system (Höganäs, Sweden); crude fat was determined gravimetrical by organic solvent extraction (Method 920.39) using a Soxtec 2055 Foss Soxhlet extractor (Höganäs, Sweden); ash content was assayed gravimetrically by calcination in an oven at 600 °C; crude fiber was quantified by successive hydrolysis in alkaline and acidic environments (Method 962.09); neutral detergent fiber (NDF) and acid detergent fiber (ADF) contents were determined by Van Soest extraction using a Raw Fiber Extractor FIWE 6 (Velp Scientifica, Usmate, Italy) [30,31]; and finally, carbohydrate content was determined as a nitrogen-free extract (NFE), as follows: NFE (%) = dry matter% − (crude protein% + crude fat% + crude ash% + crude fiber%) [32].

### 2.6. Fatty Acid Profile

Fatty acids (FAs) from total lipid extract were determined as FA methyl esters (FAMEs) at 0 and 72 h based on the ISO/TS 17764-2:2008 [33] standard method [31] by GC-FID using a Perkin Elmer-Clarus 500 (Waltham, MA, USA) gas chromatograph equipped with a TRACE TR-Fame (Thermo Electron, Waltham, MA, USA), 60 m × 0.25 mm × 0.25 μm, capillary column.

### 2.7. Reducing Sugars

The concentration of reducing sugars at 0, 24, 48, and 72 h was measured by a modified method of Miller's work [34] using DNSA reagent (10 g/L 3,5-dinitrosalicylic acid, 2 g/L phenol, 10 g/L NaOH, and 200 g/L sodium potassium tartrate, 0.5 g/L $Na_2SO_3$). Samples were prepared by extracting the reducing sugars from 1 g of fermented feeds or controls in 25 mL distilled water in a rotatory shaker at 220 rpm and room temperature for 1 h following a 10 min centrifugation at 10,000 rpm and 4 °C. In a test tube, 1 mL of extract and 2 mL DNSA reagent were vortexed and incubated at 100 °C for exactly 5 min. After incubation, the test tubes were cooled in a water bath, and 9 mL of distilled water was added and recentrifuged. Subsequently, the samples were read on Eppendorf BioSpectrometer (Hamburg, Germany) at 540 nm in comparison to a reagent blank. A calibration curve was constructed across a range of 0.1–2 mg/mL ($R^2$ = 0.9982) using glucose as standard.

### 2.8. Soluble Proteins and Peptides, Extraction, and Quantification

### 2.8.1. Determination of Soluble Proteins

Soluble proteins and peptides in neutral solution (pH 7.0), salt solution (0.5 M NaCl), and alkaline environment (pH 10.0) at 0, 24, 48, and 72 h were determined as an indicator of the effectiveness of the fermentation process, providing direct insights on enzymatic activities, extraction efficiency, and protein hydrolysis. The extract was prepared by adding 1 g of fermented feeds or controls into 25 mL of distilled water for neutral extraction, into 25 mL 0.5 M NaCl solution for salt extraction, and for alkaline extraction, the mixture in 25 mL of distilled water was vortexed for 1 min and brought to pH 10.0 using a 3 M NaOH solution. Extraction was performed in a rotatory shaker at 220 rpm and room temperature for 1 h following a 10 min centrifugation at 10,000 rpm and 4 °C, and diluted up to 100-fold.

The soluble protein content of the supernatant of the extract was determined using Pierce BCA Protein Assay kit (Thermo Scientific, Waltham, MA, USA) with bovine serum albumin as standard.

### 2.8.2. Determination of Peptides

Soluble peptide yield was determined as TCA-soluble peptides; 1 mL of 20% trichloroacetic acid (TCA) was added to an equal volume of extract and incubated for 30 min at 4 °C for protein precipitation, followed by centrifugation [35], and diluted up to 100-fold. Peptide content was quantitated using a bicinchoninic acid (BCA) assay kit.

### 2.9. In Vitro Protein Digestibility (IVPD) and Microorganism Viability under Simulated Conditions

The in vitro gastrointestinal (GI) digestion simulation model previously described [36] was employed with some modifications. The GI simulation model consisted of 3 steps that resemble the conditions within a digestive system, namely oral, gastric, and intestinal phases, all of which each fermented (after 72 h) and control (after 0 h) sample underwent. Prior to the experiment, a test was performed to determinate the exact volume of the HCL 3 M/NaOH 3 M needed to adjust the pH of the oral, gastric, and intestinal stages for fermented and control samples as each one has a unique buffering ability. The volume was recorded and the results were adjusted according to it. Simulated solutions containing electrolytes were prepared according to the protocol, pre-warmed to 37 °C, and adjusted to their specific pH in order to mimic physiological conditions. Shortly before the experiment, enzymes and bile salts were added and mixed by employing a magnetic stirrer.

Oral stage: A 2.5 g sample was mixed with 2.5 mL of distilled water and a 5 mL simulated oral solution containing 750 U α-amylase for 2 min at pH 7.0.

Gastric stage: The oral bolus was mixed with 10 mL of simulated gastric solution containing 40,000 U pepsin for 2 h at pH 3.0.

Intestinal phase: The gastric chyme was mixed with 20 mL simulated intestinal solution containing 32 mg/mL pancreatin [37] and 24 mg/mL bile salts for 2 h at pH 7.0 [38].

Samples were taken at different points of the digestion: gastric 1 h (G1), gastric 2 h (G2), intestinal 1 h (I1), and intestinal 2 h (I2). Distinct digestion flasks were prepared per each timepoint. At the end of the prescribed time, samples were taken and immediately diluted and plated (using the method stated at Section 2.4.1) for the viability assay. For IVPD, samples were centrifuged at 4 °C and the soluble peptide content was assayed (Section 2.8.2). IVPD was calculated using the following formula [39]:

$$\text{Protein digestibility (\%)} = B/A \times 100$$

where B is the soluble peptide content after each digestion stage and A is the total protein content. Independently, a reagent blank for each timepoint was prepared and its peptide content was subtracted from the total concentration:

$$\text{Survivability rate (\%)} = B/A \times 100$$

where B is the log cfu/g viability after the I2 stage and A is the initial log cfu/g.

### 2.10. Statistical Analysis

The results were calculated on dry basis (DW). The data are the means of triplicate experiments ± standard deviations. Statistical analysis was performed using Minitab v.21.2 (State College, PA, USA) where differences between the means were evaluated by a one-way ANOVA followed by Tukey's test. For analyzing the proximate composition and FA profile, paired *t*-tests were used. Differences were considered significant at $p < 0.05$. Graphics were prepared using Prism-GraphPad v. 9.1.2 (San Diego, CA, USA).

## 3. Results

### 3.1. Effect of Fermentation on Proximate Composition and Fatty Acid Profile

Fermentation with *Bacillus licheniformis* has a notable effect on the nutritional composition of the samples as presented in Table 1. Variations in dry weight across samples are attributed to the drying conditions utilized rather than being solely a consequence of the fermentation process itself. Crude protein decreased significantly ($p < 0.05$) for all samples, the most significant being for HSC ($-5.40\%$). The increases in crude fat for HSC and FSC ($1.04\%$ respectively $1.73\%$) and decrease ($p > 0.05$) for PSC ($-3.53\%$) highlights the substrate specificity of this process along with the capacity for de novo lipid biosynthesis from carbohydrates whose lipidic profile will be further explored through FA profile analysis. Nitrogen-free extracts expressed as total carbohydrates significantly decreased ($p < 0.05$) for FSC, suggesting the bacterial capacity to ferment specific carbohydrates presented in flaxseed. Small increases were observed for HSC and were more accentuated for PSC relative to its initial content. Crude fiber does not differ significantly between fermented and unfermented ($p > 0.05$) samples for HSC, and small increases were accounted for PSC and FSC. NDF and ADF significantly increased for HSC and FSC ($p < 0.05$), whereas PSC showed small increases despite ADF ($4.46\%$), and a notable decrease in NDF was observed ($-10.25$).

**Table 1.** Changes in proximate composition after fermentation.

| Parameter | HSC T 0 h | HSC T 72 h | PSC T 0 h | PSC T 72 h | FSC T 0 h | FSC T 72 h |
|---|---|---|---|---|---|---|
| DM (%) | 93.58 ± 0.38 | 95.82 ± 0.1 | 90.91 ± 0.1 | 95.17 ± 0.03 | 90.54 ± 0.06 | 96.44 ± 0.05 |
| CP (% DM) | 38.00 ± 0.22 | 35.95 ± 0.08 | 45.27 ± 0.1 | 44.65 ± 0.17 | 33.35 ± 0.13 | 32.53 ± 0.07 |
| Crude fat (% DM) | 8.44 ± 0.11 | 8.53 ± 0.08 | 15.60 ± 0.09 | 15.05 ± 0.1 | 18.82 ± 0.09 | 19.15 ± 0.14 |
| Carbohydrates (% DM) | 17.10 ± 0.05 | 18.24 ± 0.27 | 1.55 ± 0.58 | 3.45 ± 0.47 | 34.07 ± 0.149 | 31.21 ± 0.56 |
| CF (% DM) | 30.23 ± 0.09 | 30.14 ± 0.04 | 31.18 ± 0.29 | 32.27 ± 0.15 | 10.44 ± 0.23 | 11.78 ± 0.22 |
| NDF (% DM) | 43.98 ± 0.13 | 47.16 ± 0.1 | 45.28 ± 0.28 | 40.64 ± 0.33 | 27.32 ± 0.17 | 28.22 ± 0.24 |
| ADF (% DM) | 29.96 ± 0.25 | 33.83 ± 0.12 | 29.60 ± 0.17 | 30.92 ± 0.19 | 13.99 ± 0.21 | 16.82 ± 0.13 |
| Ash (% DM) | 6.20 ± 0.06 | 7.11 ± 0.06 | 6.37 ± 0.09 | 4.56 ± 0.05 | 4.11 ± 0.09 | 4.48 ± 0.06 |

DM = dry matter; CP = crude protein; CF = crude fiber; NDF = neutral detergent fiber; ADF = acid detergent fiber. Each value represents the mean for three replications ± standard deviation.

The effect of solid-state fermentation of OSC on the FA metabolism of *Bacillus licheniformis* is displayed in Table 2. The results revealed notable alterations in the levels of saturated, monounsaturated, and polyunsaturated FAs across the substrates. A significant increase in total saturated FAs was observed in FSC with 8.59% ($p > 0.05$), contrasting with a decrease in PSC with 6.61%, while no significant changes were noted in HSC. Palmitic acid, which serves as a readily available energy source, decreased with 12.94% and 7.51% for PSC and FSC, respectively, whereas a notable increase of 7.41% was observed in HSC. Levels of monounsaturated FAs increased significantly with 3.59% and 6.91% for PSC and FSC substrates ($p < 0.05$), but decreased for HSC, with 6.20%. Total polyunsaturated fatty acids (PUFAs) show variation, where small significant increases were observed for HSC and PSC, with 0.69% and 1.21%, respectively, and a notable decrease in FSC fermentation with 3.64% ($p < 0.05\%$). Fermentation does not induce noticeable changes in the total ratio of omega-6 to omega-3 FAs (n-6/n-3 ratio). However, individual fatty acids such as linoleic acid and omega-6 decreased with 0.33% and 2.04% for HSC and FSC, increasing with 3.61% for FSC after 72 h. Alpha linolenic acid, an omega-3 PUFA, is also affected by fermentation, decreasing with 6.15% for HSC, 22.88% for PSC, and 3.51% for FSC, respectively. Arachidonic acid, which serves as a precursor for pro-inflammatory eicosanoids, exhibited noteworthy changes, increasing approx. 5-fold in HSC and decreasing below undetectable levels for PSC.

**Table 2.** Changes in fatty acid profile after fermentation.

| FA (g/100 g) | HSC T 0 h | HSC F 72 h | PSC T 0 h | PSC F 72 h | FSC T 0 h | FSC F 72 h |
|---|---|---|---|---|---|---|
| Myristic C14:0 | 0.04 ± 0.01 | 0.06 ± 0.01 | 0.12 ± 0.02 | 0.11 ± 0.01 | 0.09 ± 0.02 | 0.05 ± 0.01 |
| Pentadecanoic C15:0 | 0.04 ± 0.01 | 0.11 ± 0.02 | 0.04 ± 0.01 | 0.19 ± 0.02 | ND | 0.22 ± 0.02 |
| Palmitic C16:0 | 8.61 ± 0.05 | 9.29 ± 0.04 | 14.53 ± 0.09 | 12.65 ± 0.15 | 7.67 ± 0.13 | 7.51 ± 0.19 |
| Heptadecanoic C17:0 | 0.72 ± 0.04 | ND | 0.03 ± 0.0 | 0.05 ± 0.0 | ND | 0.08 ± 0.01 |
| Stearic C18:0 | 2.66 ± 0.04 | 3.05 ± 0.06 | 5.03 ± 0.07 | 5.45 ± 0.06 | 3.32 ± 0.08 | 3.99 ± 0.2 |
| Arachic C20:0 | 1.85 ± 0.03 | 1.75 ± 0.05 | ND | ND | ND | ND |
| **TOTAL SATURATED** | 13.97 ± 0.08 | 14.28 ± 0.18 | 19.76 ± 0.19 | 18.45 ± 0.22 | 11.09 ± 0.24 | 12.04 ± 0.38 |
| Pentadecenoic C15:1 | 0.12 ± 0.02 | ND | 0.07 ± 0.02 | ND | 0.05 ± 0 | ND |
| Palmitoleic C16:1 | 0.19 ± 0.02 | 0.12 ± 0.01 | 0.11 ± 0.03 | 0.05 ± 0.0 | 0.08 ± 0.01 | 0.06 ± 0.01 |
| Heptadecenoic C17:1 | 0.81 ± 0.07 | ND | ND | 0.05 ± 0.0 | ND | 0.73 ± 0.07 |
| Oleic C18:1 | 11.49 ± 0.05 | 11.71 ± 0.1 | 28.4 ± 0.3 | 29.51 ± 0.13 | 21.9 ± 0.37 | 22.77 ± 0.08 |
| **TOTAL MONO-UNSATURATED** | 12.61 ± 0.16 | 11.84 ± 0.12 | 28.59 ± 0.31 | 29.61 ± 0.13 | 22.04 ± 0.38 | 23.56 ± 0.13 |
| Linoleic Cis C18:2n6 | 54.13 ± 0.07 | 53.96 ± 0.08 | 48.48 ± 0.41 | 50.23 ± 0.24 | 15.05 ± 0.08 | 14.74 ± 0.06 |
| Linolenic γ C18:3n6 | 0.09 ± 0.02 | ND | ND | ND | 0.21 ± 0.03 | ND |
| α Linolenic C18:3n3 | 17.4 ± 0.04 | 16.33 ± 0.06 | 0.39 ± 0.04 | 0.3 ± 0.04 | 51.14 ± 0.07 | 49.34 ± 0.67 |
| Octadecatetraenoic C18:4n3 | 0.66 ± 0.03 | 1.7 ± 0.08 | 0.24 ± 0.02 | 0.33 ± 0.06 | 0.16 ± 0.02 | 0.24 ± 0.02 |
| Eicosadienoic C20:2n6 | 0.33 ± 0.03 | 0 ± 0 | 0.38 ± 0.04 | 0.14 ± 0.04 | 0.07 ± 0.0 | ND |
| Eicosatrienoic C20:3n6 | ND | ND | ND | ND | ND | ND |
| Arachidonic C20:4n6 | 0.07 ± 0.0 | 0.47 ± 0.04 | 0.17 ± 0 | ND | ND | ND |
| Eicosapentaenoic C20:5n3 | 0.02 ± 0.0 | ND | 0.22 ± 0.04 | ND | ND | ND |
| Docosapentaenoic C22:5n3 | ND | 0.12 ± 0.03 | ND | ND | ND | ND |
| Docosadienoic C22:2n6 | 0.22 ± 0.02 | 0.68 ± 0.04 | 0.16 ± 0.02 | 0.14 ± 0 | ND | ND |
| **Other Fatty Acids** | 0.49 ± 0.02 | 0.68 ± 0.06 | 0.6 ± 0.08 | ND | 0.11 ± 0.0 | ND |
| **TOTAL POLY-UNSATURATED** | 73.44 ± 0.22 | 73.95 ± 0.41 | 50.81 ± 0.54 | 51.43 ± 0.42 | 66.76 ± 0.21 | 64.33 ± 0.72 |
| **n-6/n-3 ratio** | 0.75 | 0.75 | 0.98 | 0.99 | 0.23 | 0.23 |

ND, not detected; each value represents the mean of three replications ± standard deviation (SD). Results are expressed as g FAME/100 g oil.

### 3.2. Effect of Fermentation pH and Reducing Sugars

The initial pH levels of the studied substrates were slightly acidic, at 5.9, 6.5, and 6.6 for FSC, HSC, and PSC, respectively, as shown in Figure 1.

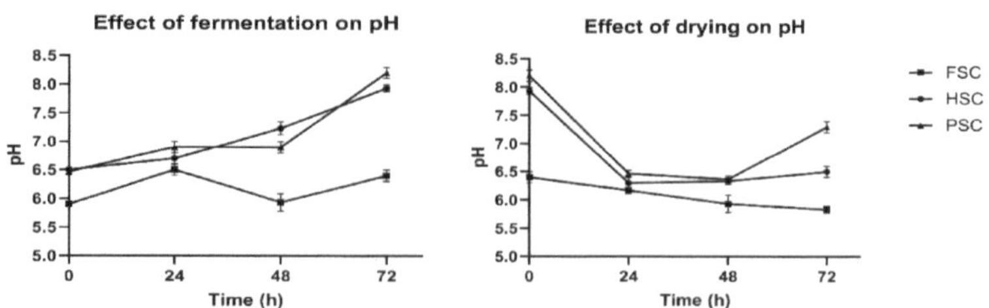


**Figure 1.** Variation of pH during fermentation and after the drying process.

As fermentation progressed, there was a shift toward alkalinity, with final pH levels reaching 7.9 and 8.2 for HSC and PSC. In contrast, the pH remained slightly acidic for FSC, measuring 6.4 at the end of fermentation. For HSC, the pH increased during fermentation, showing an alkalinization trend, while for PSC, alkalinization started only after 48 h of fermentation; finally, for FSC, a slower acidification was observed in the first 48 h of fermentation, followed by a slightly rise in pH toward 6.4. After drying, the pH returned to its initial slightly acidic state for HSC and FSC, suggesting the volatilization of alkaline compounds that increased the pH during fermentation. On the other hand, PSC maintained a slight elevation above the baseline pH, measured at 7.3.

As shown in Figure 2, for FSC, reducing sugar content initially decreased from 1.33% in the first 24 h to 1.09%, followed by a subsequent increase to 1.46% at 72 h. For HSC, sugars were predominantly consumed, showing a decline from 1.09% to 0.55%. On the other hand, the PSC exhibited an increase in reducing sugar concentration, rising from 0.36% to 0.86% at 24 h, followed by a further increase to 0.89% at the end of fermentation.

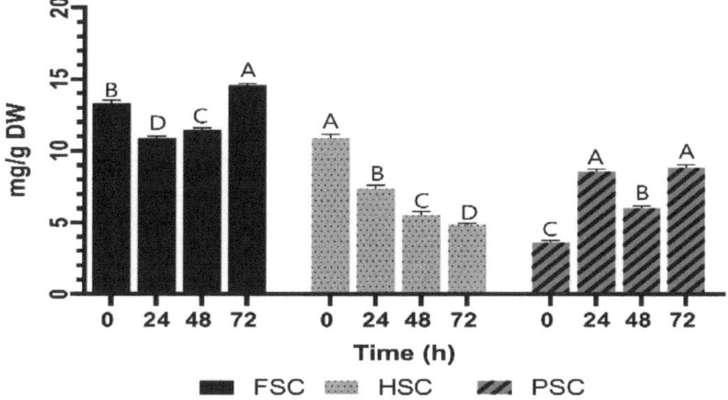


**Figure 2.** Reducing sugar concentrations at different timepoints. The use of distinct uppercase in the graph corresponds to the results of Tukey's post hoc comparison that highlight differences in reducing sugar concentrations across the four timepoints.

### 3.3. Effect of Fermentation on Growth Dynamics and Spore Formation

In this case, *Bacillus licheniformis* growth and spore formation exhibited consistency through all examined samples, as demonstrated in Figure 3. Initial count increased in the first 24 h of fermentation from initial 8.57 log cfu/g up to 11.66, 10.97, and 10.17 log cfu/g for HSC, PSC, and FSC, respectively ($p < 0.05$), following a declining phase until the end of fermentation, which was accompanied by an increased spore formation (not determined for T0). Viability reached the lowest point at 72 h of fermentation for all samples, measuring 9.80, 9.47, and 9.32 log cfu/g for HSC, PSC, and FSC ($p < 0.05$). Sporulation rate (%) relative to total viable counts progressed over time (24 h, 48 h, and 72 h), with a rising trend, as follows: from 39.21% and 71.78% to 84.92% for HSC; from 52.99% and 72.19% to 86.36% for PSC; and from 43.60% and 51.61% to 82.26% for FSC ($p < 0.05$). As HSC and PSC displayed an approx. similar pattern, FSC rose from 43.60% to 51.61% in the first 48 h and rapidly surged to 82.26%.

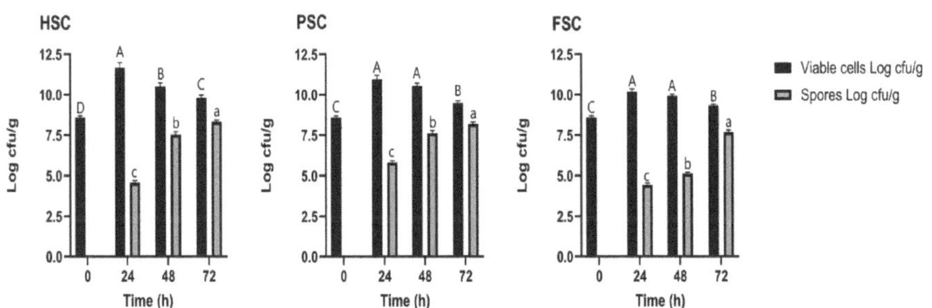



**Figure 3.** Viability and spore concentrations at different intervals. Bars represent SD of means. The use of distinct uppercase and lowercase letters in the graph corresponds to the results of Tukey's post hoc comparisons. These comparisons highlight differences in viability across the four timepoints and separate differences in spore concentration over the same time intervals.

### 3.4. Effect of Fermentation on Protein Solubility and Peptide Content

As illustrated in Figure 4, this study indicates a significant improvement in water protein solubility over time due to fermentation ($p < 0.05$) for HSC and FSC, and most notably for PSC, where an approx. 4.8-fold increase was observed after 72 h. Water-soluble protein content (mg/g) increased from $15.87 \pm 0.61$ to $69.40 \pm 1.06$ for HSC, from $90.44 \pm 0.33$ to $127.64 \pm 0.67$ for FSC, and from $32.46 \pm 0.70$ to $156.56 \pm 0.57$ for PSC. Peptide yield, due to *Bacillus licheniformis'* enzymatic activity, increased in the same manner, improving peptide content (mg/g) from $10.62 \pm 0.53$ to $38.99 \pm 0.67$ for HSC, $43.73 \pm 0.33$ to $63.45 \pm 0.67$ for FSC, and $20.96 \pm 0.25$ to $45.48 \pm 0.88$ for PSC.

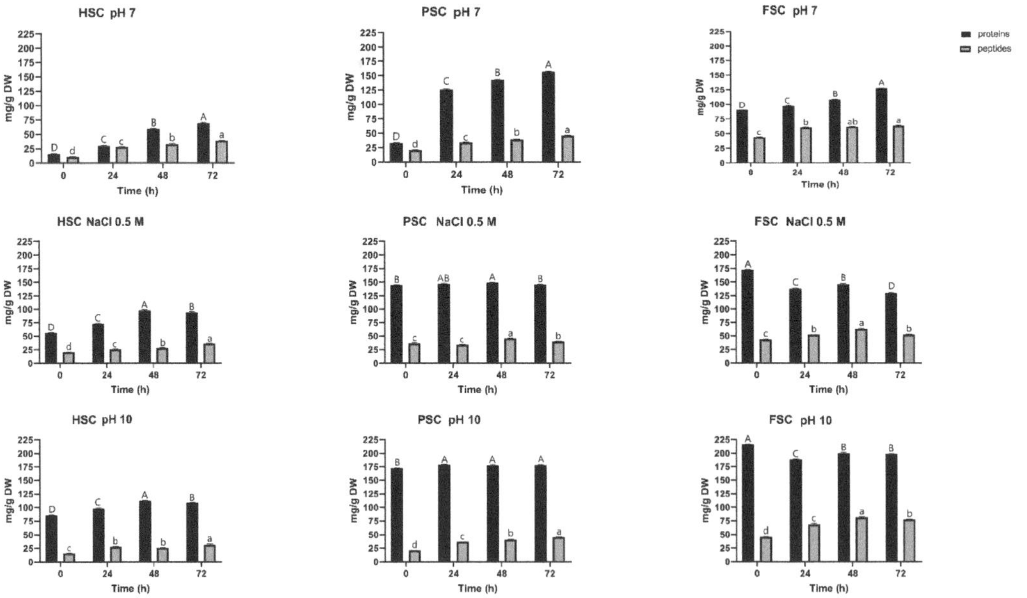



**Figure 4.** Soluble protein and peptide content during fermentation in water, NaCl, and alkaline solution. Bars represent the mean SD. The use of distinct uppercase and lowercase letters in the graph corresponds to the results of Tukey's post hoc comparisons. These comparisons highlight differences in protein solubility across the four timepoints and separate differences in peptide yield over the same time intervals.

Fermentation positively affected NaCl protein extraction for HSC, increasing the extraction yield (mg/g) from 55.44 ± 0.79 to a maximum of 97.72 ± 1.17 at 48 h. In the case of FSC, a noticeable decrease in protein extracted by salt solution from 172.39 ± 0.72 to 129.53 ± 0.94 was observed. Nevertheless, SSF was not found to be relevant for PSC protein salt extraction. Peptide content (mg/g) increased from 20.36 ± 0.46 to 35.61 ± 0.61 for HSC after 72 h. In contrast, for FSC, an increase was observed from 43.14 ± 0.81 to 62.84 ± 0.95 after 48 h of fermentation, followed by a decrease to 51.78 ± 1.05 at 72 h. Similarly with PSC, peptide yield (mg/g) increased from 36.29 ± 1.13 to 44.96 ± 0.85 mg/g at 48 h, followed by a decrease to 39.39 ± 1.02 mg/g at 72 h, probably due microbial consumption or future hydrolysis to amino acids.

Similar trends were observed in the alkaline extraction as in the salt extraction process. SSF positively influenced HSC, resulting in an increase in extraction yield (mg/g) from 85.53 ± 0.78 to 108.64 ± 0.85. Conversely, for FSC, there was a decrease from 215.79 ± 0.97 to 198.26 ± 0.44. In the case of PSC, a marginal increase of approximately 2% was noted within the initial 24 h, rising from 171.63 ± 1.24 to 178.15 ± 0.77. The lowest content of peptides in comparison with other extractions was obtained through alkaline extraction for HSC, yielding from 15.40 ± 0.71 mg/g to 31.35 ± 1.28 mg/g. PSC's final concentration was similar with water or salt extraction, ranging from 20.31 ± 0.26 to 44.84 ± 0.78 mg/g, and FSC yielded the highest amount at the end of fermentation from 45.38 ± 0.90 mg/g to 77.72 ± 0.56 mg/g, suggesting that FSC exhibited a more pronounced response to enzymatic hydrolysis compared to all other studied samples and conditions.

### 3.5. Bacillus licheniformis Survival under Simulated GI Conditions

In this study, we employed a simulation to consider multiple factors and their interactions simultaneously, rather than isolating them individually, with the viability outcome on different digestion stages, as illustrated in Figure 5. In the first hour of gastric stage, a decline in viability of about 1 log cfu/g was observed for HSC and FSC due to acidic conditions, and 0.12 log cfu/g for PSC. A modest rise in viability from G1 to G2 was observed across all samples, although statistical significance was only noted for FSC ($p < 0.05$). This indicates adaptability to the medium conditions, particularly noticeable in the case of FSC. Transitioning to the intestinal (I1) phase of digestion resulted in a rise in total count by 0.35 and 0.22 log cfu/g for HSC and PSC, respectively. In contrast, there was a small decrease of 0.11 log cfu/g observed for FSC. This decreasing trend persisted in the I2 stage and was particularly significant for HSC and FSC (0.22 log cfu/g, 0.17 log cfu/g, respectively), probably attributable to the presence of bile salts. PSC, in contrast, showed signs of reaching a stationary phase. Despite the modest microbial dynamics observed, *Bacillus* spores demonstrated resilience across the studied conditions. Survivability rate accounted for 90.2%, 101.5%, and 85.72% for HSC, PSC, and FSC when compared to initial counts before simulation.

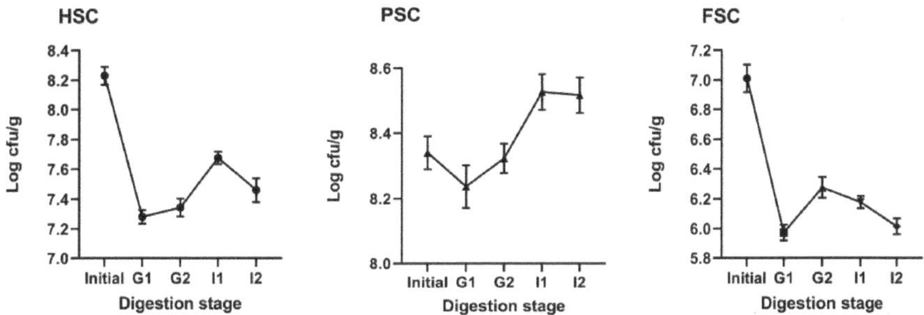


**Figure 5.** *Bacillus licheniformis* viability during GI simulation on different digestion stages for fermented (T72) samples. Bars represent the mean standard deviation.

### 3.6. In Vitro Protein Digestibility

Our approach involved measuring the short-chain peptides generated during simulated digestion using TCA-soluble peptides as a marker [39].

Additionally, we employed the BCA assay for the detection and quantification of peptides. The BCA capability to detect peptides at least three amino acids in length and a few dipeptides and amino acids [40] offers a comprehensive assessment of protein breakdown during digestion.

Peptide content of fermented and unfermented OSC, expressed as mg/g DW after each digestion stage, is shown in Figure 6. After 1 h of gastric digestion (G1), the fermented samples (T72) exhibit significantly higher digestibility rates compared to the unfermented ones (T0). Specifically, the digestibility rates for HSC, PSC, and FSC in the fermented samples are 34.73%, 27.48%, and 44.24%, respectively, whereas those in the unfermented samples are 31.28%, 15.20%, and 35.77%. These findings underscore the role of microbial enzymes in facilitating enzymatic hydrolysis under acidic conditions, improving pepsin digestion at low pH.

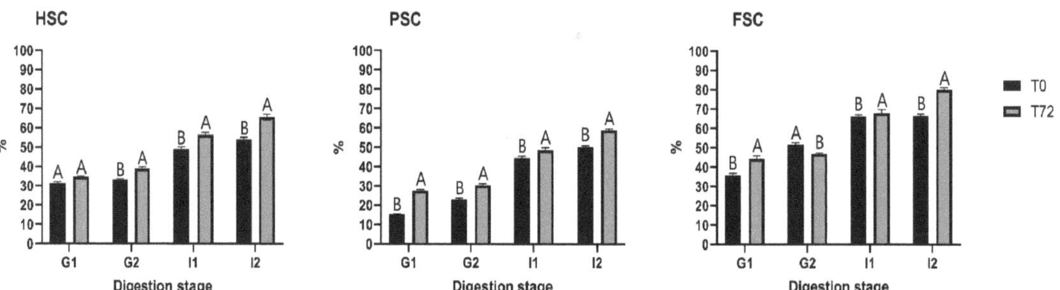


**Figure 6.** Changes in digestibility (%) during digestion stages. Bars represent the mean SD. The use of distinct uppercase letters in the graph corresponds to the results of Tukey's post hoc comparisons. These comparisons highlight differences in digestibility (%) between unfermented samples (T0) and fermented samples (T72) across the same stages of digestion.

The G2 digestibility results, representing protein breakdown after 2 h of gastric digestion, demonstrate notable improvements across all OCS, where a substantial increase was observed for fermented (T72) HSC and PSC. Following fermentation, the digestibility of HSC increases from 33.30% to 38.98%, while in PSC, it rises from 22.91% to 30.37%. In contrast, FSC demonstrates higher digestibility in the unfermented sample compared to its counterpart, with rates of 51.59% and 46.96%, respectively.

Intestinal digestion markedly increased the amount of liberated peptides, augmented by various proteases, including trypsin from pancreatin composition. SSF improved intestinal (I1 and I2) digestion for all substrates in comparison with unfermented ones, reaching in I1 a digestibility of 56.43%, 48.62%, and 68.03% for fermented (T72) HSC, PSC, and FSC, respectively, in contrast with 48.84%, 44.52%, and 66.41% for unfermented (T0) ones.

At the end of I2, the most pronounced differences were between FSC samples, of about 13.50%, reaching 80.16% in fermented ones, followed by HSC with a difference of approx. 11.60%, reaching 65.67%, and finally PSC, where fermentation improved digestion, with approx. 8.60%, reaching a final protein digestibility of 58.94%.

## 4. Discussions

In this study, we investigate the dynamic changes of macronutrients and microbial lifecycles through the complete process of the production to consumption of probiotic-enriched feeds by employing chemico-microbiological analysis and in vitro simulated conditions, which allow us to gather comprehensive data within a complex environmental context. This approach may find relevance in various areas such as animal nutrition, by-product utilization in spore production, or protein and peptide isolation.

Based on the presented results, several notable changes in the nutritional composition of the samples due to fermentation with *Bacillus licheniformis* can be observed. The loss of protein content in all samples can be regarded as a result of the formation of nitrogenous volatile substances, such as ammonia and amines produced during fermentation from protein degradation by microbes [41], that creates an alkaline environment associated with an inhibition in spoilage microorganism and a modification of organoleptic characteristics [42]. As our results show, the drying method could reverse this process, resulting in subsequent loss in volatile compounds associated with a returning of the pH to initial levels.

A potential explanation for crude fiber, NDF, and ADF increases could be related to microbial biomass accumulation through the fermentation process as cells typically contain complex carbohydrates as structural components such as cell walls. A reduction in NDF (composed of cellulose, hemicellulose, and lignin) suggest the presence of carbohydrase activity in PSC. The ability of a bacterial strain to ferment carbohydrates is crucial for various metabolic activities, including growth and reproduction. As the chosen substrates are low in readily available sugars < 1–2% [43–45], the strain is likely to rely on its repertoire of carbohydrase to hydrolyze complex carbohydrates for its carbon metabolism. In this context, different dynamic patterns of production and consumption of reducing sugars can be attributed to substrate composition.

Bacillus sp. possess the ability to alter the composition of FAs within their cells and surrounding environment; this modification of FA patterns is crucial for various cellular functions such as membrane formation and adaptation to different environments. Notably, the process of de novo biosynthesis of FAs has been demonstrated to occur during sporulation [46]. In studies investigating the metabolic adaptation of *Bacillus licheniformis* to stressors such as thermal stress [47] or varying growth stages [48], alterations in cell membrane composition may occur to regulate fluidity. In our study, the dynamic synthesis of specific fatty acids, whether up-regulated or down-regulated, could be attributed to the various stressors encountered during fermentation in the OCS medium. These findings underscore the complex alterations induced by fermentation in the FA profile of OSC, which could have implications for their nutritional value and suitability for various applications in animal feed formulations. However, more studies are required to fully elucidate the influence of different environmental conditions on FA profile modulation by *Bacillus* sp.

When encountering a new life medium, endospore formation is a crucial survival strategy for certain bacteria. It is triggered by various environmental cues (temperature, pH, aeration, presence of certain minerals, carbon or nitrogen concentrations), and one such trigger occurs at the end of the exponential growth phase when nutritional deprivation is recognized, such as during a transition from a nutrient-rich to a nutrient-poor environment. Another factor is population density, as the culture grows, there is an accumulation of a secreted peptide known as the competence and sporulation factor, which acts as an autoinducer for quorum-sensing mechanisms that trigger sporulation [49–51]. Our results suggest that changing culture conditions from inoculum to SSF using HSC, PSC, and FSC create an environmental stress that directly induces the expression of the sporulation mechanism [52,53].

A delay in sporulation in FSC at the 48 h mark suggests that it may harbor substrate-specific nutrients that delay sporulation without increasing in total viable count. Furthermore, a longer fermentation time is required to asses, as Gray et al. [54] suggested, whether the remaining vegetative cells will eventually sporulate or persist in a slow-growing state, rendering them resilient to environmental conditions. Interestingly, based on plated colonies and microscopy observations, new colony morphologies emerged after 3 days of fermentation, typically seen in *Bacillus* spp. biofilms where strains differentiate in different populations [52], suggesting that OSC promotes the apparition of a new subpopulation of genetically identical cells but with different gene expressions [51]. Spore production has been attempted on various subproducts through SSF. Zhao et al. [55] achieved a spore count of 11 log cfu/g of *Bacillus licheniformis* on wheat bran and straw powder, while Chistyakov et al. [56] reported a notable production of 11.95 log cfu/g *Bacillus amyloliquefaciens* on soybean meal, showing that agricultural by-products are an economical alternative for the production of bacterial spores.

For a functional probiotic-based feed to effectively exert its activity, the probiotic microorganisms must endure harsh or challenging environmental conditions from the GI tract, including low pH, high temperatures, elevated levels of bile salts, and nutrient deprivation [57]. Considering the resilience of Bacillus spores in withstanding gastrointestinal-simulated conditions as shown in this study, an intriguing question arises regarding their capacity to transition into vegetative states and sustain their life cycle and potentially colonizing the intestine. The observed increase in viability from G1 to G2 during gastric simulation suggests the strain's capability to proliferate and adapt to initial gastric conditions. However, a subsequent decline in CFU during intestinal simulation indicates a shift into a new adaptation phase for the bacteria, where a lesser reduction viability suggests that PSC has a beneficial food matrix effect over *Bacillus licheniformis* [58]. However, further investigation is needed to clarify if the spore germination versus the remaining vegetative cells is responsible for these dynamics.

In this work, we also conducted a preliminary evaluation on the potential of employing a short fermentation of protein-rich substrates for enhancing protein and peptide recovery. In general, seed storage proteins can be classified based on solubility; they include water-soluble albumins, salt-soluble globulins, hydrophobic prolamins which are commonly soluble in 70% ethanol, and glutenin primarily extracted using acidic solvents or alkali [59,60]; however, some quantity of protein may remain unextracted if not soluble in these solvents [61]. The protein fractions derived from HSC primarily comprise edestin (globulin), which constitutes approximately 60–80%, while albumin around 25% [62]. PSC consists of 87% extractable proteins (glutelin 49%, globulin 20.4%, albumin 13.5%, and prolamin 4.3%) [63] and FSC (albumin 38.1%, glutelin 33.9%, and globulin 27.9%) [64]. An industry-relevant extraction is considered a salt extraction (known as micellization), which is considered a milder extraction associated with less structural and conformation-changes [61]. It affects solubility by altering the electrostatic interactions (ionic strength) promoting aggregation [62]. Alkali extraction is widely used at an industrial level; it is suggested that the use of alkali can disrupt disulfide cross-linking in proteins and to ionize neutral and acidic amino acids, therefore increasing solubility [62]. Even though alkali has high efficiency, it favors the racemization of the amino acids L-isomers to D-analogues in a concentration- and exposure time-dependent manner that could lead to toxic and antagonistic actions in animals, thus reducing feeds' nutritional quality [65]. Our results suggest that peptide yield was improved for all OSCs by employing water, alkaline or salt extraction, probably due to enzymes like proteases and cellulases that are widely used in order to improve hydrolysis degree, solubility, and functionality [66]. However, the extraction method and the optimization of strain catalytic activities could further improve the extraction of protein and peptides for OSC. In other studies, hemp seed meal protein solubility increased from 6.07% to 15.15% after *Bacillus subtilis* fermentation [67]. In PSC, a remarkable peptide content of approx. 190 mg/g was achieved using *R.* oligosporus, due to high fungal proteolysis activity [68]. In flaxseed proteins, fermentation with *Bacillus altitudinis* was successfully employed to generate bioactive peptides with antioxidant and antibacterial activity [69].

The evaluation of protein digestibility is crucial for nutritional assessment in terms of the quality and bioavailability of feed products. Therefore, fermentation enhances in vitro protein digestibility possibly by *Bacillus licheniformis* proteolytic activity that can target plant-based proteins, thereby promoting hydrolysis, protein modification [70–72], and remediation of anti-nutritional factors such as trypsin inhibitors [19]. Some studies using lactic bacteria achieved 65–79% IVPD in hemp [73]. Alternatively, indigenous microbial communities' microbiota were used for PSC fermentation where protein digestibility was 68.7% after 7 days of fermentation [74]. Isolation of flaxseed protein can give a digestibility coefficient of 68% [75], while in other cases, fermentation with *Aspergillus oryzae* was directly correlated with proteolytic activity, thereby enhancing digestibility to values between 70 and 85% after 96 h [76].

## 5. Conclusions

Fermentation is a dynamic process with substrate specificity orchestrated by the degradation and synthesis of new compounds. While proximate composition provides a broad overview, it may not fully capture the complex chemical transformations occurring during fermentation. We did not find *Bacillus licheniformis* strain ATCC 21424 capable of making extreme changes in proximate composition or fatty acid profile; however, it seems capable of markedly improving water protein solubility and salt solution solubility in the case of HSC, accompanied by the generation of higher peptide yields. The SSF of oilseed cakes induces sporulation, demonstrating that it is suitable for producing probiotic-enriched feeds that can withstand harsh environmental conditions from simulated GI tracts along with a noticeable improvement in protein digestibility. Furthermore, our study underscores the potential of laboratory-domesticated strains like ATCC 21424 in improving feed digestibility and functionality through SSF. Future research focusing on strain adaptation to fermentation media holds promise for enhancing the capabilities of such strains.

**Author Contributions:** Conceptualization, D.R. and E.V.; methodology, D.R. and M.D.; software, M.D.; validation, E.V.; formal analysis, D.R.; investigation, D.R.; resources, G.C.; data curation, M.D. and G.C.; writing—original draft preparation, D.R.; writing—review and editing, D.R., M.D. and G.C.; visualization, M.D.; supervision, G.C. and E.V.; project administration, G.C.; funding acquisition, G.C. All authors have read and agreed to the published version of the manuscript.

**Funding:** This research was funded by the Romanian Ministry of Agriculture and Rural Development (Project ADER 8.1.7) and Ministry of Research, Innovation, and Digitalization (Project PN23-20.04.01 and Grant PFE 8/2021).

**Institutional Review Board Statement:** Not applicable.

**Informed Consent Statement:** Not applicable.

**Data Availability Statement:** Data is contained within the article.

**Conflicts of Interest:** The authors declare no conflict of interest.

## Abbreviations

| | |
|---|---|
| HSC | hemp seed cake |
| PSC | pumpkin seed cake |
| FSC | flaxseed cake |

## References

1. Singh, R.; Langyan, S.; Sangwan, S.; Rohtagi, B.; Khandelwal, A.; Shrivastava, M. Protein for human consumption from oilseed cakes: A Review. *Front. Sustain. Food Syst.* **2022**, *6*, 856401. [CrossRef]
2. Dozier, W.A.; Hess, J.B. Soybean meal quality and analytical techniques. In *Soybean and Nutrition*; InTech Open: London, UK, 2011; pp. 111–124.
3. Keller, M.; Reidy, B.; Scheurer, A.; Eggerschwiler, L.; Morel, I.; Giller, K. Soybean meal can be replaced by faba beans, pumpkin seed cake, spirulina or be completely omitted in a forage-based diet for fattening bulls to achieve comparable performance, carcass and meat quality. *Animals* **2021**, *11*, 1588. [CrossRef] [PubMed]
4. Ancuța, P.; Sonia, A. Oil Press-Cakes and Meals Valorization through Circular Economy Approaches: A Review. *Appl. Sci.* **2020**, *10*, 7432. [CrossRef]
5. Lekha, P.K.; Lonsane, B.K. Production and application of tannin acyl hydrolase: State of the art. *Adv. Appl. Microbiol.* **1997**, *44*, 215–260. [CrossRef] [PubMed]
6. Lizardi-Jiménez, M.A.; Hernández-Martínez, R. Solid state fermentation (SSF): Diversity of applications to valorize waste and biomass. *3 Biotech* **2017**, *7*, 44. [CrossRef] [PubMed]
7. Cerda, A.; Artola, A.; Barrena, R.; Font, X.; Gea, T.; Sánchez, A. Innovative production of bioproducts from organic waste through solid-state fermentation. *Front. Sustain. Food Syst.* **2019**, *3*, 63. [CrossRef]

8. Soares, V.F.; Castilho, L.R.; Bon, E.P.; Freire, D.M. High-yield *Bacillus subtilis* protease production by solid-state fermentation. *Appl. Biochem. Biotechnol.* **2005**, *121–124*, 311–319. [CrossRef] [PubMed]
9. Jatuwong, K.; Kumla, J.; Suwannarach, N.; Matsui, K.; Lumyong, S. Bioprocessing of agricultural residues as substrates and optimal conditions for phytase production of Chestnut Mushroom, *Pholiota adiposa*, in solid state fermentation. *J. Fungi* **2020**, *6*, 384. [CrossRef]
10. Sadh, P.K.; Kumar, S.; Chawla, P.; Duhan, J.S. Fermentation: A boon for production of bioactive compounds by processing of food industries wastes (by-products). *Molecules* **2018**, *23*, 2560. [CrossRef]
11. Olukomaiya, O.; Fernando, C.; Mereddy, R.; Li, X.; Sultanbawa, Y. Solid-state fermented plant protein sources in the diets of broiler chickens: A review. *Anim. Nutr.* **2019**, *5*, 319–330. [CrossRef]
12. Manan, M.A.; Webb, C. Design aspects of solid state fermentation as applied to microbial bioprocessing. *J. Appl. Biotechnol. Bioeng.* **2017**, *4*, 511–532. [CrossRef]
13. Ramirez-Olea, H.; Reyes-Ballesteros, B.; Chavez-Santoscoy, R.A. Potential application of the probiotic *Bacillus licheniformis* as an adjuvant in the treatment of diseases in humans and animals: A systematic review. *Front. Microbiol.* **2022**, *26*, 993451. [CrossRef]
14. Bernardeau, M.; Lehtinen, M.J.; Forssten, S.D.; Nurminen, P. Importance of the gastrointestinal life cycle of *Bacillus* for probiotic functionality. *J. Food Sci. Technol.* **2017**, *54*, 2570–2584. [CrossRef]
15. Solanki, P.; Putatunda, C.; Kumar, A.; Bhatia, R.; Walia, A. Microbial proteases: Ubiquitous enzymes with innumerable uses. *3 Biotech* **2021**, *11*, 428. [CrossRef] [PubMed]
16. Dai, C.; Yan, P.; Xu, X.; Huang, L.; Dabbour, M.; Benjamin, K.M.; He, R.; Ma, H. Effect of single and two-stage fermentation on the antioxidative activity of soybean meal, and the structural and interfacial characteristics of its protein. *LWT* **2023**, *183*, 114938. [CrossRef]
17. Liu, D.; Guo, Y.; Ma, H. Production of value-added peptides from agro-industrial residues by solid-state fermentation with a new thermophilic protease-producing strain. *Food Biosci.* **2023**, *53*, 102534. [CrossRef]
18. Wu, P.; Guo, Y.; Golly, M.K.; Ma, H.; He, R.; Luo, S.; Zhang, C.; Zhang, L.; Zhu, J. Feasibility study on direct fermentation of soybean meal by Bacillus stearothermophilus under non-sterile conditions. *JSFAAE* **2018**, *99*, 3291–3298. [CrossRef] [PubMed]
19. Dai, C.; Hou, Y.; Xu, H.; Huang, L.; Dabbour, M.; Mintah, B.K.; He, R.; Ma, H. Effect of solid-state fermentation by three different Bacillus species on composition and protein structure of soybean meal. *J. Sci. Food Agric.* **2022**, *102*, 557–566. [CrossRef]
20. Dai, C.; Ma, H.; He, R.; Huang, L.; Zhu, S.; Ding, Q.; Luo, L. Improvement of nutritional value and bioactivity of soybean meal by solid-state fermentation with Bacillus subtilis. *LWT* **2017**, *86*, 1–7. [CrossRef]
21. González Pereyra, M.L.; Di Giacomo, A.L.; Lara, A.L.; Martínez, M.P.; Cavaglieri, L.; Gonz, M.L. Aflatoxin-degrading *Bacillus* sp. strains degrade zearalenone and produce proteases, amylases and cellulases of agro-industrial interest. *Toxicon* **2020**, *180*, 43–48. [CrossRef]
22. Leeuwendaal, N.K.; Stanton, C.; O'Toole, P.W.; Beresford, T.P. Fermented Foods, Health and the Gut Microbiome. *Nutrients* **2022**, *14*, 1527. [CrossRef] [PubMed] [PubMed Central]
23. Xu, S.; Lin, Y.; Zeng, D.; Zhou, M.; Zeng, Y.; Wang, H.; Yi, Z.; Zhu, H.; Pan, K.; Jing, B.; et al. *Bacillus licheniformis* normalize the ileum microbiota of chickens infected with necrotic enteritis. *Sci. Rep.* **2018**, *8*, 1744. [CrossRef] [PubMed]
24. Lin, E.-R.; Cheng, Y.-H.; Hsiao, F.S.-H.; Proskura, W.S.; Dybus, A.; Yu, Y.-H. Optimization of solid-state fermentation conditions of Bacillus licheniformis and its effects on Clostridium perfringens-induced necrotic enteritis in broilers. *Rev. Bras. Zootec.* **2019**, *48*, e20170298. [CrossRef]
25. Yang, L.; Zeng, X.; Qiao, S. Advances in research on solid-state fermented feed and its utilization: The pioneer of private customization for intestinal microorganisms. *Anim. Nutr.* **2021**, *7*, 905–916. [CrossRef] [PubMed]
26. Hung, D.-Y.; Cheng, Y.-H.; Chen, W.-J.; Hua, K.-F.; Pietruszka, A.; Dybus, A.; Lin, C.-S.; Yu, Y.-H. Bacillus licheniformis-Fermented Products Reduce Diarrhea Incidence and Alter the Fecal Microbiota Community in Weaning Piglets. *Animals* **2019**, *9*, 1145. [CrossRef]
27. Yi, W.; Liu, Y.; Fu, S.; Zhuo, J.; Wang, J.; Shan, T. Dietary novel alkaline protease from *Bacillus licheniformis* improves broiler meat nutritional value and modulates intestinal microbiota and metabolites. *Anim Microbiome* **2024**, *6*, 1. [CrossRef] [PubMed]
28. Li, C.; Zhao, K.; Ma, L.; Zhao, J.; Zhao, Z.M. Effects of drying strategies on sporulation and titer of microbial ecological agents with *Bacillus subtilis*. *Front. Nutr.* **2022**, *9*, 1025248. [CrossRef]
29. AOAC. *Association of Official Agricultural Chemists*, 17th ed.; Official Methods of Analysis of AOAC International: Gaithersburg, MD, USA, 2000.
30. Vlaicu, P.A.; Untea, A.E.; Turcu, R.P.; Saracila, M.; Panaite, T.D.; Cornescu, G.M. Nutritional composition and bioactive compounds of basil, thyme and sage plant additives and their functionality on broiler thigh meat quality. *Foods* **2022**, *11*, 1105. [CrossRef]
31. Häbeanu, M.; Lefter, N.A.; Toma, S.M.; Dumitru, M.; Cismileanu, A.; Surdu, I.; Gheorghe, A.; Dragomir, C.; Untea, A. Changes in ileum and cecum volatile fatty acids and their relationship with microflora and enteric methane in pigs fed different fiber levels. *Agriculture* **2022**, *12*, 451. [CrossRef]
32. Ciurescu, G.; Vasilachi, A.; Idriceanuand, L.; Dumitru, M. Effects of corn replacement by sorghum in broiler chickens diets on performance, blood chemistry, and meat quality. *Ital. J. Anim. Sci.* **2023**, *12*, 537–547. [CrossRef]
33. *ISO/TS 17764-2:2008*; Animal Feeding Stuffs. ISO: Geneva, Switzerland, 2008.
34. Miller, G.L. Use of Dinitrosalicylic Acid Reagent for Determination of Reducing Sugar. *Anal. Chem.* **1959**, *31*, 426–428. [CrossRef]

35. Chen, N.; Zhao, M.; Sun, W. Effect of protein oxidation on the in vitro digestibility of soy protein isolate. *Food Chem.* **2013**, *141*, 3224–3229. [CrossRef] [PubMed]
36. Brodkorb, A.; Egger, L.; Alminger, M.; Alvito, P.; Assunção, R.; Ballance, S.; Bohn, T.; Bourlieu-Lacanal, C.; Boutrou, R.; Carrière, F. INFOGEST static in vitro simulation of gastrointestinal food digestion. *Nat. Protoc.* **2019**, *14*, 991–1014. [CrossRef]
37. Marttinen, M.; Anjum, M.; Saarinen, M.T.; Ahonen, I.; Lehtinen, M.J.; Nurminen, P.; Laitila, A. Enhancing bioaccessibility of plant protein using probiotics: An in vitro study. *Nutrients* **2023**, *15*, 3905. [CrossRef]
38. Coelho, M.C.; Ribeiro, T.B.; Oliveira, C.; Batista, P.; Castro, P.; Monforte, A.R.; Rodrigues, A.S.; Teixeira, J.; Pintado, M. *In Vitro* gastrointestinal digestion impact on the bioaccessibility and antioxidant capacity of bioactive compounds from tomato flours obtained after conventional and ohmic heating extraction. *Foods* **2021**, *10*, 554. [CrossRef] [PubMed]
39. Ketnawa, S.; Ogawa, Y. In vitro protein digestibility and biochemical characteristics of soaked, boiled and fermented soybeans. *Sci. Rep.* **2021**, *11*, 14257. [CrossRef] [PubMed]
40. Hall, A.E.; Moraru, C.I. Effect of High pressure processing and heat treatment on *in vitro* digestibility and trypsin inhibitor activity in lentil and faba bean protein concentrates. *LWT* **2011**, *152*, 112342. [CrossRef]
41. Allagheny, N.; Obanu, Z.A.; Campbell-Platt, G.; Owens, J.D. Control of ammonia formation during Bacillus subtilis fermentation of legumes. *Int. J. Food Microbiol.* **1996**, *29*, 321–333. [CrossRef] [PubMed]
42. Akanni, G.B.; De Kock, H.L.; Naudé, Y.; Buys, E.M. Volatile compounds produced by *Bacillus* species alkaline fermentation of bambara groundnut (*Vigna subterranean* (L.) Verdc) into a dawadawa-type African food condiment using headspace solid-phase microextraction and GC × GC–TOFMS. *Int. J. Food Prop.* **2018**, *21*, 929–941. [CrossRef]
43. Bernacchia, R.; Preti, R.; Vinci, G. Chemical composition and health benefits of flaxseed. *Austin. J. Nutri. Food Sci.* **2014**, *2*, 1045.
44. Habib, A.; Biswas, S.; Siddique, A.H.; Mohammad, M.; Uddin, B.; Hasan, S.; MMH, K.; Uddin, M.; Islam, M.; Hasan, M.; et al. Nutritional and lipid composition analysis of pumpkin seed (*Cucurbita maxima* Linn.). *J. Nutr. Food Sci.* **2015**, *5*, 4. [CrossRef]
45. Montero, L.; Ballesteros-Vivas, D.; Gonzalez-Barrios, A.F.; Sánchez-Camargo, A.D.P. Hemp seeds: Nutritional value, associated bioactivities and the potential food applications in the colombian context. *Front. Nutr.* **2023**, *11*, 1039180. [CrossRef] [PubMed]
46. Diomandé, S.E.; Guinebretière, M.H.; Broussolle, V.; Brillard, J. Role of fatty acids in *Bacillus* environmental adaptation. *Front. Microbiol.* **2015**, *6*, 813. [CrossRef] [PubMed] [PubMed Central]
47. Dong, Z.; Chen, X.; Cai, K.; Chen, Z.; Wang, H.; Jin, P.; Liu, X.; Permaul, K.; Singh, S.; Wang, Z.-X. Exploring the Metabolomic Responses of Bacillus licheniformis to Temperature Stress by Gas Chromatography/Mass Spectrometry. *J. Microbiol. Biotechnol.* **2018**, *28*, 473–481. [CrossRef]
48. Lopes, C.; Barbosa, J.; Maciel, E.; Costa, E.; Alves, E.; Ricardo, F.; Domingues, P.; Mendo, S.; Domingues, M.R.M. Decoding the Fatty Acid Profile of Bacillus li-cheniformis I89 and Its Adaptation to Different Growth Conditions to Investigate Possible Biotechnological Applications. *Lipids* **2019**, *54*, 245–253. [CrossRef] [PubMed]
49. Miller, M.B.; Bassler, B.L. Quorum sensing in bacteria. *Annu. Rev. Microbiol.* **2001**, *55*, 165–199. [CrossRef]
50. Logan, N.A.; Vos, P.D. *Bacillus*. In *Bergey's Manual of Systematics of Archaea and Bacteria*; John Wiley & Sons, Inc.: Hoboken, NJ, USA, 2015; pp. 1–163. [CrossRef]
51. Vlamakis, H.; Aguilar, C.; Losick, R.; Kolter, R. Control of cell fate by the formation of an architecturally complex bacterial community. *Genes Dev.* **2008**, *22*, 945–953. [CrossRef]
52. Tasaki, S.; Nakayama, M.; Shoji, W. Morphologies of *Bacillus subtilis* communities responding to environmental variation. *DGD* **2017**, *59*, 369–378. [CrossRef] [PubMed]
53. Elisashvili, V.; Kachlishvili, E.; Chikindas, M.L. Recent advances in the physiology of spore formation for *Bacillus* probiotic production. *Probiotics Antimicrob. Prot.* **2019**, *11*, 731–747. [CrossRef]
54. Gray, D.A.; Dugar, G.; Gamba, P.; Strahl, H.; Jonker, M.J.; Hamoen, L.W. Extreme slow growth as alternative strategy to survive deep starvation in bacteria. *Nat. Commun.* **2019**, *10*, 890. [CrossRef]
55. Zhao, S.; Deng, L.; Hu, N.; Zhao, B.; Liang, Y. Cost-effective production of *Bacillus licheniformis* using simple netting bag solid bioreactor. *World J. Microb. Biotechnol.* **2008**, *24*, 2859–2863. [CrossRef]
56. Chistyakov, V.; Mlnikov, V.; Chikindas, M.L.; Khutsishvili, M.; Chagelishvili, A.; Bren, A.; Kostina, N.; Cavera, V.; Elisashvili, V. Poultry-beneficial solid-state *Bacillus amyloliquefaciens* B-1895 fermented soybean formulation. *Biosci. Microbiota Food Health* **2015**, *34*, 25–28. [CrossRef] [PubMed]
57. Golnari, M.; Bahrami, N.; Milanian, Z.; Rabbani Khorasgani, M.; Asadollahi, M.A.; Shafiei, R.; Fatemi, S.S.A. Isolation and characterization of novel *Bacillus* strains with superior probiotic potential: Comparative analysis and safety evaluation. *Sci. Rep.* **2024**, *14*, 1457. [CrossRef] [PubMed]
58. Flach, J.; van der Waal, M.B.; van den Nieuwboer, M.; Claassen, E.; Larsen, O.F.A. The underexposed role of food matrices in probiotic products: Reviewing the relationship between carrier matrices and product parameters. *Crit. Rev. Food Sci. Nutr.* **2018**, *58*, 2570–2584. [CrossRef] [PubMed]
59. Branlard, G.; Bancel, E. Protein extraction from cereal seeds. *Plant Proteom.* **2007**, *355*, 15–26. [CrossRef]
60. Osborne, T.B. *The Vegetable Proteins*; Longmans: London, UK, 1924. Available online: https://wellcomecollection.org/works/p8 hecz9k/items?canvas=32 (accessed on 14 February 2024).
61. Hadnađev, M.S.; Hadnađev, D.T.; Pojić, M.M.; Šarić, B.M.; Mišan, A.Č.; Jovanov, P.T.; Sakač, M.B. Progress in vegetable proteins isolation techniques: A review. *Food Feed. Res.* **2017**, *44*, 11–21. [CrossRef]

62. Gouseti, O.; Larsen, M.E.; Amin, A.; Bakalis, S.; Petersen, I.L.; Lametsch, R.; Jensen, P.E. Applications of enzyme technology to enhance transition to plant proteins: A Review. *Foods* **2023**, *12*, 2518. [CrossRef]
63. Chen, H.; Xu, B.; Wang, Y.; Li, W.; He, D.; Zhang, Y.; Xing, X. Emerging natural hemp seed proteins and their functions for nutraceutical applications. *Food Sci. Hum. Wellness* **2023**, *12*, 929–941. [CrossRef]
64. Pham, T.T.; Tran, T.T.T.; Ton, N.M.N.; Le, V.V.M. Effects of pH and salt concentration on functional properties of pumpkin seed protein fractions. *J. Food Process. Preserv.* **2016**, *41*, e13073. [CrossRef]
65. Friedman, M.; Gumbmann, M.R.; Masters, P.M. Protein-alkali reactions: Chemistry, toxicology, and nutritional consequences. *Nutr. Toxicol. Asp. Food Saf.* **1984**, *177*, 367–412. [CrossRef]
66. Del Mar Contreras, M.; Lama-Muñoz, A.; Manuel Gutiérrez-Pérez, J.; Espínola, F.; Moya, M.; Castro, E. Protein extraction from agri-food residues for integration in biorefinery: Potential techniques and current status. *Bioresour. Technol.* **2019**, *280*, 459–477. [CrossRef] [PubMed]
67. Zhang, M.; Li, T.; Guo, G.; Liu, Z.; Hao, N. Production of Nattokinase from Hemp Seed Meal by Solid-State Fermentation and Improvement of Its Nutritional Quality. *Fermentation* **2023**, *9*, 469. [CrossRef]
68. Starzyńska-Janiszewska, A.; Duliński, R.; Stodolak, B. Fermentation with Edible Rhizopus Strains to Enhance the Bioactive Potential of Hull-Less Pumpkin Oil Cake. *Molecules* **2020**, *25*, 5782. [CrossRef] [PubMed]
69. Hwang, C.-F.; Chen, Y.-A.; Luo, C.; Chiang, W.-D. Antioxidant and antibacterial activities of peptide fractions from flaxseed protein hydro-lysed by protease from Bacillus altitudinis HK02. *Int. J. Food Sci. Technol.* **2016**, *51*, 681–689. [CrossRef]
70. Li, X.; Cheng, Y.; Yi, C.; Hua, Y.; Yang, C.; Cui, S. Effect of ionic strength on the heat-induced soy protein aggregation and the phase separation of soy protein aggregate/dextran mixtures. *Food Hydrocoll.* **2009**, *23*, 1015–1023. [CrossRef]
71. Song, P.; Zhang, X.; Wang, S.; Xu, W.; Wang, F.; Fu, R.; Wei, F. Microbial proteases and their applications. *Front. Microbiol.* **2023**, *14*, 1236368. [CrossRef]
72. Ayad, A. Characterization and Properties of Flaxseed Protein Fractions. 2010. Available online: https://www.google.com.hk/url?sa=t&source=web&rct=j&opi=89978449&url=https://escholarship.mcgill.ca/downloads/hh63sw24h.pdf&ved=2ahUKEwiZp5L_is2FAxVsk1YBHT30BvEQFnoECBwQAQ&usg=AOvVaw2eSSYehMgfvPYYv7M3kLSb (accessed on 14 February 2024).
73. Pontonio, E.; Verni, M.; Dingeo, C.; Diaz-de-Cerio, E.; Pinto, D.; Rizzello, C.G. Impact of Enzymatic and Microbial Bioprocessing on Antioxidant Properties of Hemp (*Cannabis sativa* L.). *Antioxidants* **2020**, *9*, 1258. [CrossRef] [PubMed]
74. Giami, S.Y. Effect of fermentation on the seed proteins, nitrogenous constituents, antinutrients and nutritional quality of fluted pumpkin (*Telfairia occidentalis* Hook). *Food Chem.* **2004**, *88*, 397–404. [CrossRef]
75. Marambe, H.K.; Shand, P.J.; Wanasundara, J.P.D. In vitrodigestibility of flaxseed (*Linum usitatissimum* L.) protein: Effect of seed mucilage, oil and thermal processing. *Int. J. Food Sci. Technol.* **2012**, *48*, 628–635. [CrossRef]
76. Wu, M.; Wang, L.; Li, D.; Wang, Y.; Mao, Z.; Chen, X.D. The Digestibility and Thermal Properties of Fermented Flaxseed Protein. *Int. J. Food Eng.* **2012**, *8*. [CrossRef]

*Article*

# Comparison of the Effect of Drying Treatments on the Physicochemical Parameters, Oxidative Stability, and Microbiological Status of Yellow Mealworm (*Tenebrio molitor* L.) Flours as an Alternative Protein Source

Desislava Vlahova-Vangelova [1,*], Desislav Balev [1], Nikolay Kolev [1], Stefan Dragoev [1], Evgeni Petkov [2] and Teodora Popova [2,*]

[1] Department of Meat and Fish Technology, University of Food Technologies, 26 Maritsa Blvd, 4002 Plovdiv, Bulgaria; d_balev@uft-plovdiv.bg (D.B.); n_kolev@uft-plovdiv.bg (N.K.); s_dragoev@uft-plovdiv.bg (S.D.)

[2] Agricultural Academy, Institute of Animal Science-Kostinbrod, Pochivka St, 2232 Kostinbrod, Bulgaria; e_petkov@ias.bg

* Correspondence: d_vangelova@uft-plovdiv.bg (D.V.-V.); t_popova@ias.bg (T.P.)

**Citation:** Vlahova-Vangelova, D.; Balev, D.; Kolev, N.; Dragoev, S.; Petkov, E.; Popova, T. Comparison of the Effect of Drying Treatments on the Physicochemical Parameters, Oxidative Stability, and Microbiological Status of Yellow Mealworm (*Tenebrio molitor* L.) Flours as an Alternative Protein Source. *Agriculture* **2024**, *14*, 436. https://doi.org/10.3390/agriculture14030436

Academic Editors: Petru Alexandru Vlaicu, Arabela Elena Untea and Mihaela Saracila

Received: 11 February 2024
Revised: 5 March 2024
Accepted: 6 March 2024
Published: 7 March 2024

**Abstract:** The increasing production of edible insects on an industrial scale makes it crucial to implement appropriate technologies after harvesting to process safe and high quality insect products. The aim of this work was to compare the impact of different drying treatments used in the production of flour from *Tenebrio molitor* larvae. The larvae were subjected to freeze-drying (FD), conventional drying (CD), microwave drying (MWD), microwave drying without freezing prior blanching (MWDL), and microwave drying with addition of 0.1% butylated hydroxytoluene (BHT) during the blanching of the larvae (MWDA). The studied parameters included water activity ($a_w$), instrumental colour, chemical composition, lipid oxidative processes, antioxidant activity, as well as microbiological status. The freeze-drying and conventional drying of the larvae reduced the $a_w$ of the derived flours ($p < 0.0001$); however, their nutritional profile revealed lower protein ($p < 0.0001$) and considerably higher fat content ($p < 0.0001$) compared to the flours after microwave treatments. The conventional drying and microwave treatment with BHT induced significantly darker colour ($p < 0.0001$) in comparison to the other methods. Despite the advantages of the microwave drying as a fast and energy efficient method, it displayed some negative effects associated with low lipid stability such as higher acid value (AV) and secondary products of lipid oxidation (TBARS) ($p < 0.0001$). This was also observed in the MWDA flour, indicating a certain pro-oxidative effect of the BHT. Regardless of the drying method, all the flours had a low microbial load.

**Keywords:** yellow mealworm; drying treatments; flours; nutrients; lipid oxidation; antioxidant activity; safety

## 1. Introduction

The agriculture industry is alarmed by a future shortage of food products due to the increasing global population [1]. In recent decades, there has been a continuous increase in the demand for proteins worldwide [2]. It is known that insects are a complete source of proteins, vitamins, and minerals of organic origin. According to Wageningen University, insect protein can replace 10% of the animal protein in the animal feeds and 20% in human diets in 2025 [3]. Insects have high reproductive capacity. Furthermore, the low resource utilization (land and water) and feed for their cultivation makes them a suitable alternative to meat and dairy products, against the backdrop of growing global needs for nutritious food [4]. While in many countries in Asia, Africa and Latin America insects are used as a major protein source, in Western societies people still resist eating them and they have greater potential as animal feed rather than human [5]. Researchers have studied the aversion to insects to find ways to overcome it and have found that the species of insects chosen for consumption is considered a significant factor [6]. For example, some species are considered more acceptable and visually appealing than others [7]. Yellow mealworms (the larvae of *Tenebrio molitor*) are consumed as a part of a regular diet or for medicinal purposes in some countries outside the EU. The yellow mealworms were the first insects approved by EFSA for human consumption [8], considering the consumption of mealworms as generally safe, as the larvae can be consumed "both as a whole, dried insect and in powder form" [9]. Consumers in Europe are more inclined to consume insects when the body parts are not visible [10]. Therefore, the development of new food products including insects is expected to attract consumer interest. With the increase in production of edible insects on an industrial scale, it is crucial to implement appropriate technologies after their harvesting and processing. Thus, the safety, preservation, quality improvement, fractionation, and storage of the insects and insect products will be guaranteed.

In its report, the World Bank announced inflation in domestic food prices ranging from 5 to 30% in the majority of countries worldwide [2]. This, in turn, necessitates the search for cheaper and more efficient methods for food processing and production. Conventional food and feed processing may include several individual operations. The pathways for insect processing may vary depending on the nature of the initial material and the desired end product. In recent years, new food processing technologies (e.g., high-pressure processing, ultrasound, or microwave waves) have shown potential as alternatives or synergies to conventional technologies. The main challenges in insect processing are the development of efficient, environmentally friendly, and low-cost processing technologies, waste minimization, utilization, and incorporation of by-products. Drying is the most common technology used to extend the shelf life of foods and feeds. Dehydration enhances the microbiological stability, reduces oxidation, and possibly improves the colour and texture of end products. From traditional sun drying to innovative technologies such as freeze-drying or microwave drying, these methods have been studied to increase shelf life and safety. Despite their effectiveness, some pre-treatments are recommended before drying to eliminate or reduce the overall microbiological load and to inactivate enzymes, initiators of lipid and protein oxidation. Blanching is the most commonly used method, both at a traditional and industrial level, as it has safety advantages, reduces lipid oxidation, and inactivates enzymes, contributing to extending shelf life. Microwave drying, in combination with various types of freezing, has been studied by Vlahova-Vangelova [11]. Microwave drying takes less time to dry insects compared to freezing and conventional drying. Finally, drying affects the final quality of the proteins and lipids and the extraction of other metabolites with potential health benefits. Despite the research on the methods of processing insects for their further use in foods and feeds, the information on suitable modes and methods of drying, as well as the qualities of the derived products, including flours, still remains rather scarce. This work was designed to compare the impact of different drying treatments applied in the production of flour from *Tenebrio molitor* larvae as an alternative protein source. This will enable the clarification of their advantages or disadvantages in regards to maximal preservation of the nutritional parameters of the insect flours and selection of the best method for further processing and implementation in the strategies of animal and human nutrition.

## 2. Materials and Methods

### 2.1. Larvae Rearing

The mealworm larvae were reared in plastic boxes (600 × 400 × 120 mm), at an insect farm located in Petrich, Bulgaria (Vi Bi Ef PRO Ltd.) in controlled microclimate (T = 25 ± 2 °C, RH = 50–70%). The insects were fed on wheat bran up to day 70. Fruits and vegetables were also added 2–3 times a week. Three days prior to harvest, the mealworms were not fed. The proximate composition of the larvae after harvest and before any treatment is presented in Table 1:

**Table 1.** Proximate composition of the larvae after harvesting.

| Protein, % | Fat, % | Ashes, % | Moisture, % | Carbohydrates, % |
|---|---|---|---|---|
| 14.85 | 1.20 | 1.02 | 77.30 | 5.63 |

## 2.2. Drying Treatments

After harvesting, the larvae were separated into five groups (1 kg each). Four of the groups were frozen at $-18\ ^\circ$C and kept at this temperature for 24 h. All the larvae were blanched (100 $^\circ$C, 45 s) and cooled in cold water prior to drying (Figure 1). For the purpose of the study, five drying treatments were applied:

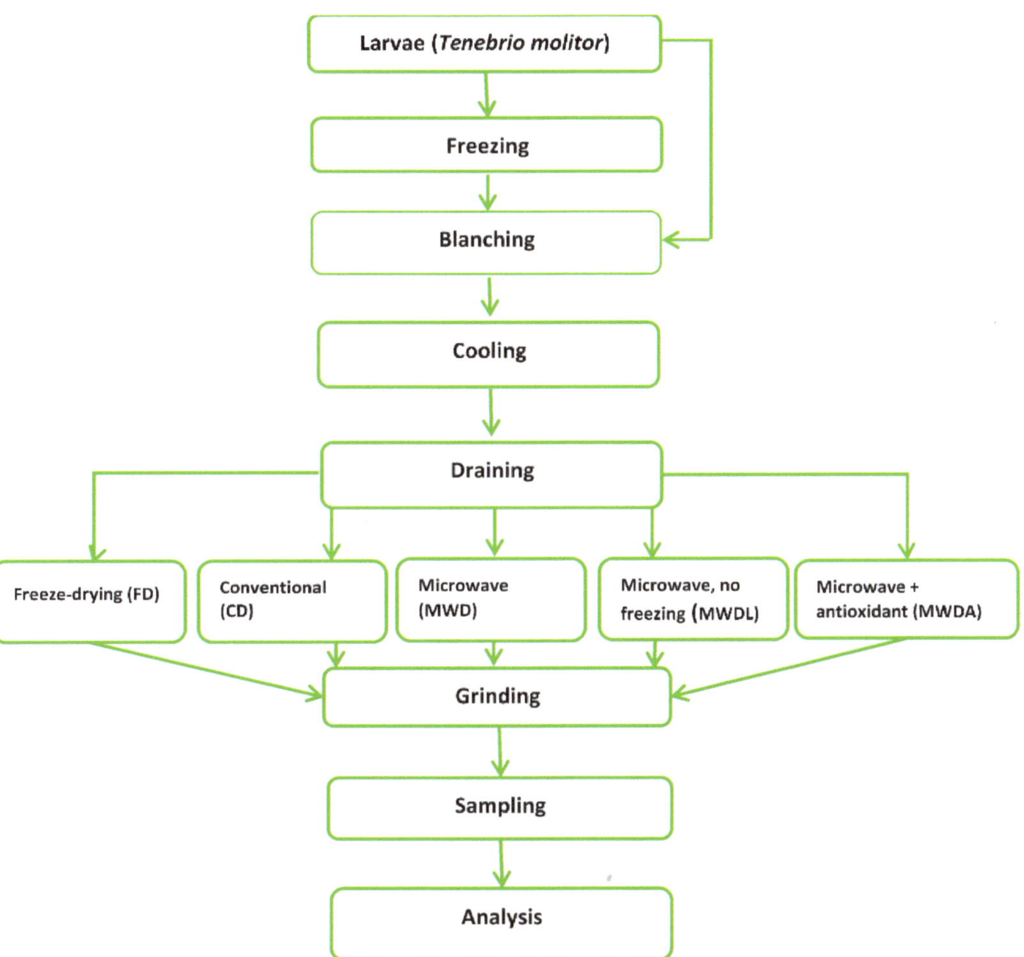


**Figure 1.** Scheme of the drying treatments applied to *Tenebrio molitor* larvae.

Freeze-drying (FD)—the larvae were freeze-dried (T= $-45$–$-40\ ^\circ$C; 0.400 mBar) for 12–18 h using laboratory freeze dryer CRYODOS—50 (Telstar Industrial, S.L., Terrassa, Spain).

Conventional drying (CD)—the conventional drying was carried out in a hot air rack oven (KC-100/200, Zalmed, Warsaw, Poland), for 6 h (T = 60 $^\circ$C, RH = 17%, air speed 1 m/s).

Microwave drying (MWD)—the frozen and blanched larvae were dried in microwave (MWD307/WH, Whirlpool, Benton Harbor, MI, USA) at 825 W for 7 min.

Microwave drying without freezing (MWDL)—the live directly blanched larvae were dried in the microwave oven at 825 W for 7 min.

Microwave drying + antioxidant (MWDA)—after freezing, the larvae were blanched in water containing 0.1% BHT and dried in the microwave at 825 W for 7 min.

After drying, the insect flours were prepared by grinding the larvae using a Nutribullet blender (NB-WL046A-02, Capital Brands, Los Angeles, CA, USA). Each flour was divided into three samples of approximately 200 g. The samples were stored in sealed plastic bags until analysis.

### 2.3. Water Activity ($a_w$) and Instrumental Colour Measurements

Water activity was determined using an $a_w$ meter LabSwift-aw (Novasina AG, Lachen, Switzerland) at 20 °C.

Colour measurements of the flours were conducted using a Konica Minolta CR-400 chromameter (Konica Minolta Holding, Inc., Ewing, NJ, USA) by measuring the lightness (L*), redness (a*), and yellowness (b*). The chromameter had the following settings: aperture = 8 mm, standard observer 2°, and illuminant D65. The instrument was calibrated using a standard white plate (Y = 94.3, x = 0.3134 and y = 0.3197).

### 2.4. Proximate Composition

The total nitrogen was determined by the AOAC [12], and the protein content was calculated by nitrogen-to-protein conversion factor 4.76, as proposed by Janssen et al. [13] to compensate for the presence of chitin-derived nitrogen.

The fat content was determined after extraction performed by Soxhlet apparatus [14].

Total ash content was measured after incineration of the insect flour [15].

The moisture content of the flours was calculated after drying at 104–105 °C using a KERN MLS 65 3A—moisture analyser (Kern & Sohn Gmbh, Ballingen, Germany) until constant weight [16]. The carbohydrate content was calculated using the following equation [17]:

$$\text{Carbohydrates} = 100 - \text{Protein} - \text{Fat} - \text{Ash} - \text{Moisture, \%.}$$

### 2.5. Acid Value (AV)

The degree of lipolysis was described by the acid value of the extracted lipids and measured following the method of Kardash and Tur'yan [18]. The extraction of lipids was performed according to the method of Bligh and Dyer [19]. After extraction, 1 g of lipid was dissolved in 20 $cm^3$ neutral alcohol–ether mixture with added phenolphthalein. The mixture was titrated with 0.01N KOH. The volume of the used KOH was recorded. The acid value was calculated as AV = (V × F × 5.6104)/m, mg KOH/g, where:

V—volume of KOH used for titration, g.

F—factor of 0.1 nKOH = 0.996.

m—the weight of the sample, g.

### 2.6. Peroxide Value (PV)

Peroxide value was determined according to Shantha and Decker [20]. Briefly, the extracted lipid (0.1 g) was mixed in a glass tube with 50 μL iron (II) solution, 50 μL $NH_4$ SCN (300 mg/mL), and $CHCl_3$:$CH_3OH$ (3:5, *v/v*) for a final volume of 10 mL. The samples were incubated for 5 min at room temperature and the absorbance was determined by spectrophotometer at 507 nm against a blank, containing all the reagents except the sample. The PV was calculated through a standard curve set up using $Fe^{3+}$ chloride standard solution (10 μg/mL). The results were presented as $meqO_2$/kg lipid.

### 2.7. TBARS Content

The content of the thiobarbituric acid reactive substances was determined as described by Botsoglou et al. [21] with slight modifications. Ten grams of flour were homogenized with 50 mL NaCl (0.9%), and left for 5 min. Furthermore, 50 mL of trichoroacetic acid (10%) were added and the samples were filtered through (Filtrax, Grade 391). The filtered samples (4 mL) were then mixed with 1 mL 2-thiobabituric acid (1%) and were incubated at 70 °C for 30 min. After cooling to room temperature, the absorbance of the samples was determined at 532 nm against a blank, containing distilled water instead of the sample. TBARS concentrations were calculated using 1, 1, 3, 3 tetraethoxypropane as standard. The results were expressed as mg MDA/kg product or TBARS units.

### 2.8. Antioxidant Activity

The antioxidant activity of the flours was measured through DPPH assay according to the method of Brand-Williams et al. [22], modified by Dinkova et al. [23]: 250 μL of the sample extract (in methanol $1/10$, $w/v$) was mixed with 2250 μL of methanolic solution of DPPH ($6 \times 10^{-5}$ M) in a UV-macro cuvette. The cuvettes were sealed and left in the dark for 15 min at room temperature. Absorbance was measured at 515 nm against a blank of pure methanol. The results were expressed as μmol TE (Trolox Eq)/100 g sample.

### 2.9. Microbiological Assay

All of the preparation and decimal dilutions of sample suspensions were conducted according to ISO 6887-4:2017 [24]. The microbiological status was presented by the Total Plate Count (TPC), Coliforms count, and *E. coli* count, following the procedure of ISO 4833-1:2013/Amd 1:2022 [25].

All of the analyses were performed in triplicates.

### 2.10. Statistical Evaluation

The statistical evaluation was performed through one way ANOVA procedure and post-hoc comparisons (Tukey HSD, $p < 0.05$) using JMP v. 7 statistical software [26].

## 3. Results

### 3.1. Water Activity and Instrumental Colour

The drying treatments significantly affected the $a_w$ and they differed among all the flours. The lowest value of this parameters was measured in the CD flour, whereas the MWDA flour exhibited the highest $a_w$ (Table 2). The freezing of the larvae prior to blanching for the microwave treatment decreased the water activity, and as a result the MWD flour had lower aw when compared to MWDL flour.

**Table 2.** Water activity and instrumental colour parameters of the flours derived from *Tenebrio molitor* larvae according to the drying treatment.

| Treatment | $a_w$ | L* | a* | b* |
|:---:|:---:|:---:|:---:|:---:|
| FD | 0.35 [d] | 52.76 [b] | 6.55 [c] | 11.79 [b] |
| CD | 0.14 [e] | 49.09 [c] | 6.17 [d] | 9.95 [c] |
| MWD | 0.50 [c] | 51.74 [b] | 7.29 [a] | 12.94 [a] |
| MWDL | 0.54 [b] | 54.70 [a] | 6.63 [c] | 13.27 [a] |
| MWDA | 0.63 [a] | 49.54 [c] | 7.00 [b] | 13.37 [a] |
| SEM | 0.01 | 0.44 | 0.05 | 0.34 |
| *p* | <0.0001 | <0.0001 | <0.0001 | <0.0001 |

FD, freeze-drying; CD, conventional drying; MWD, microwave drying; MWDL, microwave drying without freezing prior blanching; MWDA, microwave drying + 0.1% BHT; SEM, standard error of the mean. The means connected with different letters in a row are significantly different ($p < 0.05$).

The drying treatments had a significant effect on the lightness of the flours (Table 2). The darkest colour was observed in the CD and MWDA flours, whereas the MWDL flour was the lightest. The freeze-drying and microwave drying of the insect larvae also resulted in the lighter colour of the FD and MWD flours, when compared to the oven dried and MWDA flours. In addition to the lower L*, the CD flour also had lower a* and b*, when compared to the flours derived after the other studied methods of drying. Higher a* values were observed in the MWD and MWDA flour. These flours, together with the MWDL flour, exhibited higher b* values in comparison to the FD and CD samples.

### 3.2. Proximate Composition

The freeze-drying and conventional drying of the *Tenebrio molitor* larvae led to lower protein content of the derived flours (Table 3). The highest protein content was measured in the MWDL followed by MWD flours. Contrary to protein content, the fat percentage was significantly higher in FD and CD flours when compared to the rest. FD, CD, and MWDA flours had ash content within the range of 4.16–4.23%, which was significantly higher than MWD and MWDL.

**Table 3.** Proximate composition of the flours derived from *Tenebrio molitor* larvae according to the drying treatment.

| Treatment | Protein, % | Fat, % | Ash, % | Moisture, % | Carbohydrate, % |
|---|---|---|---|---|---|
| FD | 41.21 [d] | 20.82 [a] | 4.17 [a] | 6.07 | 27.73 [c] |
| CD | 41.42 [d] | 21.11 [a] | 4.23 [a] | 5.62 | 27.62 [c] |
| MWD | 47.14 [b] | 12.71 [b] | 2.87 [c] | 5.71 | 31.57 [ab] |
| MWDL | 49.99 [a] | 11.35 [b] | 3.83 [b] | 5.18 | 29.65 [bc] |
| MWDA | 44.97 [c] | 11.42 [b] | 4.17 [a] | 6.29 | 33.15 [a] |
| SEM | 0.75 | 0.55 | 0.10 | 0.46 | 0.97 |
| *p* | <0.0001 | <0.0001 | <0.0001 | 0.1034 | 0.0001 |

FD, freeze-drying; CD, conventional drying; MWD, microwave drying; MWDL, microwave drying without freezing prior blanching; MWDA, microwave drying + 0.1% BHT; SEM, standard error of the mean. The means connected with different letters in a row are significantly different ($p < 0.05$).

The latter differed in regard to the ash content as MWD displayed lower values of this parameter in comparison to MWDL flour. With regard to the carbohydrates, their percentage was lower in the FD and CD than in MWD and MWDA flours, while the carbohydrate content of the MWDL flour had an intermediate position.

### 3.3. Lipid Stability

The hydrolysis in lipids of the derived flours were presented through the acid value (Table 4).

**Table 4.** Acid value (AV), peroxide value (PV), and TBARS value in the lipids of flours derived from *Tenebrio molitor* larvae according to the drying treatment.

| Treatment | AV, mgKOH/g | PV, meqO$_2$/kg | TBARS, mgMDA/kg |
|---|---|---|---|
| FD | 1.94 [c] | 1.27 [a] | 0.05 [c] |
| CD | 1.69 [d] | 1.22 [ab] | 0.06 [c] |
| MWD | 2.58 [b] | 0.97 [bc] | 0.12 [a] |
| MWDL | 2.55 [b] | 0.91 [c] | 0.10 [b] |
| MWDA | 3.10 [a] | 1.16 [abc] | 0.10 [b] |
| SEM | 0.04 | 0.10 | 0.005 |
| *p* | <0.0001 | 0.0080 | <0.0001 |

FD, freeze-drying; CD, conventional drying; MWD, microwave drying; MWDL, microwave drying without freezing prior blanching; MWDA, microwave drying + 0.1% BHT; SEM, standard error of the mean. The means connected with different letters in a row are significantly different ($p < 0.05$).

Lower AV was found in the FD and CD flours, while the microwave drying in all three modes stimulated the lipid hydrolysis.

The content of the primary products of lipid oxidation also depended on the drying treatment. The highest levels of peroxides were measured in the flours derived after freeze-drying and conventional drying of the *Tenebrio molitor* larvae. Microwave drying; however, was associated with lower content of peroxides, except for the MWDA flour.

The analysis of the secondary products of lipid oxidation showed that FD and CD flours had significantly lower TBARS levels when compared to all the flours derived from microwave dried insects.

### 3.4. Antioxidant Activity

Not surprisingly, MWDA displayed the highest radical scavenging activity (Figure 2). It was followed by the MWDL and MWD flours. The lowest radical scavenging activity was observed in the FD and CD samples.

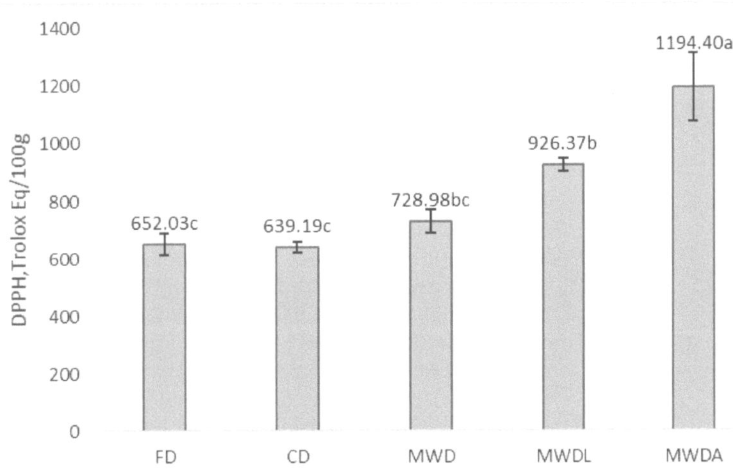


**Figure 2.** Antioxidant activity of the flours derived from *Tenebrio molitor* larvae according to the drying treatment. FD, CD, MWD, MWDL, and MWDA represent freeze-drying, conventional drying, microwave drying, microwave drying without freezing prior blanching, and microwave drying + 0.1% BHT, respectively. The means connected with different letters in a row are significantly different ($p < 0.05$).

### 3.5. Microbiological Status

The total plate count was higher in the FD flours in comparison to all the other flours in this study; however, the differences were not significant (Table 5).

**Table 5.** Microbiological status of the flours derived from *Tenebrio molitor* larvae according to the drying treatments.

| Treatment | TPC, $\log_{10}$ CFU/g | Coliforms, $\log_{10}$ CFU/g | Yeasts Moulds, $\log_{10}$ CFU/g | *E. coli*, $\log_{10}$ CFU/g |
|---|---|---|---|---|
| FD | 4.52 | 2.93 | N.D. | N.D. |
| CD | 3.91 | 3.19 | N.D. | N.D. |
| MWD | 3.93 | 2.97 | N.D. | N.D. |
| MWDL | 3.97 | 2.87 | N.D. | N.D. |
| MWDA | 3.99 | 2.90 | N.D. | N.D. |
| SEM | 0.62 | 0.50 | | |
| *p* | 0.7279 | 0.9411 | | |

FD, freeze-drying; CD, conventional drying; MWD, microwave drying; MWDL, microwave drying without freezing prior blanching; MWDA, microwave drying + 0.1% BHT; SEM, standard error of the mean; TPC, total plate count; N.D.—not detected.

The coliform content varied within 2.87–3.19 $\log_{10}$CFU/g, with no statistical difference among the drying treatments. Yeasts, moulds, and *E. coli* were not detected.

## 4. Discussion

Drying of the larvae is of crucial importance for the further processing of the insects so that the nutritional quality and safety are guaranteed. Some studies have already reported changes in the quality characteristics of edible insects or insect flours due to different drying methods [11,15,27–29]. The results of this study confirmed the effect of freeze-drying, oven drying, and microwave drying of the larvae on the water activity of the flours that was observed by Krönke et al. [28] and Vlahova-Vangelova et al. [11]. The freeze-drying and conventional drying considerably reduced the water activity of the flours, while it was higher after microwave drying. Substances with higher water activity tend to support microbial growth; most bacteria usually require $a_w$ at least 0.91 [30], while some moulds and yeasts need $a_w$ in the range of 0.61–0.65 [31]. Colour is one of the most important quality attributes. It is a crucial factor influencing the consumer's decision when buying a new product, but also can indicate chemical changes that occur in food [32]. Colour can change during drying due to chemical and biochemical reactions [33]. In our study, the lowest $L^*$ were observed for the flours derived after conventional drying and microwave drying of the larvae blanched in the presence of BHT. The other microwave treatments led to a lighter colour, compared to conventional drying and freeze-drying. In their study, Trukhanova et al. [34] observed lighter colour of the microwave dried larvae of *Tenebrio molitor*, when compared to the larvae dried through convection, which is in line with our results. The lower $L^*$ values observed in the CD flour can be attributed to Maillard reaction, due to the longer drying process (6 h). Usually, this type of browning occurs in foods that have high lipid content and where reactions between products of lipid oxidation with amino acids, amines, and proteins occur [32]. Maillard reaction and lipid oxidation are interrelated and should be considered simultaneously in regard to colour changes [35]. The microwave treated flours also exhibited elevated content of secondary products of lipid oxidation, which was considerably higher in comparison to FD and CD flour. Furthermore, the significantly darker colour of the MWDA flour suggests a certain pro-oxidative effect of the BHT in the amount used to treat the larvae for the present experiment. Some phenolic compounds might act as pro-oxidants depending on the conditions of the environment (such as high pH, presence of oxygen molecules, and high concentration of transition metals) [36].

There was considerable variation in the chemical components of the flours according to the drying treatment. Generally, the flours derived from microwave dried larvae had higher protein content when compared to the FD and CD flours. Additionally, the flours derived from larvae that were not frozen prior to blanching exhibited the highest protein content of all the flours. This indicates the negative effect of freezing, most likely associated with the denaturation of proteins in the other flours in the study [37]. The higher protein content in the MWD, MWDL, and MWDA flours corresponded to their lower fat percentage when compared to the FD and CD flours. The latter did not differ in regard to the protein and fat content. Our results coincided with those of Selaledi and Mabelele [1], who did not observe any differences in the protein content of *Tenebrio molitor* larvae when freeze-dried and oven dried. In contrast to our data, Krönke et al. [38] found that rack oven dried mealworm larvae had higher protein and fat content than freeze-dried larvae. The authors also reported dramatic increase in moisture after freeze-drying, whereas in our study, we failed to observe any significant difference in the moisture content among the flours. The moisture content only tended to be higher in the MWDA flour and corresponded to its highest $a_w$.

When comparing the different drying treatments applied in this study, we can conclude that the conventional drying is associated with lowest degree of free fatty acid accumulation, whereas the microwave drying stimulated the hydrolytic changes in the lipid fraction of the insect flours. Other studies also reported a similar effect of the microwave treatment on the lipid stability in foods, associated with a considerable degree of hydrolysis [39,40]. The freezing prior to blanching of the *Tenebrio molitor* larvae did not affect the lipid hydrolysis in the flours derived from the microwave treated larvae. The comparison of the flours derived after the different microwave treatments showed that in the BHT addition there was a 20% increase in the free fatty acid content, again indicating the potential pro-oxidative effect of BHT. According to Pérez-Torres et al. [41], the high concentrations of strong antioxidants such as BHT in the food matrices exert a pro-oxidative effect. Hence, the higher degree of hydrolysis of the lipids in the MWDA flour might be due to the higher dose of BHT. The elevated AV of the flours derived after microwave treatment corresponded to their higher TBARS values. Usually, with advancing of lipid oxidation, triacylglycerols are converted into fatty acids and glycerol thus increasing the acid value, which is in line with our results. On the other hand, we observed a decrease in the PV of the MWD and MWDL flours, corresponding to the elevated levels of TBARS in these flours. The lipid peroxides are unstable and susceptible to decomposition [42] and formation of other products, such as alcohols, ketones, and aldehydes. The freeze-drying and conventional drying were associated with a significant increase in the peroxide content in the insect flours and the same was observed in the MWDA samples. In all the samples regardless of the drying treatment, PV remained below 2 meq/$O_2$, which is considered as a threshold recommended by EFSA [43].

Despite the higher degree of lipid oxidation, the flours derived of microwave dried larvae displayed higher antioxidant activity in comparison to the FD and CD flours. Not surprisingly, the MDWA flour exhibited the highest antioxidant activity. As reported in previous studies, the antioxidant content correlates positively with the radical scavenging activity [44]. On the other hand, the higher antioxidant activity measured in the MWD, MWDL and MWDA flours contradicts to the higher degree of lipid oxidation. It can be suggested that the products from lipid oxidation have reacted with DPPH, presenting falsely increased antioxidant activity [45].

Microbiological status is crucial for the safety of the foods. In our study, we observed a low microbial load of the flours. In their study, Bußler et al. [46] observed a high microbial contamination in *Tenebrio molitor* larvae before processing (8.1 log CFU/g) that reduced to 4.3 log CFU in the high protein fraction. Kluder et al. [47] found a low microbial load in whole *Tenebrio molitor* larvae after boiling and roasting (<1.7 log CFU/g); however, after crushing, the TVC was considerably augmented (2.5–4.8 log CFU/g). The analysis of the results showed that the total plate count measured in the FD flours was higher than the CD, MWD, MWDL, and MWDA flours. Caparros Megido et al. [48] reported data about different treatments on the microbial load in edible insects. Similar to our results, the authors measured the total aerobic count in freeze-dried mealworm and house crickets as 4.47 and 4.05 log CFU/g, respectively, significantly higher than the microbial load in the sterilized insects. Messina et al. [49] presented a microbiological profile of powders prepared using mealworm and house crickets after prolonged storage. They identified seven microbial populations in the *Tenebrio molitor* powder, consisting of LAB cocci, enterococci, pseudomonads, CPS, and members of the Bacillaceae family, while yeasts and moulds were not detected. The flours presented in this study showed insignificant differences in regard to the coliforms. Yeast, moulds, and *E. coli* were not detected, showing the high level of microbial safety of the applied drying treatments.

## 5. Conclusions

The drying treatments caused considerable changes in the qualities of the flours prepared from *Tenebrio molitor* larvae. The flours obtained by freeze-dried and conventionally dried larvae had lower water activity and showed considerably higher fat content than the microwave treated flours; however, their nutritional profile was less favourable with lower protein. The conventionally dried flour, as well as the flour derived from microwave treated insects with addition of BHT, exhibited a darker colour that can probably be attributed to the Maillard reaction and lipid oxidation. Although known as a fast and energy efficient method, in this study, the microwave drying displayed some negative effects associated with a higher degree of lipid oxidation in the flours. Increased oxidation was also observed after BHT treatment, indicating that the dose of the antioxidant used in the study (0.1%) is not recommended for obtaining a good quality in the flours treated with microwaves. Showing both positive and negative effects of the applied treatments, this study contributes to the development of optimal drying modes for preparation of insect flours. However, further experiments are needed to determine the most appropriate drying technique that will help to preserve the best nutritional qualities of the flours while restricting the processes that will deteriorate them.

**Author Contributions:** Conceptualization, D.V.-V. and D.B.; methodology, D.V.-V. and N.K.; formal analysis, N.K., D.V.-V., D.B. and S.D.; investigation, D.V.-V., N.K., D.B., S.D., T.P. and E.P.; resources D.V.-V. and D.B.; data curation, N.K., T.P. and E.P.; writing—original draft preparation, D.V.-V., T.P. and N.K.; writing—review and editing, S.D. and D.B.; supervision, S.D.; project administration, D.V.-V. and D.B.; funding acquisition, D.V.-V. All authors have read and agreed to the published version of the manuscript.

**Funding:** This research was funded by Bulgarian National Science Fund, Ministry of Education and Science in Bulgaria (Grant number KP-06-PN-76/7).

**Institutional Review Board Statement:** The insects were reared according to the requirements of Commission Regulation (EU) No 10/2011 of 14 January 2011.

**Data Availability Statement:** The data presented in this study are available upon request from the corresponding authors.

**Conflicts of Interest:** The authors declare no conflicts of interest.

## References

1. Selaledi, L.; Mabelebele, M. The influence of drying methods on the chemical composition and body color of yellow mealworm (*Tenebrio molitor* L.). *Insects* **2021**, *12*, 333. [CrossRef]
2. Food Security Update. World Bank Response to Rising Food Insecurity. The World Bank. Available online: https://thedocs.worldbank.org/en/doc/40ebbf38f5a6b68bfc11e5273e1405d4-0090012022/related/Food-Security-Update-XCV-11-9-23.pdf (accessed on 11 December 2023).
3. Insects as Replacement of Animal Protein Gets a Boost. Available online: https://www.wur.nl/en/newsarticle/insects-as-replacement-of-animal-protein-gets-a-boost.htm (accessed on 21 January 2024).
4. Smarzyński, K.; Sarbak, P.; Musiał, S.; Jeżowski, P.; Piątek, M.; Kowalczewski, P.Ł. Nutritional analysis and evaluation of the consumer acceptance of pork pâté enriched with cricket powder—Preliminary study. *Open Agric.* **2019**, *4*, 159–163. [CrossRef]
5. Kim, T.K.; Yong, H.I.; Kim, Y.B.; Kim, H.W.; Choi, Y.S. Edible Insects as a protein source: A review of public perception, processing technology, and research trends. *Food Sci. Anim. Resour.* **2019**, *39*, 521–540. [CrossRef] [PubMed]
6. Schäeufele, I.; Albores, E.B.; Hamm, U. The role of species for the acceptance of edible insects: Evidence from a consumer survey. *Br. Food J.* **2019**, *121*, 2190–2204. [CrossRef]
7. Fischer, A.R.H.; Steenbekkers, B. All insects are equal, but some insects are more equal than others. *Br. Food J.* **2018**, *120*, 852–863. [CrossRef] [PubMed]
8. Turck, D.; Castenmiller, J.; De Henauw, S.; Hirsch Ernst, K.; Kearney, J.; Maciuk, A.; Knutsen, H. Safety of dried yellow mealworm (*Tenebrio molitor* larva) as a novel food pursuant to Regulation (EU) 2015/2283. *EFSA J.* **2021**, *19*, 6343. [CrossRef]
9. Novel Food: Insects–Food of the Future?! Available online: https://www.eurofins.de/food-analysis/food-news/food-testing-news/novel-food_insects-as-food-of-the-future/ (accessed on 30 January 2024).
10. Hartmann, C.; Shi, J.; Giusto, A.; Siegrist, M. The psychology of eating insects: A cross-cultural comparison between Germany and China. *Food Qual. Prefer.* **2015**, *44*, 148–156. [CrossRef]
11. Vlahova-Vangelova, D.; Balev, D.; Kolev, N.; Stoyanov, V. Effect of drying regimes on the quality and safety of alternative protein source (*Tenebrio molitor* L.). *Acta Sci. Pol. Technol. Aliment.* **2023**, *22*, 217–225. [CrossRef]
12. AOAC. *Official Methods of Analysis of AOAC International: AOAC 984.13 1994 Protein (Crude) in Animal Feed and Pet Food*; AOAC: Rockville, MD, USA, 1996.
13. Janssen, R.; Vincken, J.; van den Broek, L.; Fogliano, V.; Lakemond, C. Nitrogen to protein conversion factors for three edible insects: *Tenebrio molitor, Alphitobiu diaperinus*, and *Hermetia illucens. J. Agric. Food Chem.* **2017**, *65*, 2275–2278. [CrossRef]
14. *ISO 1444:1996*; Meat and Meat Products—Determination of Free Fat Content. ISO: Geneva, Switzerland, 1996.
15. Lenaerts, S.; Van Der Borght, M.; Callens, A.; Van Campenhout, L. Suitability of microwave drying for mealworms (*Tenebrio molitor*) as alternative to freeze drying: Impact on nutritional quality and colour. *Food Chem.* **2018**, *254*, 129–136. [CrossRef]

16. Vandeweyer, D.; Lenaerts, S.; Callens, A.; Van Campenhout, L. Effect of blanching followed by refrigerated storage or industrial microwave drying on the microbial load of yellow mealworm larvae (*Tenebrio molitor*). *Food Control* **2017**, *71*, 311–314. [CrossRef]
17. Joymak, W.; Ngamukote, S.; Chantarasinlapin, P.; Adisakwattana, S. Unripe papaya by-product: From food wastes to functional ingredients in pancakes. *Foods* **2021**, *10*, 615. [CrossRef]
18. Kardash, E.; Tur'yan, Y.I. Acid value determination in vegetable oils by indirect titration in aqueous-alcohol media. *Croat. Chem. Acta* **2005**, *78*, 99–103.
19. Bligh, E.G.; Dyer, W.J. A rapid method of total lipid extraction and purification. *Can. J. Biochem. Physiol.* **1959**, *37*, 911–917. [CrossRef]
20. Shantha, N.C.; Decker, E.A. Rapid, sensitive, iron-based spectrophotometric methods for determination of peroxide values of food lipids. *J. AOAC Int.* **1994**, *7*, 421–424. [CrossRef]
21. Botsoglou, N.A.; Fletouris, D.J.; Papageorgiou, G.E.; Vassilopoulos, V.N.; Mantis, A.J.; Trakatellis, A.G. Rapid, sensitive, and specific thiobarbituric acid method for measuring lipid peroxidation in animal tissue, food, and feedstuff samples. *J. Agric. Food Chem.* **1994**, *42*, 1931–1937. [CrossRef]
22. Brand-Williams, W.; Cuvelier, M.E.; Berset, C. Use of a free radical method to evaluate antioxidant activity. *LWT* **1995**, *28*, 25–30. [CrossRef]
23. Dinkova, R.; Heffels, P.; Shikov, V.; Weber, F.; Schieber, A.; Mihalev, K. Effect of enzyme-assisted extraction on the chilled storage stability of bilberry (*Vaccinium myrtillus* L.) anthocyanins in skin extracts and freshly pressed juices. *Food Res. Int.* **2014**, *65*, 35–41. [CrossRef]
24. *ISO 6887 4:2017*; Microbiology of the Food Chain—Preparation of Test Samples, Initial Suspension and Decimal Dilutions for Microbiological Examination—Part 4: Specific Rules for the Preparation of Miscellaneous Products. ISO: Geneva, Switzerland, 2017.
25. *ISO 4833 1:2013/Amd 1:2022*; Microbiology of the Food Chain—Horizontal Method for the Enumeration of Microorganisms—Part 1: Colony Count at 30 °C by the Pour Plate Technique—Amendment 1: Clarification of Scope. ISO: Geneva, Switzerland, 2013.
26. *JMP, Version 7*; SAS Institute Inc.: Cary, NC, USA, 2007.
27. Fombong, F.T.; Van Der Borght, M.; Vanden Broeck, J. Influence of freeze-drying and oven-drying post blanching on the nutrient composition of the edible insect *Ruspolia differens*. *Insects* **2017**, *8*, 102. [CrossRef] [PubMed]
28. Kröncke, N.; Böschen, V.; Woyzichovski, J.; Demtröder, S.; Benning, R. Comparison of suitable drying processes for mealworms (*Tenebrio molitor*). *Innov. Food Sci. Emerg. Technol.* **2018**, *50*, 20–25. [CrossRef]
29. Yisa, N.K.; Osuga, I.M.; Subramanian, S.; Ekesi, S.; Emmambux, M.N.; Duodu, K.G. Effect of drying methods on the nutrient content, protein and lipid quality of edible insects from East Africa. *J. Insects Food Feed* **2022**, *9*, 647–659. [CrossRef]
30. Allen, L.V., Jr. Quality control: Water activity considerations for beyond-use dates. *Int. J. Pharm. Compd.* **2018**, *22*, 288–293. [PubMed]
31. Tapia, M.S.; Alzamora, S.M.; Chirife, J. Effects of Water Activity (aw) on Microbial Stability: As a Hurdle in Food Preservation. In *Water Activity in Foods: Fundamentals and Applications*; Barbosa-Cánovas, G.V., Fontana, A.J., Jr., Schmidt, S.J., Labuza, T.P., Eds.; Blackwell Publishing Ltd.: Oxford, UK, 2008; pp. 239–271. [CrossRef]
32. Hidalgo, F.J.; Zamora, R. The role of lipids in nonenzymatic browning. *Grasas Aceites* **2000**, *51*, 35–49. [CrossRef]
33. Bonazzi, C.; Dumoulin, E. Quality changes in food materials as influenced by drying processes. In *Modern Drying Technology*; Tsotsas, E., Mujumdar, A.S., Eds.; Wiley-VCH Verlag GmbH & Co. KGaA: Weinheim, Germany, 2011; pp. 1–20.
34. Trukhanova, K.A.; Mechtaeva, E.V.; Novikova, M.V.; Sorokoumov, P.N.; Ryabukhin, D.S. Influence of drying and pretreatment methods on certain parameters of yellow mealworm larvae (*Tenebrio molitor*). *Theory Pract. Meat Process.* **2022**, *7*, 247–257. [CrossRef]
35. Zamora, R.; Hidalgo, F.J. Coordinate contribution of lipid oxidation and Maillard reaction to the nonenzymatic food browning. *Crit. Rev. Food Sci. Nutr.* **2005**, *45*, 49–59. [CrossRef]
36. Cotoras, M.; Vivanco, H.; Melo, R.; Aguirre, M.; Silva, E.; Mendoza, L. In vitro and in vivo evaluation of the antioxidant and prooxidant activity of phenolic compounds obtained from grape (*Vitis vinifera*) pomace. *Molecules* **2014**, *19*, 21154. [CrossRef] [PubMed]
37. Lee, S.; Jo, K.; Jeong, H.G.; Choi, Y.-S.; Kyoung, H.; Jung, S. Freezing-induced denaturation of myofibrillar proteins in frozen meat. *Crit. Rev. Food Sci. Nutr.* **2024**, *64*, 1385–1402. [CrossRef] [PubMed]
38. Kröncke, N.; Grebenteuch, S.; Keil, C.; Demtröder, S.; Kroh, L.; Thünemann, A.F.; Benning, R.; Haase, H. Effect of different drying methods on nutrient quality of the yellow mealworm (*Tenebrio molitor* L.). *Insects* **2019**, *10*, 84. [CrossRef]
39. Chandrasekaran, S.; Ramanathan, S.; Basak, T. Microwave food processing—A review. *Food Res. Int.* **2013**, *52*, 243–261. [CrossRef]
40. Jiang, H.; Liu, Z.; Wang, S. Microwave processing: Effects and impacts on food components. *Crit. Rev. Food Sci. Nutr.* **2018**, *58*, 2476–2489. [CrossRef]
41. Pérez-Torres, I.; Guarner-Lans, V.; Rubio-Ruiz, M.E. Reductive stress in inflammation-associated diseases and the pro-oxidant effect of antioxidant agents. *Int. J. Mol. Sci.* **2017**, *18*, 2098. [CrossRef] [PubMed]
42. Talbot, G. The stability and shelf life of fats and oils. In *The Stability and Shelf Life of Food*; Subramaniam, P., Ed.; Elsevier: Amsterdam, The Netherlands, 2016; pp. 461–503. [CrossRef]
43. European Food Safety Authority (EFSA). Scientific opinion on fish oil for human consumption. Food hygiene, including rancidity. *EFSA J.* **2010**, *8*, 1874. [CrossRef]

44. Kumar, S.; Sandhir, R.; Ojha, S. Evaluation of antioxidant activity and total phenol in different varieties of *Lantana camara* leaves. *BMC Res. Notes* **2014**, *7*, 560. [CrossRef] [PubMed]
45. Jeong, M.K.; Yeo, J.D.; Jang, E.Y.; Kim, M.-J.; Lee, J.H. Aldehydes from oxidized lipids can react with 2,2-diphenyl-1-picrylhydrazyl (DPPH) free radicals in isooctane systems. *J. Am. Oil Chem. Soc.* **2012**, *89*, 1831–1838. [CrossRef]
46. Bußler, S.; Rumpold, B.A.; Jander, E.; Rawel, H.M.; Schlüter, O.K. Recovery and techno-functionality of flours and proteins from two edible insect species: Meal worm (*Tenebrio molitor*) and black soldier fly (*Hermetia illucens*) larvae. *Heliyon* **2016**, *2*, e00218. [CrossRef]
47. Klunder, H.C.; Wolkers-Rooijackers, J.; Korpela, J.M.; Nout, M.J.R. Microbiological aspects of processing and storage of edible insects. *Food Control* **2012**, *26*, 628–631. [CrossRef]
48. Caparros Megido, R.; Desmedt, S.; Blecker, C.; Béra, F.; Haubruge, É.; Alabi, T.; Francis, F. Microbiological load of edible insects found in Belgium. *Insects* **2017**, *8*, 12. [CrossRef]
49. Messina, C.M.; Gaglio, R.; Morghese, M.; Tolone, M.; Arena, R.; Moschetti, G.; Santulli, A.; Francesca, N.; Settanni, L. Microbiological profile and bioactive properties of insect powders used in food and feed formulations. *Foods* **2019**, *8*, 400. [CrossRef]

*Article*

# Monitoring of Chemical and Fermentative Characteristics during Different Treatments of Grape Pomace Silage

Tea Sokač Cvetnić [1], Veronika Gunjević [1,2], Anja Damjanović [1], Anita Pušek [1,3], Ana Jurinjak Tušek [1], Tamara Jakovljević [4,*], Ivana Radojčić Redovniković [1] and Darko Uher [2]

[1]    Faculty of Food Technology and Biotechnology, University of Zagreb, Pierotti St. 6, 10 000 Zagreb, Croatia; tsokac@pbf.hr (T.S.C.); vgunjevic@pbf.hr (V.G.); adamjanovic@pbf.hr (A.D.); apusek@pbf.hr (A.P.); ana.tusek.jurinjak@pbf.unizg.hr (A.J.T.); irredovnikovic@pbf.hr (I.R.R.)
[2]    Faculty of Agriculture, University of Zagreb, Svetošimunska 25, 10 000 Zagreb, Croatia; duher@agr.hr
[3]    Salvus d.o.o., Toplička Street 100, 49 240 Donja Stubica, Croatia
[4]    Croatian Forest Research Institute, Cvjetno Naselje 41, 10 450 Jastrebarsko, Croatia
*    Correspondence: tamaraj@sumins.hr

**Abstract:** Grape pomace is a fibrous food with satisfactory quantities of residual sugars. It meets the desirable characteristics for conservation in the form of silage for later use in animal feed, mainly for ruminant herbivores. Fresh grape pomace was subdivided into three treatment groups: grape pomace as a control, grape pomace treated with an inoculum of lactic acid bacteria, and grape pomace treated with zeolite. The treatments were performed in micro-silos over 90 days. There was a significant change ($p < 0.05$) in the chemical characteristics, content of biologically active compounds, and fermentative characteristics during the silage of all treatments. After 30, 60 and 90 days of ensiling, silages treated with inoculum and zeolite had better fermentation quality indicated by significantly ($p < 0.05$) lower pH and ammonia-nitrogen contents compared with those of the control. Also, the additives have decreased the total polyphenols and tannins for 97% in average which confirmed that lactic acid bacteria and zeolite positively effect on the degradation of polyphenols and tannins in grape pomace silage. The Flieg score was calculated and the values were above 80% what refers to excellent silage. In conclusion, our results suggest that inoculant and zeolite supplementation improves the quality of grape pomace silage for later use in animal feed.

**Keywords:** fermentation quality; grape pomace; lactic acid bacteria; silage; zeolite

**Citation:** Sokač Cvetnić, T.; Gunjević, V.; Damjanović, A.; Pušek, A.; Jurinjak Tušek, A.; Jakovljević, T.; Radojčić Redovniković, I.; Uher, D. Monitoring of Chemical and Fermentative Characteristics during Different Treatments of Grape Pomace Silage. *Agriculture* **2023**, *13*, 2264. https://doi.org/10.3390/agriculture13122264

Academic Editors: Petru Alexandru Vlaicu, Arabela Elena Untea and Mihaela Saracila

Received: 27 October 2023
Revised: 6 December 2023
Accepted: 7 December 2023
Published: 12 December 2023

## 1. Introduction

The wine industry produces a large amount of unprocessed plant waste, known as grape pomace. Approximately 25% of the total mass of grapes used in wine production ends up as biowaste, which is, globally, around 9 million tons of waste every year [1]. In Croatia, the wine industry produces 15,000 tons of solid waste annually. A wide distribution of the leading grape variety in Croatia, called Graševina, is responsible for producing the largest share of wine waste [2]. Grape pomace consists of skins, seeds, and stems, and is considered a valuable by-product because grapes are rich in biologically active compounds. Improperly disposed grape pomace presents an environmental risk in terms of soil and water pollution. To reduce the amount of waste, a small portion of pomace is used for oil production or polyphenol extraction, or is processed into animal feed [3]. However, grapes are a seasonal fruit and large quantities of pomace are generated in a short period, making it difficult to store in a way that ensures the maintenance of desirable chemical characteristics. To preserve the stability of fresh plant material, the biomass can be dried; for example, vacuum drying is proven to be an effective method for conserving fresh biomass. However, vacuum drying requires additional equipment, which increases the final cost of storage, especially on a large industrial scale [2]. For the said reasons, ensilage is another method that can be applied in such cases [4]. Ensilage is a fermentative process used for the conservation of by-products. Lactic acid bacteria facilitate the fermentation of moist biomass under anaerobic conditions. As a result of converting water-soluble carbohydrates into organic acids, the pH value rapidly drops, which inhibits the growth of undesirable micro-organisms, such as yeast and mold. Therefore, the further decomposition and deterioration of the biomass are prevented and long-term storage under those conditions is provided [5]. When grape pomace is ensiled, it can produce high-quality feed for ruminant animals, such as cows or sheep. The fermentation process can produce a valuable by-product called biogas, which can be used as a renewable energy source. This can be economically beneficial for wineries that have a lot of disposed grape pomace. In addition to its potential as a feed source, the ensiling of grape pomace offers ecological advantages. Ensiling contributes to a reduction in agricultural waste by minimizing the amount of pomace left in the fields. To prevent greenhouse gas production during silage, it is advisable to monitor lactate degradation and butyrate formation by clostridia, as the production of methane can adversely affect the silage quality. The quantities of greenhouse gas emissions also depend on factors such as the forage type, dry matter concentration, and silage additives [6].

Grape pomace is especially desirable for ensiling because of its already low pH value and high content of polyphenols and lipids, which prevent protein degradation [7]. Polyphenolic compounds have already attracted a large market interest as antioxidants that can be used as additives for various applications and that exert numerous health-promoting effects on humans [8]. In contrast, polyphenol effects have been only recently studied in animal diets, and effects such as an enhanced oxidative stability of meat, the modulation of intestinal microbiota, and a reduction in synthetic antioxidant additions such as vitamin E have been reported [9]. Monitoring the total polyphenolic content during ensilage is essential for the quality control of the process. Polyphenols, being natural antioxidants, play a crucial role in enhancing the overall quality and nutritional value of silage. Variations in the polyphenolic content can serve as indicators of potential issues, such as spoilage during storage or excessive heating in the process, due to their degradation [10]. Moreover, polyphenols aid in optimizing the nutritional quality of the silage for specific animal diets [9]. Tannins are naturally occurring phenolic compounds that may exhibit positive effects on human health and ruminant performance. Nevertheless, concentrations higher than 50 g kg$^{-1}$ dry matter (DM) generally decrease feed intake and reduce nutrient availability [11]. Since white grapes contain much higher concentrations of tannins, decreasing their concentration during silage is desirable. Tannins are capable of binding proteins, minerals, and carbohydrates, which can negatively affect the nutritional value of the ensiled biomass [12]. The main goal of ensilage is to improve the microbiological and nutritional quality of by-products [13].

Various types of additives can be applied early in the process to ensure that fermentation occurs appropriately and to improve the silage quality. These additives can be chemical or microbiological (also known as starter cultures or inoculants), with microbiological additives being more commonly used [14]. Chemical additives such as acids and salts improve the aerobic stability of the process. Urea can also be beneficial by increasing the protein content and reducing the fiber in silage, making it more nutritious [13]. The inoculant is a product composed of strains of one or more species of micro-organisms, and it must be viable at the time of use. The main positive effects of the use of these micro-organisms in silages are a reduction in dry matter losses, an increase in the production of microbial metabolites of interest, the inhibition of undesirable micro-organisms, and an improvement in the microbiological and nutritional quality [14]. The benefits of inoculated silage include the inhibition of pathogenic micro-organisms, lower concentrations of toxins and undesirable compounds, interactions with ruminal micro-organisms, and changes in rumen fermentation (e.g., a reduction in methane production) [15]. Recently, much attention has been directed towards zeolites—additives from the group of mycotoxin adsorbents that considerably reduce the toxic effects of the metabolic products of mildew, which are present in many feeds. The results of previous research indicated that zeolite had a positive effect on the quality of corn, alfalfa, red clover, and ryegrass silages as well as the silages of sugar beet pulp, in the way that they bound moisture and improved the activity of bacteria in the silages [16]. Irrespective of the ensilage approach applied, the main goal of ensilage is to improve the microbiological and nutritional quality of by-products [13].

The objective of this study was to evaluate the impact of different additives on the quality of grape pomace silage. Briefly, Graševina grape pomace was ensiled over 90 days in three different experiments: grape pomace as a control, grape pomace with an inoculum of lactic acid bacteria, and grape pomace with zeolite. After 30, 60, and 90 days of ensiling, chemical analyses of the grape pomace were performed. The following parameters were determined: the dry matter (DM), water-soluble carbohydrate (WSC), crude protein (CP), neutral detergent fiber (NDF), acid detergent fiber (ADF), total polyphenol (TP), and total tannin (TT) content. Furthermore, the fermentative characteristics of the grape pomace were determined, such as the pH, silage acids (lactic, acetic, and butyric acids), the ammonia–nitrogen ($NH_3$-N) content, and the Flieg score, which describes the silage quality.

## 2. Materials and Methods

### 2.1. Materials

Graševina grape pomace (*Vitis vinifera*) was obtained from the Croatian native grape cultivar Kutjevo d.d. in September 2021. All solvents were purchased from Kemika d.d. (Zagreb, Croatia) and were of HPLC grade.

### 2.2. Silage of Grape Pomace

After harvest and wine production, the grape pomace was used for ensilage. Three different treatments were carried out: grape pomace as a control, grape pomace with an inoculum of lactic acid bacteria (0.0002% $w/w$) (SIL-ALL 4 × 4+ FVA, Nutritech, Auckland, New Zealand), and grape pomace with zeolites (0.2% $w/w$) (zeolite clinoptilolite, Heil-tropfen, London, UK). The commercial inoculum comprised 4 bacterial strains (*Lactobacillus plantarum*, *Pediococcus acidilactici*, *Pedioccus pentosaceus*, and *Propionibacterium acidipropionici*), 4 enzymes (α-amylase, cellulase, β-glucanase, and xylanase), sucrose, and silica. The additives were mixed homogenously with the grape pomace.

In detail, each treatment was carried out in three replicates of 0.5 kg each, and they were packed into air-tight polyethylene bags. A total of 27 bags (3 treatments × 3 ensiling days × 3 replicates) were stored at a temperature of 18–20 °C. At 30, 60, and 90 days of ensiling, 9 bags (3 treatments × 3 replicates) were opened and the chemical composition and fermentative characteristics were analyzed. Initial fresh grape pomace samples were taken before ensiling and the same analysis was carried out.

### 2.3. Extract Preparation and Sample Analysis

The amount of total dry matter and the residual moisture content of Graševina grape pomace were determined using a conventional dryer (Instrumentaria d.d., Sesvete, Croatia) at $103 \pm 2$ °C with a pressure $\leq 100$ mmHg (13.3 kPa) [2]. The WSCs were extracted in distilled water in a ratio of 1:10. The extraction was performed for 2 h in a water bath at 70 °C, and after the extraction, the mixture was filtered through a filter paper. The WSC content in the supernatant was determined using the phenol–sulfuric method described by Nielsen [17]. The total nitrogen (TN) was determined using an elemental analyzer with a spectrophotometer (LaboMed UV-VIS, Los Angeles, CA, USA), and the crude protein content was calculated via multiplying the TN by 6.25. The NDF and ADF were determined according to Van Soest et al. [18]. The extraction of phenolic compounds was carried out using an 80% ($w/w$) methanol solution according to the method described by Palma and Barroso [19], with some modifications. After the extraction, the mixture was filtered through a filter paper and the total phenolic (TP) content was determined using the Folin–Ciocalteu method [20]. The absorbance was measured at 760 nm (Specord 50 PLUS UV/VIS spectrophotometer, Analytik, Jena, Germany) and the results were expressed as the grams of gallic acid equivalent per kilogram of dry matter (g GAE kg$^{-1}$ DM). The extraction of tannins was carried out according to the method described by Hagerman et al. [21] using 70% ($w/w$) acetone. After the extraction, the mixture was filtered through a filter paper and the total tannin (TT) content was determined using the Bate-Smith method as described by Ribéreau-Gayon et al. [22]. The absorbance was measured at 550 nm (Specord 50 PLUS UV/VIS spectrophotometer, Analytik, Jena, Germany) and the results were expressed as the tannin mass over the mass of dry matter (g kg$^{-1}$). For the determination of pH, the extract was prepared as described by Ni at el. [23] and the pH was measured using a digital pH meter (Mettler Toledo, Greifensee, Switzerland). The organic acid concentrations were determined using high-performance liquid chromatography (HPLC) [2]. The NH$_3$-N content was analyzed using Nessler's method as described by Merck [24]. All the analyses were performed in triplicate. The parameter of silage quality, the Flieg score, was calculated using the following equation [25]:

$$\text{Flieg score} = [220 + (\text{DM} - 15)] - 40 \cdot \text{pH} \tag{1}$$

*2.4. Statistical Analysis*

The statistical analyses were performed using Statistica 13.0 (Tibco Software Inc., Palo Alto, Santa Clara, CA, USA). Differences between the means of the chemical characteristics and the fermentative characteristics during the grape pomace silage were tested using a two-way analysis of variance (two-way ANOVA) at the significance level of $p < 0.05$, followed by Tukey's honestly significant difference (HSD) test. The SS (sum of squares) in total, the SS among subjects, the SS between subjects A and B, the MS (mean square) for subjects A and B, the df (degrees of freedom) for subjects A and B, and the F-ratio for subjects 1 and 2 were the parameters used for the two-way ANOVA. A two-way ANOVA is a statistical method that tests the group means of two or more factors. This type of analysis examines groups that have been divided into multiple categories based on the values of both independent variables. The formulas for two-way ANOVA are given in Table 1:

**Table 1.** Formulas for two-way ANOVA.

| Source of Variation | d.f. | SS | MS | $F_0$ |
|---|---|---|---|---|
| Factor A (between groups) | a − 1 | $SSA = \sum\limits_{i=1}^{a} n_i(\overline{y_i} - \overline{y})^2$ | $MSA = \frac{SSA}{(a-1)}$ | $\frac{MSA}{MSE}$ |
| Factor B (between groups) | b − 1 | $SSB = \sum\limits_{j=1}^{b} n_i\left(\overline{y_j} - \overline{y}\right)^2$ | $MSB = \frac{SSB}{(b-1)}$ | $\frac{MSB}{MSE}$ |
| Error (within groups) | (a − 1)(b − 1) | $SSE = SST - SSA - SSB$ | $MSE = \frac{SSE}{(a-1)(b-1)}$ | |
| Total | N − 1 | $SST = \sum\limits_{i=1}^{a}\sum\limits_{j=1}^{n}\left(y_{ij} - \overline{y}\right)^2$ | | |

n = 3.

## 3. Results and Discussion

*3.1. Effect of Additives and Storage Time on the Chemical Characteristics of Grape Pomace Silage*

The main characteristics of fresh grape pomace and the effect of additives and storage time on the chemical composition of grape pomace silage are presented in Table 2. In general, there was a significant change in the chemical characteristics after 30 days of the silage of grape pomace in all the treatments in comparison to fresh grape pomace; with a longer silage period, the chemical characteristics did not change considerably, and the values were consistent until the end of the process.

One of the factors that affected the ensiling process was the DM content of the plant. According to the literature, to obtain good-quality silage, the interval needs to be between 280 and 400 g kg$^{-1}$ to avoid high nutrient losses [4]. As these authors have reported, dry matter values within this range contributed to a reduction in nutrient losses, which can be leached together with excess moisture. The grape pomace used in our experiment for ensiling showed an adequate dry matter content, and at the end of all treatments, the values were higher than 400 g kg$^{-1}$, which indicated no significant nutrient loss. Also, an increase in the dry matter content was noticed in the treatment with zeolite, and similar results have been reported by Herremans et al. [16]. In this previous study, they found that an application of zeolite (1%) increased the content of dry matter in ryegrass and red clover silage compared to a control. Good fermentation probably results in total dry matter losses of less than 10 to 12%, and poor fermentation coupled with poor storage conditions results in total dry matter losses of greater than 20%. Losses of dry matter may also come from runoff, oxidation, and the loss of volatile organic compounds [26]. In this work, the control and grape pomace with an inoculum had dry matter losses of 8.75% and 12.00%, respectively, which indicates good fermentation. On the other hand, the experiment with zeolite showed an increase in dry matter of 6.58%, which was expected and is in line with the literature.

**Table 2.** Chemical composition of fresh and ensiled grape pomace. Results are presented as average value ± standard errors.

| Variable | Day of Ensiling | Control | Grape Pomace + Lactic Acid Bacteria | Grape Pomace + Zeolite |
|---|---|---|---|---|
| DM $(g\ kg^{-1})$ | 0 | $465.50 \pm 3.05$ [A,a] | $465.50 \pm 3.05$ [A,a] | $465.50 \pm 3.05$ [A,a] |
| | 30 | $485.72 \pm 4.79$ [A,a] | $431.67 \pm 9.57$ [A,a] | $513.44 \pm 6.14$ [A,a] |
| | 60 | $463.03 \pm 15.01$ [A,a] | $428.68 \pm 12.23$ [A,a] | $494.59 \pm 0.79$ [A,a] |
| | 90 | $424.74 \pm 27.76$ [B,a] | $409.62 \pm 0.31$ [B,b] | $498.33 \pm 9.52$ [A,a] |
| WSC $(g\ kg^{-1}\ DM)$ | 0 | $54.83 \pm 0.47$ [A,a] | $54.83 \pm 0.47$ [A,a] | $54.83 \pm 0.47$ [A,a] |
| | 30 | $48.13 \pm 0.02$ [A,b] | $23.40 \pm 0.01$ [B,b] | $34.05 \pm 0.01$ [C,b] |
| | 60 | $11.89 \pm 0.01$ [A,c] | $10.13 \pm 0.01$ [B,c] | $11.76 \pm 0.02$ [A,c] |
| | 90 | $13.40 \pm 0.01$ [A,d] | $11.78 \pm 0.01$ [B,d] | $7.88 \pm 0.07$ [C,d] |
| CP $(g\ kg^{-1}\ DM)$ | 0 | $96.25 \pm 0.02$ [A,a] | $96.25 \pm 0.02$ [A,a] | $96.25 \pm 0.02$ [A,a] |
| | 30 | $105.60 \pm 0.02$ [A,b] | $93.70 \pm 0.02$ [B,b] | $91.80 \pm 0.02$ [C,b] |
| | 60 | $98.10 \pm 0.02$ [A,c] | $98.10 \pm 0.02$ [A,c] | $98.10 \pm 0.02$ [A,c] |
| | 90 | $100.00 \pm 0.02$ [B,d] | $102.50 \pm 0.02$ [A,d] | $99.40 \pm 0.02$ [C,d] |
| NDF $(g\ kg^{-1}\ DM)$ | 0 | $451.70 \pm 0.54$ [A,d] | $451.70 \pm 0.54$ [A,d] | $451.70 \pm 0.54$ [A,d] |
| | 30 | $454.80 \pm 0.54$ [B,c] | $511.90 \pm 0.54$ [A,c] | $511.20 \pm 0.54$ [A,c] |
| | 60 | $572.00 \pm 0.54$ [B,b] | $579.90 \pm 0.54$ [A,a] | $559.90 \pm 0.54$ [C,b] |
| | 90 | $593.00 \pm 0.54$ [A,a] | $572.80 \pm 0.54$ [C,b] | $585.90 \pm 0.54$ [B,a] |
| ADF $(g\ kg^{-1}\ DM)$ | 0 | $395.40 \pm 0.65$ [A,c] | $395.4 \pm 0.65$ [A,d] | $395.40 \pm 0.65$ [A,d] |
| | 30 | $392.20 \pm 0.65$ [C,c] | $445.6 \pm 0.65$ [B,c] | $467.10 \pm 0.65$ [A,c] |
| | 60 | $527.20 \pm 0.65$ [B,b] | $554.2 \pm 0.65$ [A,a] | $529.70 \pm 0.65$ [B,b] |
| | 90 | $573.10 \pm 0.65$ [A,a] | $547.1 \pm 0.65$ [C,b] | $553.10 \pm 0.65$ [B,a] |

DM: dry matter; WSC: water-soluble carbohydrate; CP: crude protein; NDF: neutral detergent fiber; ADF: acid detergent fiber. (n = 3). [A–C] The same superscript capital letters within a row denote no significant differences ($p > 0.05$) between the values obtained for the different treatments regarding the control sample according to Tukey's ANOVA (real $p$-values are given in Supplementary Table S1). [a–d] The same superscript lowercase letters within a column denote no significant differences ($p > 0.05$) between values obtained for different days of storage according to Tukey's ANOVA (real $p$-values are given in Supplementary Table S1).

The concentration of WSCs during the silage was influenced by factors such as the temperature, moisture content, fertilizers, additives, and time of ensiling. In the ensiling process, carbohydrates are the key source of fermentable substrates. In an anaerobic environment, water-soluble carbohydrates are converted into organic acids with the help of lactic acid bacteria [27]. The initial value of WSCs was 54.83 g kg$^{-1}$ DM, which corresponds to the values reported in the literature [28]. As expected, the WSC concentrations decreased rapidly in all the grape pomace silages over the first 30 days of ensiling (Table 2), especially in the experiment with lactic acid bacteria. The higher WSC content in the control silage (48.13 g kg$^{-1}$ DM) and lower content in the inoculant and zeolite treatments 30 days after the start of the silage could be explained by an inhibition of microbial growth by the tannins in the grape pomace silage. After 90 days of ensiling, the final WSC contents were 13.40, 11.78, and 7.88 g kg$^{-1}$ DM for the control, inoculum, and zeolite treatments, respectively. The decrease in the WSC content was mainly due to oxygen consumption by plant cells in the silages in the early stage of ensiling, followed by the fermentation of water-soluble carbohydrates by micro-organisms into lactic acid; as a consequence, the pH decreased [29]. Furthermore, the CP content in fresh grape pomace before the silage process was 96.25 g kg$^{-1}$ DM, and the approximate values have been reported by some authors [10,28]. There were no significant changes in the CP values after 90 days of ensilage. Similar results were obtained by Massaro Junior et al. [4] and Fitri et al. [30]. Herremans et al. [16] found that an application of zeolite (1%) increased the content of crude protein in red clover silage compared to the control.

The nutritional value of grape pomace silage is significantly lower than that of other forages, resulting from a high proportion of structural carbohydrates. Generally, there is great variation in the ADF and NDF values of grape pomace in the literature. For the ADF, the values are between 312.00 and 593.10 g kg$^{-1}$ DM, and for the NDF, the values are between 439.70 and 631.00 g kg$^{-1}$ DM [4,28], which correspond to the values obtained in this experiment. A detergent fiber analysis showed that the content of ADF and NDF in grape pomace silages increased after 90 days of the silage process compared to the content before the ensilage. Alipour et al. [31] and Spanghero et al. [32] also found an increase in the ADF and NDF content in grape pomace 30 and 45 days after the start of ensilage. Certain inoculants contain bacteria that could secrete specific enzymes, mainly cellulases and xylanases, that may contribute to degrading these structures and increasing fiber digestion [33].

### 3.2. Effect of Additives and Storage Time of Grape Pomace Silage on Biologically Active Compounds

Polyphenols and tannins are biologically active compounds that may provide positive impacts on human health due to their anti-tumor, anti-inflammatory, anti-viral, anti-fungal, anti-bacterial, anti-parasitic, and cardio-protective effects [8]. In past decades, the polyphenol valorization of agro-food industry by-products emerged in order to provide added value to unexploited by-products. In addition, polyphenol's global market is ever-growing, since these compounds may be used as additives to improve a product's potential health benefits across a wide range of industries, such as the cosmetic, pharmaceutical, and food industries [7,32]. On the other hand, an elevated polyphenol content in silage may have a negative impact on animal production. Polyphenols may inhibit the activity of certain digestive enzymes and modulate the gut microbiota, thus negatively affecting digestion. Nevertheless, polyphenols have also been reported to improve animal gut health due to their antioxidant and anti-inflammatory properties [34]. Due to all the aforesaid results, their preservation during ensiling is crucial. Therefore, during grape pomace ensiling, the TP and TT contents were closely monitored.

The TP content in fresh Graševina grape pomace was 58.80 g kg$^{-1}$ DM (Figure 1a). Antonić et al. [3] reported that the TP content in white grape pomace varies from 2.80 to 87.00 g kg$^{-1}$ DM. The content of polyphenols decreased rapidly in all the grape pomace silages over the first 30 days of ensiling. On average, 82.24% of the polyphenols were degraded during the first 30 days. At the end of the experiment, 97.67% of the polyphenols had been degraded, on average, in all the monitored treatments. Winkler et al. [35], on the other hand, noted that 50% of the TPs in grape pomace ensiled for 56 days was degraded. Of the three silage treatments, after 90 days, the lowest TP degradation was seen in the treatment with zeolite (96.84%), while the control and inoculum treatments exhibited lower TP retention (98.21 and 97.96% of TPs degraded, respectively). The degradation of polyphenols has also been reported in studies by Alipour and Rouzbehan [31] and Winkler et al. [35]. This degradation generally occurs due to polyphenol polymerization and oxidation [35].

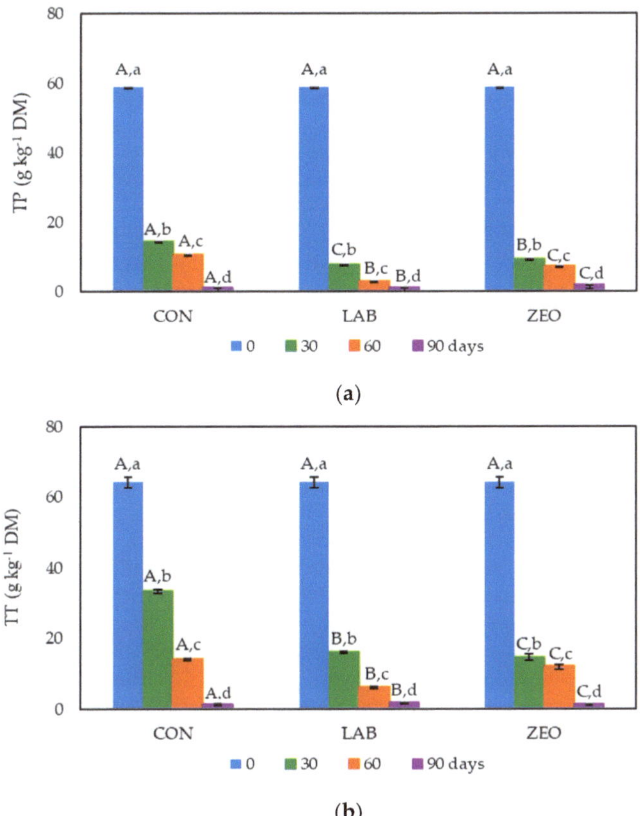

**Figure 1.** TP (**a**) and TT (**b**) content during 90 days of ensiling. Results are expressed as average values ± standard errors (CON-control; LAB-inoculum of lactic acid bacteria; ZEO-zeolite); n = 3. [A–C] The same superscript capital letters within a row denote no significant differences ($p > 0.05$) between the values obtained for the different treatments regarding the control sample according to Tukey's ANOVA. [a–d] The same superscript lowercase letters within a column denote no significant differences ($p > 0.05$) between values obtained for different days of storage according to Tukey's ANOVA.

The TT content was monitored over the 90 days of grape pomace ensiling (Figure 1b). In fresh grape pomace, the TT content was 64.16 g kg$^{-1}$ DM. A similar TT content in white grape pomace has already been reported in the literature [36]. As in the case of TPs, the most considerable number of tannins were degraded during the first 30 days of ensiling (66.71%, on average). This was slightly less evident in the control, where it was 47.92%. Nevertheless, after 90 days, the control treatment, together with the zeolite treatment, had the lowest TT content (1.24 and 1.19 mg g$^{-1}$ DM, respectively). At the end of ensiling, only 2.15% of the tannins, on average, were preserved. Condensed tannin degradation during grape pomace ensiling has also been reported by Alipour and Rouzbehan [31], whereby after 30 days of ensiling, only 27.56% of the tannins remained in the grape pomace silage. The presented results (Figure 1) indicate that ensiling is not a suitable method for polyphenol and tannin preservation. Therefore, other methods for biologically active compound conservation should be applied.

Sokač et al. [2] reported a high stability for polyphenols and tannins during grape pomace drying. The authors noted that grape pomace drying at 70 °C ensured the highest level of secondary metabolite retention. Therefore, drying could be an acceptable alternative to ensure secondary metabolite stability in grape pomace. Despite the above-mentioned polyphenol and tannin degradation in grape pomace silage, ensiling has been reported to increase the digestibility of grape pomace, and thus, improve its feed value [9]. Grape pomace is generally considered to be a low-energy source when compared to the common forage used in animal diets. Nevertheless, ensiling may improve its digestibility, ensuring a year-round feed source [37]. The application of grape pomace as feed does not bring an extremely high added value to the whole wine-making process; however, it decreases the amount of wasted by-product and ensures its complete usage.

*3.3. Effect of Additives and Storage Time on the Fermentative Characteristics of Grape Pomace Silage*

Table 2 illustrates the dynamics of the fermentation quality of grape pomace silage during ensiling. Generally, the fermentation quality of silages improved by a low pH, a high lactic acid content, and a low level of ammonia-N can be achieved through the treatment of silage with different additives, including inoculants, chemical matter, and enzymes [12,21,26]. The pH of fresh grape pomace was 5.88, and after 90 days of the ensiling process, the value decreased. At the end of the silage, the pH value in the control treatment and the treatment with LAB was 4.06, and in the treatment with zeolite, it was 4.27. The decrease in the pH was expected due to the fermentation of water-soluble carbohydrates by lactic acid bacteria (LAB), and as a result, organic acids (mainly lactic acid) were produced [29]. The pH value of silage is an important index for evaluating the success of silage; well-fermented silage should have a pH of 3.80~4.20 [38]. The amount of acid required to decrease the original pH of 6 to a stable pH depends on the contents of the silage dry matter, the water-soluble carbohydrate content, and the crude protein content. Furthermore, forages with a high dry matter content are fermented at a slower rate than forages with low dry matter because of low water activity [39]. There are different opinions on the suitability of pH values for silage: some authors define a pH value of 4.2 as the upper threshold for a positive assessment of silage [40], while other researchers state that the final pH is not important; what matters is the decreasing rate, as this parameter is more important for inhibiting a secondary fermentation occurrence [4].

Lactic acid should be the primary acid in good silages. This acid is stronger than the other acids present in silage (acetic, propionic, and butyric acids), and as mentioned before, it is usually responsible for the decrease in the pH value. Lactic acid should represent at least 65 to 70% of the total silage acids in a good silage [26]. In this study, the lactic acid content before the silage of grape pomace was 28.35 g kg$^{-1}$ DM, and after 30 days, the values in all the treatments decreased. Moreover, after 60 days of silage, an increase in lactic acid was noted and, finally, it decreased until the end of the process (Table 3). The content of lactic acid was higher in the treatment with zeolite than in the control and inoculum treatments. Also, Đorđević et al. [41] found in their research that the application of zeolite increased the content of lactic acid in corn, alfalfa, and perennial ryegrass silage compared to a control.

When lactic acid decreases, acetic acid increases (Table 3), which can be explained by the conversion of lactic acid to acetic acid with a prolonged ensilage time, in accordance with Ni et al. [23]. Der Bedrosian et al. [42] reported that some strains of lactic acid bacteria can utilize lactic acid anaerobically when sugar is limiting, which reduces the lactic acid concentration and increases the acetic acid concentration. Acetic acid is the acid with the second highest concentration in silage, usually ranging from 1 to 3%. The acetic acid content in silage is also affected by the dry matter content [38], while the acetic acid content decreases as the dry matter content increases, as also confirmed by the results of Juráček et al. [43]. Acetic acid is a promoter of aerobic stability during the ensiling process [44] and an effective inhibitor of fungi [45].

**Table 3.** Fermentative characteristics of the grape pomace silage. Results are presented as average value ± standard errors.

| Variable | Day of Ensiling | Control | Grape Pomace + Lactic Acid Bacteria | Grape Pomace + Zeolite |
|---|---|---|---|---|
| pH | 0 | $5.88 \pm 0.14$ [A,a] | $5.88 \pm 0.08$ [A,a] | $5.88 \pm 0.05$ [A,a] |
| | 30 | $4.01 \pm 0.05$ [A,d] | $3.95 \pm 0.03$ [A,d] | $4.09 \pm 0.03$ [A,d] |
| | 60 | $4.12 \pm 0.04$ [A,b] | $4.12 \pm 0.03$ [A,b] | $4.20 \pm 0.01$ [A,c] |
| | 90 | $4.06 \pm 0.05$ [A,c] | $4.06 \pm 0.02$ [A,c] | $4.27 \pm 0.01$ [A,b] |
| Lactic acid (g kg$^{-1}$ DM) | 0 | $6.09 \pm 0.07$ [A,b] | $6.09 \pm 0.07$ [A,a] | $6.09 \pm 0.04$ [C,a] |
| | 30 | $0.18 \pm 0.18$ [B,d] | $6.98 \pm 0.20$ [A,a] | $5.08 \pm 0.15$ [B,a] |
| | 60 | $11.19 \pm 0.33$ [B,a] | $6.80 \pm 0.18$ [C,a] | $11.54 \pm 0.26$ [A,b] |
| | 90 | $4.58 \pm 0.03$ [B,c] | $0.17 \pm 0.29$ [C,b] | $7.86 \pm 0.01$ [B,d] |
| Acetic acid (g kg$^{-1}$ DM) | 0 | $15.39 \pm 0.10$ [A,c] | $15.39 \pm 0.10$ [A,c] | $15.39 \pm 0.10$ [A,a] |
| | 30 | $22.25 \pm 0.29$ [A,a] | $15.72 \pm 0.13$ [B,c] | $14.63 \pm 0.02$ [C,a] |
| | 60 | $11.79 \pm 0.31$ [B,d] | $16.53 \pm 0.87$ [A,b] | $8.68 \pm 0.02$ [C,b] |
| | 90 | $19.19 \pm 0.11$ [B,b] | $24.40 \pm 0.37$ [A,a] | $14.53 \pm 0.02$ [C,a] |
| Butyric acid (g kg$^{-1}$ DM) | 0 | - | - | - |
| | 30 | - | - | - |
| | 60 | - | - | - |
| | 90 | - | - | - |
| NH$_3$-N (g kg$^{-1}$ DM) | 0 | $0.00 \pm 0.04$ [A,c] | $0.00 \pm 0.04$ [A,c] | $0.00 \pm 0.04$ [A,c] |
| | 30 | $1.19 \pm 0.02$ [A,a] | $0.90 \pm 0.02$ [B,a] | $0.84 \pm 0.01$ [C,a] |
| | 60 | $0.92 \pm 0.01$ [A,b] | $0.63 \pm 0.01$ [C,b] | $0.79 \pm 0.01$ [B,b] |
| | 90 | $0.97 \pm 0.01$ [A,b] | $0.99 \pm 0.02$ [A,a] | $0.94 \pm 0.02$ [A,a] |

($n = 3$). [A-C] The same superscript capital letters within a row denote no significant differences ($p > 0.05$) between the values obtained for the different treatments regarding the control sample according to Tukey's ANOVA (real $p$-values are given in Supplementary Table S2). [a-d] The same superscript lowercase letters within a column denote no significant differences ($p > 0.05$) between values obtained for different days of storage according to Tukey's ANOVA (real $p$-values are given in Supplementary Table S2).

The butyric acid content was not determined in any of the grape pomace silage treatments. Furthermore, butyric acid is undesirable in silage due to its inhibitory effect on lactic acid bacteria and yeast growth. However, the presence of butyric acid in a range between 0.1% and 0.6% would not affect the silage quality [46]. Belém et al. [28] reported that the low values observed for butyric acid may be related to the low ammoniacal nitrogen in the silage, indicating low *Clostridium* spp. activity and a high silage quality. Besides the presence of butyric acid and lower-than-normal concentrations of lactic acid, clostridial silages are often characterized by a higher-than-normal pH and higher-than-normal concentrations of acetic acid and NH$_3$-N [38].

Furthermore, the NH$_3$-N concentration is a reliable indicator of protein degradation [25,44]. During the ensiling process, NH$_3$-N accumulation can be explained by the activity of plant enzymes and fermentation by *Clostridium* and *Enterobacter* [47]. Usually, silage with high concentrations of NH$_3$-N coupled with butyric acid may also have significant concentrations of other undesirable end products, such as amines, that may reduce animal performance [38]. Compared to the control, a significant reduction in ammonia was observed in the silages with the inoculant or zeolite opened after 30 and 60 days (Table 3). This may have been the result of the lower pH values, the activity of homofermentative lactic acid bacteria, or lactic acid production. Some authors [5,46] have also found that an inoculation with lactic acid bacteria or the use of zeolites significantly decreased the NH$_3$-N levels compared with the controls. At the end of the silage, the lowest values of NH$_3$-N were found for the treatment with zeolite in comparison with the control and the treatment with an inoculum (Table 3), which agrees with the literature [14,39].

Finally, the silage quality can be expressed by the Fleig score, which is the relationship between the dry matter content and the pH value of the silage [48]. The Flieg point was determined for all the treatments, and the value for the fresh grape pomace was 61.85. After the first 30 days of ensiling, the Flieg score increased significantly and the maximum values were achieved; these values were 141.74 for the control, 133.33 for the treatment with the inoculum of lactic acid bacteria, and 144.08 for the treatment with zeolite. At the end of the ensiling, the values decreased by 7.90%, on average, in comparison to the values determined 30 days after the process. The Flieg score had a value >80 for excellent silage, 61–80 for good, 41–60 for medium, 21–40 for weak, and 0–20 for poor silage [48,49]. Considering the mentioned scale and the calculated values for the Flieg score, in all the treatments, the silage was of a great quality. Similar results to those obtained in this work were reported by Zehra Saricicek et al. [25] for corn silage.

## 4. Conclusions

The present study illustrates the chemical and fermentative characteristics during grape pomace silage carried out in different treatments: control, with an inoculum of lactic acid bacteria, and with zeolite. After 30 days, the silages treated with additives had a better fermentation quality, as indicated by a significantly ($p < 0.05$) lower pH and $NH_3$-N content and a higher lactic acid content compared with those of the control. Thus, our results suggest that inoculant and zeolite supplementation can improve the quality of grape pomace silage for its use in animal feed, mainly for ruminant herbivores.

**Supplementary Materials:** The following supporting information can be downloaded at: https://www.mdpi.com/article/10.3390/agriculture13122264/s1, Table S1. *p*-values of two-way ANOVA of chemical composition of the fresh and the ensiled grape pomace; Table S2. *p*-values of two-way ANOVA of fermentative characteristics of the grape pomace silage.

**Author Contributions:** Conceptualization, I.R.R. and V.G.; methodology, I.R.R., D.U. and V.G.; software, V.G., A.J.T. and T.S.C.; validation, I.R.R., D.U., A.J.T. and T.J.; formal analysis, V.G., A.D., T.S.C. and A.P.; investigation, V.G., A.D., T.S.C. and A.P.; resources, I.R.R.; data curation, V.G., A.D., T.S.C. and A.P.; writing—original draft preparation, V.G., T.S.C., A.D. and A.P.; writing—review and editing, V.G., T.S.C., A.D., A.P., A.J.T., T.J., I.R.R. and D.U.; visualization, V.G., T.S.C., A.D. and A.P.; supervision, I.R.R. and T.J.; project administration, I.R.R.; funding acquisition, I.R.R. All authors have read and agreed to the published version of the manuscript.

**Funding:** This work was supported by the European Union through the European Regional Development Fund, Competitiveness and Cohesion 2014–2020 (KK.01.1.1.07.0007).

**Institutional Review Board Statement:** Not applicable.

**Data Availability Statement:** The authors confirm that the data supporting the findings of this study are available within the article and Supplementary Materials.

**Acknowledgments:** The authors warmly acknowledge Kutjevo d.d. (Kutjevo, Croatia) for providing the raw Graševina grape pomace biomass.

**Conflicts of Interest:** The co-author Pušek worked at the faculty at the time of performing the research. However, there are no potential commercial interests because this firm has nothing with the silage or any other similar work. The authors declare no conflict of interest.

## References

1. Sirohi, R.; Tarafdar, A.; Singh, S.; Negi, T.; Gaur, V.K.; Gnansounou, E.; Bharathiraja, B. Green Processing and Biotechnological Potential of Grape Pomace: Current Trends and Opportunities for Sustainable Biorefinery. *Bioresour. Technol.* **2020**, *314*, 123771. [CrossRef] [PubMed]
2. Sokač, T.; Gunjević, V.; Pušek, A.; Tušek, A.J.; Dujmić, F.; Brnčić, M.; Ganić, K.K.; Jakovljević, T.; Uher, D.; Mitrić, G.; et al. Comparison of Drying Methods and Their Effect on the Stability of Graševina Grape Pomace Biologically Active Compounds. *Foods* **2022**, *11*, 112. [CrossRef] [PubMed]
3. Antonić, B.; Jančíková, S.; Dordević, D.; Tremlová, B. Grape Pomace Valorization: A Systematic Review and Meta-Analysis. *Foods* **2020**, *9*, 1627. [CrossRef]
4. Massaro Junior, F.L.; Bumbieris Junior, V.H.; Zanin, E.; da Silva, L.d.D.F.; Galbeiro, S.; Pereira, E.S.; Neumann, M.; Mizubuti, I.Y. Effect of Storage Time and Use of Additives on the Quality of Grape Pomace Silages. *J. Food Process. Preserv.* **2020**, *44*, e14373. [CrossRef]
5. Yang, F.; Wang, Y.; Zhao, S.; Wang, Y. Lactobacillus Plantarum Inoculants Delay Spoilage of High Moisture Alfalfa Silages by Regulating Bacterial Community Composition. *Front. Microbiol.* **2020**, *11*, 1989. [CrossRef] [PubMed]

6.  Schmithausen, A.J.; Deeken, H.F.; Gerlach, K.; Trimborn, M.; Weiß, K.; Büscher, W.; Maack, G.C. Greenhouse Gas Formation during the Ensiling Process of Grass and Lucerne Silage. *J. Environ. Manag.* **2022**, *304*, 114142. [CrossRef]
7.  Ke, W.C.; Yang, F.Y.; Undersander, D.J.; Guo, X.S. Fermentation Characteristics, Aerobic Stability, Proteolysis and Lipid Composition of Alfalfa Silage Ensiled with Apple or Grape Pomace. *Anim. Feed Sci. Technol.* **2015**, *202*, 12–19. [CrossRef]
8.  Cravotto, G.; Mariatti, F.; Gunjevic, V.; Secondo, M.; Villa, M.; Parolin, J.; Cavaglià, G. Pilot Scale Cavitational Reactors and Other Enabling Technologies to Design the Industrial Recovery of Polyphenols from Agro-Food by-Products, a Technical and Economical Overview. *Foods* **2018**, *7*, 130. [CrossRef]
9.  De Bellis, P.; Maggiolino, A.; Albano, C.; De Palo, P.; Blando, F. Ensiling Grape Pomace With and Without Addition of a Lactiplantibacillus Plantarum Strain: Effect on Polyphenols and Microbiological Characteristics, In Vitro Nutrient Apparent Digestibility, and Gas Emission. *Front. Vet. Sci.* **2022**, *9*, 808293. [CrossRef]
10. Alba-Mejía, J.E.; Dohnal, V.; Domínguez-Rodríguez, G.; Středa, T.; Klíma, M.; Mlejnková, V.; Skládanka, J. Ergosterol and polyphenol contents as rapid indicators of orchardgrass silage safety. *Heliyon* **2020**, *9*, e14940. [CrossRef]
11. Peng, K.; Jin, L.; Niu, Y.D.; Huang, Q.; McAllister, T.A.; Yang, H.E.; Denise, H.; Xu, Z.; Acharya, S.; Wang, S.; et al. Condensed Tannins Affect Bacterial and Fungal Microbiomes and Mycotoxin Production during Ensiling and upon Aerobic Exposure. *Appl. Environ. Microbiol.* **2018**, *84*, e02274-17. [CrossRef]
12. Fitri, A.; Obitsu, T.; Sugino, T. Effect of Ensiling Persimmon Peel and Grape Pomace as Tannin-Rich Byproduct Feeds on Their Chemical Composition and In Vitro Rumen Fermentation. *Anim. Sci. J.* **2021**, *92*, e13524. [CrossRef]
13. Carvalho, B.F.; Sales, G.F.C.; Schwan, R.F.; Ávila, C.L.S. Criteria for Lactic Acid Bacteria Screening to Enhance Silage Quality. *J. Appl. Microbiol.* **2021**, *130*, 341–355. [CrossRef]
14. Muck, R.E.; Nadeau, E.M.G.; McAllister, T.A.; Contreras-Govea, F.E.; Santos, M.C.; Kung, L. Silage Review: Recent Advances and Future Uses of Silage Additives. *J. Dairy Sci.* **2018**, *101*, 3980–4000. [CrossRef]
15. Han, H.; Wang, C.; Huang, Z.; Zhang, Y.; Sun, L.; Xue, Y.; Guo, X. Effects of Lactic Acid Bacteria-Inoculated Corn Silage on Bacterial Communities and Metabolites of Digestive Tract of Sheep. *Fermentation* **2022**, *8*, 320. [CrossRef]
16. Herremans, S.; Decruyenaere, V.; Beckers, Y.; Froidmont, E. Silage Additives to Reduce Protein Degradation during Ensiling and Evaluation of In Vitro Ruminal Nitrogen Degradability. *Grass Forage Sci.* **2019**, *74*, 86–96. [CrossRef]
17. Nielsen, S.S. Phenol-Sulfuric Acid Method for Total Carbohydrates. In *Food Analysis Laboratory Manual*, 1st ed.; Nielsen, S.S., Ed.; Springer: Boston, MA, USA, 2010; pp. 47–53. [CrossRef]
18. Van Soest, P.J.; Robertson, J.B.; Lewis, B.A. Methods for Dietary Fiber, Neutral Detergent Fiber, and Nonstarch Polysaccharides in Relation to Animal Nutrition. *J. Dairy Sci.* **1991**, *74*, 3583–3597. [CrossRef]
19. Palma, M.; Barroso, C.G. Ultrasound-Assisted Extraction and Determination of Tartaric and Malic Acids from Grapes and Winemaking by-Products. *Anal. Chim. Acta* **2002**, *458*, 119–130. [CrossRef]
20. Singleton, V.L.; Orthofer, R.; Lamuela-Raventós, R.M. Analysis of Total Phenols and Other Oxidation Substrates and Antioxidants by Means of Folin-Ciocalteu Reagent. *Methods Enzymol.* **1999**, *299*, 152–178. [CrossRef]
21. Hagerman, A.E. Extraction of Tannin from Fresh and Preserved Leaves. *J. Chem. Ecol.* **1988**, *14*, 453–461. [CrossRef]
22. Gayon, P.R.; Stonestreet, E. Dosage Des Tanins Du Vin Rouge et Determination de Leur Structure. *Chim. Anal.* **1966**, *48*, 188–196.
23. Ni, K.; Wang, F.; Zhu, B.; Yang, J.; Zhou, G.; Pan, Y.; Tao, Y.; Zhong, J. Effects of Lactic Acid Bacteria and Molasses Additives on the Microbial Community and Fermentation Quality of Soybean Silage. *Bioresour. Technol.* **2017**, *238*, 706–715. [CrossRef]
24. Merck, E. *The Testing of Water: A Selection of Chemical Methods for Practical Use*, 10th ed.; Merck Darmstadt: Darmstadt, Germany, 1964; pp. 7–13.
25. Zehra Saricicek, B.; Yildirim, B.; Kocabas, Z.; Ozgumus Demir, E. Effect of Storage Time on Nutrient Composition and Quality Parameters of Corn Silage Mısır Silajının Besin Madde Kompozisyonu ve Silaj Kalite Parametreleri Üzerine Depolama Süresinin Etkileri. *Turkish J. Agric. Sci. Technol. Türk Tarım-Gıda Bilim ve Teknol. Derg.* **2016**, *4*, 934–939.
26. Kung, L., Jr. Understanding the Biology of Silage Preservation to Maximize Quality and Protect the Environment. In Proceedings of the California Alfa & Forage Symposium and Corn/Cereal Silage Conferenxe, Visalia, CA, USA, 1–2 December 2010; pp. 1–14.
27. Ali, M.F.; Tahir, M. An Overview on the Factors Affecting Water-Soluble Carbohydrates Concentration during Ensiling of Silage. *J. Plant Environ.* **2021**, *3*, 63–80. [CrossRef]
28. Belém, C.d.S.; de Souza, A.M.; de Lima, P.R.; de Carvalho, F.A.L.; Queiroz, M.A.Á.; da Costa, M.M. Digestibility, Fermentation and Microbiological Characteristics of Calotropis Procera Silage with Different Quantities of Grape Pomace. *Ciênc. Agrotecnologia* **2016**, *40*, 698–705. [CrossRef]
29. Dunière, L.; Sindou, J.; Chaucheyras-Durand, F.; Chevallier, I.; Thévenot-Sergentet, D. Silage Processing and Strategies to Prevent Persistence of Undesirable Microorganisms. *Anim. Feed Sci. Technol.* **2013**, *182*, 1–15. [CrossRef]
30. Fitri, A.; Obitsu, T.; Sugino, T.; Jayanegara, A. Ensiling of Total Mixed Ration Containing Persimmon Peel: Evaluation of Chemical Composition and In Vitro Rumen Fermentation Profiles. *Anim. Sci. J.* **2020**, *91*, e13403. [CrossRef]
31. Alipour, D.; Rouzbehan, Y. Effects of Ensiling Grape Pomace and Addition of Polyethylene Glycol on In Vitro Gas Production and Microbial Biomass Yield. *Anim. Feed Sci. Technol.* **2007**, *137*, 138–149. [CrossRef]
32. Spanghero, M.; Salem, A.Z.M.; Robinson, P.H. Chemical Composition, Including Secondary Metabolites, and Rumen Fermentability of Seeds and Pulp of Californian (USA) and Italian Grape Pomaces. *Anim. Feed Sci. Technol.* **2009**, *152*, 243–255. [CrossRef]

33. Adesogan, A.T.; Arriola, K.G.; Jiang, Y.; Oyebade, A.; Paula, E.M.; Pech-Cervantes, A.A.; Romero, J.J.; Ferraretto, L.F.; Vyas, D. Symposium Review: Technologies for Improving Fiber Utilization. *J. Dairy Sci.* **2019**, *102*, 5726–5755. [CrossRef]
34. Mariatti, F.; Gunjević, V.; Boffa, L.; Cravotto, G. Process Intensification Technologies for the Recovery of Valuable Compounds from Cocoa By-Products. *Innov. Food Sci. Emerg. Technol.* **2021**, *68*, 102601. [CrossRef]
35. Winkler, A.; Weber, F.; Ringseis, R.; Eder, K.; Dusel, G. Determination of Polyphenol and Crude Nutrient Content and Nutrient Digestibility of Dried and Ensiled White and Red Grape Pomace Cultivars. *Arch. Anim. Nutr.* **2015**, *69*, 187–200. [CrossRef]
36. González-Centeno, M.R.; Jourdes, M.; Femenia, A.; Simal, S.; Rosselló, C.; Teissedre, P.L. Characterization of Polyphenols and Antioxidant Potential of White Grape Pomace Byproducts (*Vitis vinifera* L.). *J. Agric. Food Chem.* **2013**, *61*, 11579–11587. [CrossRef]
37. Muhlack, R.A.; Potumarthi, R.; Jeffery, D.W. Sustainable Wineries through Waste Valorisation: A Review of Grape Marc Utilisation for Value-Added Products. *Waste Manag.* **2018**, *72*, 99–118. [CrossRef]
38. Kung, L.; Shaver, R.D.; Grant, R.J.; Schmidt, R.J. Silage Review: Interpretation of Chemical, Microbial, and Organoleptic Components of Silages. *J. Dairy Sci.* **2018**, *101*, 4020–4033. [CrossRef]
39. Rizk, C.; Mustafa, A.F.; Phillip, L.E. Effects of Inoculation of High Dry Matter Alfalfa Silage on Ensiling Characteristics, Ruminal Nutrient Degradability and Dairy Cow Performance. *J. Sci. Food Agric.* **2005**, *85*, 743–750. [CrossRef]
40. Fu, Z.; Sun, L.; Hou, M.; Hao, J.; Lu, Q.; Liu, T.; Ren, X.; Jia, Y.; Wang, Z.J.; Ge, G. Effects of Different Harvest Frequencies on Microbial Community and Metabolomic Properties of Annual Ryegrass Silage. *Front. Microbiol.* **2022**, *13*, 971449. [CrossRef]
41. Đorđević, N.; Grubić, G.; Adamović, M.; Stojanović, B. The Influence of Inoculant and Zeolite Supplementation on Quality of Silages Prepared from Whole Maize Plant, Lucerne and Perennial Ryegrass. In Proceedings of the International Congres: Food Technology, Quality and Safety, Novi Sad, Serbia, 13–15 November 2007; pp. 51–56.
42. Der Bedrosian, M.C.; Nestor, K.E.; Kung, L. The Effects of Hybrid, Maturity, and Length of Storage on the Composition and Nutritive Value of Corn Silage. *J. Dairy Sci.* **2012**, *95*, 5115–5126. [CrossRef]
43. Juráček, M.; Bíro, D.; Šimko, M.; Gálik, B.; Rolinec, M.; Hanušovský, O.; Vašeková, P.; Kolláthová, R.; Barantal, S. Fermentation Quality and Dry Matter Losses of Grape Pomace Silages with Urea Addition. *J. Int. Sci. Publ. Agric. Food* **2019**, *7*, 173–178.
44. Schmidt, R.J.; Kung, L. The Effects of Lactobacillus Buchneri with or without a Homolactic Bacterium on the Fermentation and Aerobic Stability of Corn Silages Made at Different Locations. *J. Dairy Sci.* **2010**, *93*, 1616–1624. [CrossRef]
45. Le Lay, C.; Coton, E.; Le Blay, G.; Chobert, J.M.; Haertlé, T.; Choiset, Y.; Van Long, N.N.; Meslet-Cladière, L.; Mounier, J. Identification and Quantification of Antifungal Compounds Produced by Lactic Acid Bacteria and Propionibacteria. *Int. J. Food Microbiol.* **2016**, *239*, 79–85. [CrossRef]
46. Öztürk, Y.E.; Gülümser, E.; Mut, H.; Başaran, U.; Çopur Doğrusöz, M. A Preliminary Study on Change of Mistletoe (*Viscum album* L.) Silage Quality According to Collection Time and Host Tree Species. *Turkish J. Agric. For.* **2022**, *46*, 104–112. [CrossRef]
47. Bai, J.; Ding, Z.; Ke, W.; Xu, D.; Wang, M.; Huang, W.; Zhang, Y.; Liu, F.; Guo, X. Different Lactic Acid Bacteria and Their Combinations Regulated the Fermentation Process of Ensiled Alfalfa: Ensiling Characteristics, Dynamics of Bacterial Community and Their Functional Shifts. *Microb. Biotechnol.* **2021**, *14*, 1171–1182. [CrossRef]
48. Eliş, S.; Özyazici, M.A. Determination of the Silage Quality Characteristics of Different Switchgrass (*Panicum virgatum* L.) Cultivars. *Appl. Ecol. Environ. Res.* **2019**, *17*, 15755–15773. [CrossRef]
49. Wang, M.; Gao, R.; Franco, M.; Hannaway, D.B.; Ke, W.; Ding, Z.; Yu, Z.; Guo, X. Effect of Mixing Alfalfa with Whole-Plant Corn in Different Proportions on Fermentation Characteristics and Bacterial Community of Silage. *Agriculture* **2021**, *11*, 174. [CrossRef]

*Article*

# Dietary Effects of Black-Oat-Rich Polyphenols on Production Traits, Metabolic Profile, Antioxidative Status, and Carcass Quality of Fattening Lambs

Zvonko Antunović [1], Željka Klir Šalavardić [1,*], Boro Mioč [2], Zvonimir Steiner [1], Mislav Đidara [1], Vinko Sičaja [1], Valentina Pavić [3], Lovro Mihajlović [3], Lidija Jakobek [4] and Josip Novoselec [1]

[1] Faculty of Agrobiotechnical Sciences Osijek, J.J. Strossmayer University of Osijek, V. Preloga 1, 31000 Osijek, Croatia; zantunovic@fazos.hr (Z.A.); zsteiner@fazos.hr (Z.S.); mdidara@fazos.hr (M.Đ.); vinko-55@hotmail.com (V.S.); jnovoselec@fazos.hr (J.N.)

[2] Department of Animal Science and Technology, Faculty of Agriculture, University of Zagreb, Svetošimunska Cesta 25, 10000 Zagreb, Croatia; bmioc@agr.hr

[3] Department of Biology, J.J. Strossmayer University of Osijek, Cara Hadrijana 8, 31000 Osijek, Croatia; vpavic@biologija.unios.hr (V.P.); lmihajlovic@biologija.unios.hr (L.M.)

[4] Faculty of Food and Technology Osijek, J.J. Strossmayer University of Osijek, Franje Kuhača 18, 31000 Osijek, Croatia; lidija.jakobek@ptfos.hr

* Correspondence: zklir@fazos.hr; Tel.: +385-31-554-906

**Abstract:** The study aimed to establish the dietary effects of black oat rich in polyphenols on the production traits, metabolic profile, antioxidant status, and carcass quality of fattening lambs, after weaning. In the BO group, in the feed mixture, common oats replaced the black oat compared to the CO group. The research comprehensively investigated production indicators, blood metabolic profile, antioxidant status, and lamb carcass quality. No significant differences were found in the fattening or slaughter characteristics of lamb carcasses, except for lower $pH_1$ values in BO lamb carcasses. Significant increases in RBC, HCT, and MCV levels as well as TP, ALB, and GLOB concentrations and GPx and SOD activities in the blood of BO lambs were found. The glucose and EOS content as well as the activity of the enzymes ALT and ALP were significantly lower in the blood of the BO group than in the CO group. In the liver, the DPPH activity was significantly higher in the BO lambs compared to the CO lambs. The observed changes in glucose, protein metabolism, and antioxidant status in the blood and tissues of lambs indicate that the use of polyphenol-rich black oats in the diet of lambs under stress conditions is justified.

**Keywords:** oat; black oat; production traits; metabolic profile; antioxidative status

**Citation:** Antunović, Z.; Klir Šalavardić, Ž.; Mioč, B.; Steiner, Z.; Đidara, M.; Sičaja, V.; Pavić, V.; Mihajlović, L.; Jakobek, L.; Novoselec, J. Dietary Effects of Black-Oat-Rich Polyphenols on Production Traits, Metabolic Profile, Antioxidative Status, and Carcass Quality of Fattening Lambs. *Agriculture* **2024**, *14*, 1550. https://doi.org/10.3390/agriculture14091550

Academic Editors: Mihaela Saracila, Arabela Elena Untea and Petru Alexandru Vlaicu

Received: 2 August 2024
Revised: 4 September 2024
Accepted: 5 September 2024
Published: 7 September 2024



## 1. Introduction

Oat grain is commonly used in ruminant rations, typically making up to 20% of the feed mixture [1]. Oats are well known for having a distinctive protein composition and high nutritious content. Oat proteins are mostly composed of globulins as opposed to other grain proteins; as a result, their amino acid profile is more balanced [2]. Oats are regarded as a highly valuable grain because of their high fat content and unsaturated fatty acids, which contribute to their high energy and nutritional value [3]. Due to the ban on the use of antibiotics as growth promoters in animal feed, various biostimulants and other feed additives are increasingly being used to increase resistance and have a positive effect on animal immunity. Numerous colored feedstuffs are becoming increasingly important, including coloured cereals used in animal meals. The reason for their use lies in the richness of polyphenols, which in oats are mainly phenolic acids [4], as quality components that contribute to the above. Dietary interventions with polyphenols may alter the gut microbiome response and attenuate the weaning stress related to inflammation. Further, polyphenols elicit health-favored effects through ameliorating inflammatory processes to improve digestibility and thereby exert a protective effect on animal production [5]. The health effects of the use of oats in human and animal nutrition are well known, given the richness of oats in various biological components [6–8]. These are primarily antioxidants, proteins, lipids, vitamins, minerals, soluble fiber-β-D-glucan (but also cellulose and arabinoxylan), and especially avenanthramides, a unique group of N-cinnamoyl anthranilic acid derivatives found in oats but not in other cereals [9–12]. Recent research has also demonstrated the anti-inflammatory, antiproliferative, and antipruritic properties of oat polyphenols, which may offer extra defense against colon cancer, coronary heart disease, and skin irritation in both humans and animals [13–15]. There are numerous varieties/genotypes of oats, of which the black varieties, the so-called black oats, stand out [16,17]. In the studies of Varga et al. [17], it was found that the DPPH content was increased in black-hulled oats, indicating an increased antioxidant activity associated with a higher content of phenolic compounds in black-hulled oats. According to Fontanella et al. [18], black oats can only be grown as a forage crop or as a forage crop and cereal. There are considerably more studies in the literature in which black oats are used as pasture fodder or as a component of silage, while the available literature contains very few studies on the use of black oat grains in animal nutrition. Most of these studies were carried out in Central and North America (Mexico, Brazil) with ruminants on pasture [19–21] or as a component of silage [20,22]. The aforementioned studies emphasize the positive effects of black oats in animal feed on the production characteristics and profitability of this production. When analyzing the available literature, two papers were found showing the use of black oat grains together with other cereal grains in animal feeding, in fattening lambs, when satisfactory production and slaughter indicators and economic efficiency of production were achieved [23,24].

Sheep are subjected to a variety of physiological and environmental stresses that affect their welfare, reduce their production, and increase the risk of disease [25]. The lamb experiences extreme stress during weaning. According to Lynch et al. [26], weaning is the total physical separation of mother and child as well as the nutritional switch from milk to solid foods. According to Freitas-de-Melo et al. [27], artificial weaning is regarded as one of the most stressful events in farm animals' lives. As such, it is critical to consider the elements influencing those reactions in order to create effective weaning procedures and enhance the welfare of the sheep. The level of shock manifested by reduction in post-weaning growth rate may vary depending on weaning age and weight, the intake of solid feed before weaning as well as health status of the lamb [25]. Lambs may also experience nutritional, social, and emotional changes that result in a significant stress reaction [28]. According to studies by Damián et al. [29] and Freitas-de-Melo et al. [30], abrupt weaning is linked to reduced weight gain, increased alertness and mobility, and decreased resting and feeding behavior in ewes and lambs. In the Republic of Croatia, the normal time for natural weaning of lambs is around 4 months. Therefore, earlier weaning can cause various problems/shock in lambs. Therefore, the time of weaning and quality preparation of lambs before weaning, especially regarding feeding, are very important [31,32]. It is critical to figure out how to reduce the issues listed above that arise from lambs' weaning stress. Feeding the lambs premium feed, preferably with fodder rich in polyphenols or polyphenols extracted from certain plants that offer many health benefits, is one method. In a study using weaned lambs, for instance, Xu et al. [33] discovered that tea polyphenols in the lambs' diets had antioxidant and anti-inflammatory properties comparable to those of antibiotics.

To the best of our knowledge there is no study carried out on lambs fed black oats in their diets affecting the production traits and blood parameters. Finding out how black oat polyphenols affected the metabolic profile, antioxidative status, production traits, and carcass quality of fattening lambs was the study's main goal.

## 2. Materials and Methods

The European Union Directive 2010/63/EU, the Animal Protection Act (NN 133/06, NN 37/13, and NN 1. kg125/13), the Act on the Protection of Animals Used for Scientific Purposes (NN 55/13), and other relevant legal acts on the welfare of farm animals were all followed in this research. Consequently, the Bioethical Committee for Research on Animals of the Faculty of Agrobiotechnical Science Osijek (22-03, 17 March 2022) accepted the protocol of the animal study.

### 2.1. Experiment Design and Body Weight Analysis

The study was carried out on 20 Merinoladschaf lambs aged 70 days after weaning on the Sičaja family farm in Gašinci (Osijek-Baranja County, Croatia; 45°2000500 N, 18°1805900 E), during the year 2022. The selected lambs were fed a feed mixture and meadow hay (40:60) as well as salt and water ad libitum. In addition, an acclimatization phase was carried out to accustom the lambs to the new feed, which lasted 7 days. The selected lambs were in the same sex ratio (50 M: 50 F), body weight, age, and in good physical condition and health. The lambs were divided into 2 groups of 10 lambs each. Each group was housed and ate together in an enclosure (5 m × 4 m). The study lasted 30 days, during which body weight was determined at the beginning and end of the study by weighing (Kern EOS 150K50XL animal scales (Kern & Sohn, Balingen, Germany)). A qualified technician evaluated the body condition score (BCS), which was measured using Russel's 5-point scale [34] (1 = thin to 5 = obese).

### 2.2. Analyses of Feed Mixture and Hay

As per the National Research Council [35], the feed mixtures were provided to the lambs based on their specific needs, with an expected weight of approximately 32 kg. Using a cutting mill (Microtron MB 550; Kinematica, Luzern, Switzerland) to grind the feed mixture and meadow hay to a particle size of 1 mm, samples were taken, dried at 60 °C, and their chemical composition was examined (Table 1). The standard methods of the AOAC [36] were used to analyze the basic chemical composition of the feed. Methods for the determination of proteins, ether extract, dietary fiber, and ash of feeds were described in the work of Antunović et al. [37]. The crude protein content was estimated on the basis of the nitrogen content using the Kjeldahl method. The concentrations of the ether extracts were analyzed using the extraction system B-811 (Buchy, Flawil, Switzerland). The crude fiber content was determined according to the Weende method and the NEM according to INRAE-CIRAD-AFZ [38]. Finally, the crude ash concentration was determined by burning the feed samples at 550 °C for 6 h. The extraction and determination of polyphenols is presented in the work of Jakobek et al. [39]. Samples were weighed and 0.2 g was set in a tube and 1.5 mL of 80% (*v*/*v*) methanol was added in water. Then, samples were vortexed and extracted for 15 min with an ultrasonic water bath (RK 100, Bandelin, Berlin, Germany). Afterwards, samples were centrifuged for 10 min at $6739 \times g$ (Eppendorf, Hamburg, Germany). The extract was transferred into a separate plastic tube. Total polyphenols were determined by using the Shimadzu UV-1280 spectrophotometer (Shimadzu Europe GmbH, Duisburg, Germany). Total polyphenols were expressed as mg gallic acid equivalents (GAE)/kg of sample weight. Data are presented as the mean of three parallels.

**Table 1.** Ingredients and chemical composition of the feed mixture and meadow hay for feeding lambs.

| Ingredient (%) | Group | | Meadow Hay | Black Oat |
|---|---|---|---|---|
| | **CO** | **BO** | | |
| *Ingredients composition* | | | | |
| Corn | 40.00 | 40.00 | | |
| Oat | 15.00 * | - | | |
| Black oat | - | 15.00 | | |
| Barley | 21.30 | 21.30 | | |
| Soybean meal | 8.00 | 8.00 | | |
| Soybean toasted | 12.00 | 12.00 | | |
| Limestone | 0.30 | 0.30 | | |
| Cattle salt | 0.40 | 0.40 | | |
| Mineral premix ** | 3.00 | 3.00 | | |
| *Chemical composition (g/kg DM)* | | | | |
| Dry matter | 950.00 | 945.00 | 946.70 | 950.20 |
| Crude proteins | 155.00 | 152.90 | 102.70 | 73.60 |
| Ether extract | 48.30 | 48.60 | 7.00 | 48.70 |
| Crude fibre | 22.30 | 36.90 | 353.70 | 101.30 |
| Ash | 66.90 | 60.40 | 51.50 | 28.80 |
| NEM, MJ/kg | 8.00 | 7.00 | 4.50 | - |
| Polyphenols (total; mg gallic acid equivalents (GAE)/kg) | 1631.15 | 2234.63 | 4730.30 | 1538.46 |

CO—control oat, BO—black oat; NEM-net energy for meat production; * content of total polyphenol in oat is 1053.58 mg/kg. ** Mineral-vitamin premix for lambs: 8% Ca, 5% P, 9.5% Na, 2.00% Mg, 400,000 IU vitamin A, 40,000 IU vitamin D, 500 mg vitamin E, 4000 mg Zn, 2000 mg Mn, 60 mg I, 10 mg Co, 50 mg Se.

### 2.3. Blood Sampling and Analysis

Blood samples were taken from the lambs at the start and end of the study (days 1 and 30). Sterile vacuum tubes (Vacutube®, LT Burnik, Vodice, Slovenia) were used for blood sampling from the jugular vein. The hematological indicators (leukocytes—WBC, erythrocytes—RBC, hemoglobin content—HGB, HCT—hematocrit, mean corpuscular volume—MCV, mean corpuscular hemoglobin—MCH, mean corpuscular hemoglobin concentration-MCHC and platelet count—PLT) were determined on the Sysmex pocH 100 iV hematology 3-Diff analyzer (Sysmex Europe GmbH, Hamburg, Germany). Blood smears stained with the Pappenheim procedure and fixed in air were used to calculate the relative proportion of each kind of leukocyte. An Olympus BX 51® microscope (Olympus, Tokyo, Japan) was used to differentiate the white blood count. Following blood collection, lamb serum was separated using centrifugation for 10 min at 1609.92× $g$. An Olympus AU640 analyzer (Olympus, Tokyo, Japan) was used for analysis. Mineral concentrations (calcium—Ca; inorganic phosphorus—P; magnesium—Mg), biochemical parameters (glucose, urea, total proteins, cholesterol, albumin, globulin, triglycerides, high-density lipoprotein (HDL), low-density lipoprotein (LDL), β-hydroxybutyrate (BHB), non-esterified fatty acids (NEFA) and enzyme activities (glutathione peroxidase-GPx, superoxide dismutase-SOD, alanine aminotransferase—ALT; aspartate aminotransferase—AST; gamma-glutamyl transferase—GGT and alkaline phosphatase—ALP) were determined in serum using Olympus System Reagents (Olympus Diagnostica GmbH, Lismeehan, Ireland). The globulin content was determined as a subtraction of total protein and albumin.

### 2.4. Analyses of Carcass Measures and Meat Quality

The lambs were weighed on an automated animal scale, the Kern EOS 150K50XL (Kern und Sohn, Balingen, Germany), prior to slaughter. On the Croatian market, lamb is sold as whole carcass with head and kidneys. This is why these parts were not separated from the carcass or weighed separately. After slaughter (classic method of bleeding by severing the large blood vessels in the neck external jugular vein and common carotid artery) and bleeding of 10 lambs, the skin of the lambs was removed from the carcasses and the organs of the abdominal cavity (forestomach, stomach, spleen, intestines, and liver) and the thoracic cavity (trachea with lungs and heart) were cut off. Weighing was conducted on the skin, lower limbs, internal organs, and carcasses itself. The carcasses were measured using the standard developmental technique (linear measurement) according to Antunović et al. [40]: length (carcass length 1—os pubis to the atlas; length 2—os pubis to the first rib; length 3—os pubis to the last rib); circumference of the carcass at chest; length of the hind legs (tuber calcanei to tubercule ossis ischia); and circumference of the hind legs (the widest part). Evaluation of meat quality samples of lamb meat (musculus semimembranosus) was taken from all lambs immediately after carcass processing. The $pH_1$ values and the color of the lamb carcass were determined 45 min post-mortem, while $pH_2$ was determined 24 h post-mortem. The pH of muscle (a central part) was measured with a Mettler Toledo contact pH meter (Mettler Toledo, Columbus, OH, USA). In accordance with the CIE [41] L*a*b* color system standard, the meat's color was measured using a portable Minolta Chroma Metre CR-410 (Minolta Camera Co. Ltd., Osaka, Japan). The formula for calculating dressing percentage was (pre-slaughter weight − carcass weight × 100). Water holding capacity was measured according to the method of Sierra [42].

The hue angle and colorfulness were computed using the formulas:

$$H^* = \tan^{-1} (b^*/a^*) \times (180/\pi) \tag{1}$$

$$C^* = \sqrt{(a^*2 + b^*2)} \tag{2}$$

### 2.5. Extraction of Antioxidants from Lamb Muscle, Liver, and Kidney Samples to Assess Antioxidative Status

Immediately after slaughter, muscle (musculus semimembranosus), liver, and kidney samples were taken and observable fat was detached. Homogenates were prepared in 0.05 M phosphate buffer (pH 7) using an Ultra Turrax (IKA T18 Basic, Labortechnik, Staufen, Germany) homogenizer (10 $w/v$) and centrifuged at 12,000× $g$ for 60 min at 4 °C. The equipment used was a centrifuge Hermle Z 326 K (Hermle Labortechnik GmbH, Wehingen, Germany). The extraction was performed in triplicate. The prepared extracts were stored at −80 °C in a Binder UF V 700 Ultra-low temperature freezer (BINDER GmbH, Tuttlingen, Germany) until they were analyzed. The supernatant obtained was used for DPPH radical scavenging activity and TBARS.

### 2.5.1. Free-Radical-Scavenging Activity of 2,2-Diphenyl-1-picrylhydrazyl (DPPH)

With slight adjustments, the DPPH radical scavenging assay previously reported by Qwele et al. [43] was used to measure the extracts' overall antioxidant activity. A volume of 625 µL of 0.2 mM DPPH prepared in methanol was added to 625 µL of supernatant, which had been diluted with distilled water at a 1:4 ratio. The mixture was vortexed and allowed to stand for 30 min in the dark at room temperature. A tube containing 625 µL methanol and 625 µL DPPH solution was used as a control, while methanol alone served as a blank. To establish a baseline, ascorbic acid (AA) was utilized. We took three duplicate measurements for each. DPPH scavenging activity was calculated using Equation (1), and absorbance was measured at 517 nm using a spectrophotometer (Lambda 25, Perkin Elmer, Waltham, MA, USA).

$$\text{DPPH activity} = ((Ab + As - Am)/Ab) \times 100 \tag{3}$$

where Ab is the absorbance of the 0.1 mM DPPH radical solution at $\lambda = 517$ nm, As is the absorbance of the extracts at $\lambda = 517$ nm, and Am is the absorbance of the mixture of tested extracts and DPPH radical at 517 nm.

### 2.5.2. Thiobarbituric-Acid-Reactive Substances (TBARS)

The TBARS method was used to quantify lipid peroxidation in muscle (musculus semimembranosus), liver, and kidney samples. Slight adjustments were made to the parameters published in Liu et al.'s [44] paper. In brief, a volume of 1.5 mL of 0.1% trichloroacetic acid was added to 300 μL of supernatant and centrifuged at $6000\times g$ for 5 min at 4 °C. The equipment used was a centrifuge Hermle Z 326 K (Hermle Labortechnik GmbH, Wehingen, Germany). A volume of 800 μL TBA/TCA solution reagent (0.5% 2-thiobarbituric acid in 20% trichloroacetic acid) was added to 400 μL supernatant, vortexed, and heated to 95 °C for 30 min in a boiling water bath. The mixtures were then cooled on ice for 10 min and centrifuged at $18{,}000\times g$ for 15 min at 4 °C with a centrifuge Hermle Z 326 K (Hermle Labortechnik GmbH, Wehingen, Germany). The same procedure without adding the sample was performed for the blank and the absorbance values were measured (532 nm and 600 nm) using a spectrophotometer (Lambda 25, Perkin Elmer, Waltham, MA, USA). The concentration of malondialdehyde (MDA) was calculated using the molar extinction coefficient of MDA ($156{,}000\ \mathrm{M}^{-1}\ \mathrm{cm}^{-1}$). The results were expressed as mg MDA equivalents per kg tissue sample.

### 2.6. Statistical Analysis

The normality of data distribution was checked by the Shapiro–Wilk test. The mean values of the obtained results were calculated by MEANS procedures in the statistical program TIBICO Statistica® (version 13.3.0) [45]. Differences between mean values of analyzed results were examined by ANOVA using feeding treatment as a fixed effect.

## 3. Results

### 3.1. Production Traits

There were no significant variations in the lambs' production traits based on the feeding, as indicated by Table 2. The analysis of slaughter characteristics of lamb carcasses (Table 3) and its development (Table 4) did not reveal any significant changes. Physical properties of lamb carcass are presented in Table 5 in which a decreased value of $pH_1$ musculus semimembranosus 45 min post-mortem was determined.

**Table 2.** Production traits of lambs.

| Traits | Measuring Time (Days) | Groups (Mean ± sd) | | SEM | *p*-Values |
|---|---|---|---|---|---|
| | | CO | BO | | |
| Body mass (kg) | 1 | 23.71 ± 1.58 | 24.75 ± 1.34 | 0.34 | 0.13 |
| | 30 | 33.71 ± 2.65 | 34.00 ± 2.27 | 0.54 | 0.80 |
| Daily gain (kg) | 1–30 | 0.333 ± 0.05 | 0.309 ± 0.05 | 0.01 | 0.29 |
| Feed conversion (DM/g gain) | 1–30 | 2.86 | 2.88 | | |
| Feed consumption (g/days) | 1–30 | 0.95 | 0.89 | | |
| BCS (point) | 1 | 3.53 ± 0.27 | 3.66 ± 0.23 | 0.06 | 0.26 |
| | 30 | 3.88 ± 0.23 | 4.00 ± 0.21 | 0.05 | 0.06 |

sd—standard deviation; SEM—standard error of the mean; BCS—body condition score; DM—dry matter; CO—control oat group, BO—black oat group.

**Table 3.** Slaughtering characteristics of lamb carcasses.

| Indicators | Groups (Mean ± sd) | | SEM | *p*-Value |
|---|---|---|---|---|
| | CO | BO | | |
| Pre-slaughter mass (kg) | 33.71 ± 2.65 | 34.00 ± 2.27 | 0.54 | 0.80 |
| Carcass mass (kg) | 19.06 ± 1.33 | 18.47 ± 1.03 | 0.27 | 0.28 |
| Entrails mass * (kg) | 1.50 ± 0.30 | 1.45 ± 0.26 | 0.06 | 0.71 |
| Skin and legs mass (kg) | 4.52 ± 0.40 | 4.64 ± 0.59 | 0.11 | 0.60 |

**Table 3.** *Cont.*

| Indicators | Groups (Mean ± sd) | | SEM | *p*-Value |
|---|---|---|---|---|
| | **CO** | **BO** | | |
| Forestomach and intestines mass (kg) | 7.78 ± 0.87 | 7.09 ± 0.59 | 0.18 | 0.06 |
| Carcass dressing (%) | 56.73 ± 4.54 | 54.37 ± 1.50 | 0.78 | 0.14 |

sd—standard deviation; SEM—standard error of the mean; * lungs, trachea, heart, liver; CO—control oat group, BO—black oat group.

**Table 4.** Development measures of lamb carcasses.

| Measures | Groups (Mean ± sd) | | SEM | *p*-Value |
|---|---|---|---|---|
| | **CO** | **BO** | | |
| Carcass length 1 (cm) | 73.90 ± 3.03 | 74.40 ± 3.06 | 0.67 | 0.72 |
| Carcass length 2 (cm) | 51.95 ± 2.59 | 53.40 ± 3.45 | 0.68 | 0.30 |
| Carcass length 3 (cm) | 29.00 ± 1.53 | 29.90 ± 1.70 | 0.37 | 0.23 |
| Carcass circumference (cm) | 65.85 ± 1.84 | 66.90 ± 1.58 | 0.44 | 0.06 |
| Hind legs length (cm) | 30.60 ± 4.01 | 30.15 ± 2.54 | 0.73 | 0.21 |
| Ham circumference (cm) | 52.60 ± 1.90 | 54.05 ± 1.89 | 0.44 | 0.11 |

sd—standard deviation; SEM—standard error of the mean; carcass length 1 = (os pubis–atlas); carcass length 2 = (os pubis–first rib); carcass length 3 = (os pubis–last rib); CO—control oat group, BO—black oat group.

**Table 5.** Physical properties of lamb carcasses.

| Indicators | Groups (Mean ± sd) | | SEM | *p*-Value |
|---|---|---|---|---|
| | **CO** | **BO** | | |
| pH$_1$ | 6.51 ± 0.15 | 6.37 ± 0.10 | 0.03 | 0.03 |
| pH$_2$ | 5.69 ± 0.05 | 5.66 ± 0.05 | 0.12 | 0.24 |
| WHC (%) | 25.50 ± 5.07 | 28.23 ± 3.88 | 1.03 | 0.19 |
| | | Color | | |
| L* | 44.32 ± 1.24 | 43.00 ± 1.82 | 0.37 | 0.08 |
| a* | 19.03 ± 1.41 | 19.30 ± 0.80 | 0.25 | 0.14 |
| b* | 2.47 ± 0.26 | 2.42 ± 0.27 | 0.06 | 0.68 |
| Hue angle | 7.44 ± 0.90 | 7.17 ± 0.88 | 0.20 | 0.51 |
| Chroma | 19.19 ± 1.40 | 19.45 ± 0.79 | 0.25 | 0.61 |

sd—standard deviation; SEM—standard error of the mean; CO—control oat group, BO—black oat group; WHC—water holding capacity.

### 3.2. Blood Parameters and Antioxidative Status

Table 6 displays the hematological indicators in which a significant increase in RBC, HCT, and MCV content, and a decrease in EOS was presented at the end of study in the blood of the BO group. Table 7 presents the biochemical indicators in which significant differences were observed, namely a significant increase in TP, ALB, and GLOB concentrations, but also a significant decrease in glucose concentrations at the end of the study in the blood of the BO groups compared to CO. Enzyme activities are presented in Table 8 when feeding lambs feed mixture with black oats. Significant differences were observed in GPx and SOD activities, but a significant decrease in ALT and ALP enzyme activity was also found at the end of the study in the blood of the BO groups compared to CO.

TBARS content in all tissues in BO was lowered compared to the CO group but not significant (Figure 1). Non-significant changes in the DPPH scavenging activity were found in the muscle and kidney. However, a numerical increase in the DPPH scavenging activity was found in all tissues in BO group. Figure 2 shows that BO lambs had a significantly higher DPPH scavenging activity in the liver compared to CO lambs.

**Table 6.** Hematological parameters in lambs.

| Indicators | Measuring Time (Day) | Groups (Mean ± sd) | | SEM | *p*-Values | Ref. Values [46] |
|---|---|---|---|---|---|---|
| | | CO | BO | | | |
| WBC $\times 10^9$ L | 1 | 10.51 ± 2.74 | 10.39 ± 1.85 | 0.51 | 0.91 | 5.10–15.90 |
| | 30 | 11.31 ± 3.02 | 12.12 ± 3.60 | 0.73 | 0.59 | |
| RBC $\times 10^{12}$ L | 1 | 9.74 ± 1.19 | 10.50 ± 0.88 | 0.24 | 0.14 | 9.20–13.00 |
| | 30 | 9.17 ± 3.02 | 10.45 ± 0.63 | 0.27 | 0.01 | |
| HGB (g/L) | 1 | 117.70 ± 9.90 | 120.70 ± 2.62 | 2.02 | 0.47 | 105.00–137.00 |
| | 30 | 121.80 ± 10.41 | 129.00 ± 9.84 | 2.35 | 0.13 | |
| HCT (L/L) | 1 | 0.44 ± 0.11 | 0.48 ± 0.12 | 0.03 | 0.41 | 0.28–0.47 |
| | 30 | 0.34 ± 0.06 | 0.40 ± 0.03 | 0.01 | 0.003 | |
| MCV (fL) | 1 | 46.09 ± 18.50 | 46.50 ± 15.24 | 0.69 | 0.96 | 28–41 |
| | 30 | 36.46 ± 0.06 | 38.09 ± 1.29 | 0.35 | 0.02 | |
| MCH (pg) | 1 | 12.22 ± 1.86 | 11.57 ± 0.62 | 0.31 | 0.31 | 10–13 |
| | 30 | 13.61 ± 3.02 | 12.41 ± 1.54 | 0.54 | 0.28 | |
| MCHC (g/L) | 1 | 279.00 ± 41.70 | 261.30 ± 43.92 | 9.54 | 0.37 | 332–392 |
| | 30 | 357.90 ± 101.63 | 327.80 ± 50.85 | 18.34 | 0.20 | |
| PLT $\times 10^9$ L | 1 | 680.80 ± 68.87 | 634.60 ± 155.73 | 26.74 | 0.40 | 426.00–1142.00 |
| | 30 | 610.53 ± 47.09 | 591.80 ± 129.74 | 21.35 | 0.67 | |
| Differential blood smears (%) | | | | | | |
| EOS | 1 | 4.50 ± 3.60 | 5.40 ± 8.46 | 1.42 | 0.76 | 1–8 [47] |
| | 30 | 2.60 ± 1.65 | 1.00 ± 0.67 | 0.33 | 0.01 | |
| SEG | 1 | 33.30 ± 7.82 | 29.20 ± 5.79 | 1.70 | 0.07 | 10–50 [47] |
| | 30 | 26.10 ± 6.74 | 24.40 ± 11.02 | 1.99 | 0.68 | |
| BAC | 1 | 0.20 ± 0.42 | 0.40 ± 1.71 | 0.37 | 0.06 | 0 [47] |
| | 30 | 0 | 0.10 ± 0.32 | 0.05 | 0.33 | |
| LYMPH | 1 | 56.60 ± 9.24 | 64.50 ± 11.96 | 2.50 | 0.16 | 50–75 [47] |
| | 30 | 69.50 ± 8.44 | 73.60 ± 10.78 | 2.16 | 0.36 | |
| MONO | 1 | 4.80 ± 6.63 | 0.90 ± 1.20 | 1.13 | 0.08 | 0–4 [47] |
| | 30 | 1.70 ± 2.00 | 0.70 ± 0.95 | 0.36 | 0.17 | |
| BAS | 1 | 0.59 ± 0.70 | 0.60 ± 1.08 | 0.20 | 0.99 | 0–1 [47] |
| | 30 | 0.10 ± 0.32 | 0.20 ± 0.42 | 0.08 | 0.56 | |

sd—standard deviation; SEM—standard error of the mean; CO—control oat group, BO—black oat group; WBC—white blood cells; RBC—red blood cells; HGB—hemoglobin content; HCT—hematocrit; MCV—mean corpuscular volume; MCH—mean corpuscular hemoglobin; MCHC—mean corpuscular hemoglobin concentration; PLT—platelet count; SEG—segmented neutrophils; LYMPH—lymphocytes; EOS—eosinophils; MONO—monocytes; BAS—basophils; BAC—band cells.

**Table 7.** Lambs' blood biochemical parameters.

| Indicators | Measuring Time (Day) | Groups (Mean ± sd) | | SEM | *p*-Values | Ref. Values [46] |
|---|---|---|---|---|---|---|
| | | CO | BO | | | |
| Ca (mmol/L) | 1 | 2.77 ± 0.16 | 2.89 ± 0.23 | 0.05 | 0.20 | 2.42–2.92 |
| | 30 | 2.85 ± 0.21 | 2.68 ± 0.28 | 0.08 | 0.12 | |
| P-inorganic (mmol/L) | 1 | 3.04 ± 0.60 | 3.31 ± 0.52 | 0.13 | 0.29 | 1.88–3.34 |
| | 30 | 3.05 ± 0.35 | 2.99 ± 0.38 | 0.10 | 0.08 | |
| Mg (mmol/L) | 1 | 1.55 ± 0.19 | 1.61 ± 0.22 | 0.05 | 0.52 | 0.91–1.31 |
| | 30 | 1.52 ± 0.18 | 1.42 ± 0.14 | 0.04 | 0.17 | |
| Glucose (mmol/L) | 1 | 6.62 ± 0.52 | 6.39 ± 0.58 | 0.12 | 0.37 | 2.70–4.80 |
| | 30 | 6.57 ± 0.34 | 6.06 ± 0.67 | 0.13 | 0.046 | |

**Table 7.** *Cont.*

| Indicators | Measuring Time (Day) | Groups (Mean ± sd) | | SEM | *p*-Values | Ref. Values [46] |
|---|---|---|---|---|---|---|
| | | CO | BO | | | |
| Urea (mmol/L) | 1 | 6.55 ± 1.07 | 6.80 ± 1.83 | 0.33 | 0.72 | 5.00–9.10 |
| | 30 | 9.76 ± 1.51 | 9.67 ± 0.97 | 0.28 | 0.89 | |
| Total proteins (g/L) | 1 | 60.80 ± 4.01 | 59.19 ± 4.32 | 0.93 | 0.40 | 51.00–64.00 |
| | 30 | 58.50 ± 3.19 | 65.64 ± 4.30 | 1.16 | 0.007 | |
| ALB (g/L) | 1 | 30.87 ± 2.47 | 28.97 ± 2.01 | 0.54 | 0.08 | 30.00–37.00 |
| | 30 | 28.64 ± 4.57 | 32.32 ± 1.69 | 0.86 | 0.03 | |
| GLOB (g/L) | 1 | 29.93 ± 3.63 | 30.22 ± 3.64 | 0.79 | 0.86 | 19.00–30.00 |
| | 30 | 29.82 ± 3.62 | 33.33 ± 3.33 | 0.86 | 0.04 | |
| CHOL (mmol/L) | 1 | 1.42 ± 0.46 | 1.63 ± 0.76 | 0.14 | 0.46 | 1.35–1.97 [48] |
| | 30 | 1.43 ± 0.22 | 1.29 ± 0.30 | 0.06 | 0.26 | |
| HDL (mmol/L) | 1 | 0.73 ± 0.26 | 0.87 ± 0.19 | 0.05 | 0.19 | 0.68–0.97 [49] |
| | 30 | 0.83 ± 0.09 | 0.77 ± 0.17 | 0.03 | 0.09 | |
| LDL (mmol/L) | 1 | 0.42 ± 0.26 | 0.64 ± 0.49 | 0.09 | 0.22 | 0.10–0.50 [49] |
| | 30 | 0.45 ± 0.14 | 0.39 ± 0.17 | 0.03 | 0.80 | |
| TRIG (mmol/L) | 1 | 0.43 ± 0.10 | 0.40 ± 0.07 | 0.02 | 0.48 | |
| | 30 | 0.33 ± 0.14 | 0.29 ± 0.04 | 0.02 | 0.78 | 0.00–2.00 [48] |
| NEFA (mmol/L) | 1 | 0.35 ± 0.47 | 0.17 ± 0.18 | 0.08 | 0.27 | <0.4 |
| | 30 | 0.27 ± 0.22 | 0.26 ± 0.14 | 0.04 | 0.88 | |
| BHB (mmol/L) | 1 | 0.29 ± 0.12 | 0.28 ± 0.13 | 0.03 | 0.87 | 0.2–0.7 |
| | 30 | 0.30 ± 0.10 | 0.38 ± 0.12 | 0.03 | 0.11 | |

sd—standard deviation; SEM—standard error of the mean; CO—control oat group, BO—black oat group; Ca—calcium; P—inorganic phosphorus; Mg—magnesium; ALB—albumin; GLOB—globulin; CHOL—cholesterol, TRIG—triglycerides, HDL—high-density lipoprotein; LDL—low-density lipoprotein; NEFA—non-esterified fatty acids; BHB—β-hydroxybutyrate.

**Table 8.** Blood enzyme activity in lambs.

| Indicators | Measuring Time (Day) | Groups (Mean ± sd) | | SEM | *p*-Values | Ref. Values [46] |
|---|---|---|---|---|---|---|
| | | CO | BO | | | |
| AST (U/L) | 1 | 107.88 ± 8.84 | 116.03 ± 11.98 | 2.48 | 0.10 | 83.00–140.00 |
| | 30 | 130.55 ± 16.11 | 119.27 ± 5.76 | 2.93 | 0.052 | |
| ALT (U/L) | 1 | 13.43 ± 3.24 | 15.29 ± 5.24 | 0.97 | 0.35 | 6.00–20.00 [48] |
| | 30 | 16.06 ± 2.53 | 12.55 ± 0.69 | 0.57 | 0.0006 | |
| ALP (U/L) | 1 | 441.97 ± 114.50 | 459.56 ± 98.01 | 23.28 | 0.72 | 300–500 [50] |
| | 30 | 590.11 ± 59.36 | 508.66 ± 64.23 | 16.39 | 0.009 | |
| GGT (U/L) | 1 | 81.31 ± 10.09 | 68.58 ± 19.44 | 3.67 | 0.08 | 56.00–110.00 |
| | 30 | 72.49 ± 9.50 | 70.10 ± 10.30 | 2.14 | 0.59 | |
| GPx (U/L) | 1 | 524.02 ± 37.77 | 524.73 ± 38.98 | 8.35 | 0.97 | >600 [50] |
| | 30 | 510.27 ± 36.74 | 676.73 ±109.68 | 25.85 | 0.0005 | |
| SOD (U/mL) | 1 | 0.39 ± 0.10 | 0.43 ± 0.11 | 0.04 | 0.366 | 0.39–0.67 [50] |
| | 30 | 0.54 ± 0.08 | 0.63 ± 0.10 | 0.03 | 0.047 | |

sd—standard deviation; SEM—standard error of the mean; CO—control oat group, BO—black oat group; (LDL), AST—aspartate aminotransferase; ALT—alanine aminotransferase; ALP—alkaline phosphatase; GGT—gamma-glutamyl transferase; GPx—glutathione peroxidase; SOD—superoxide dismutase.

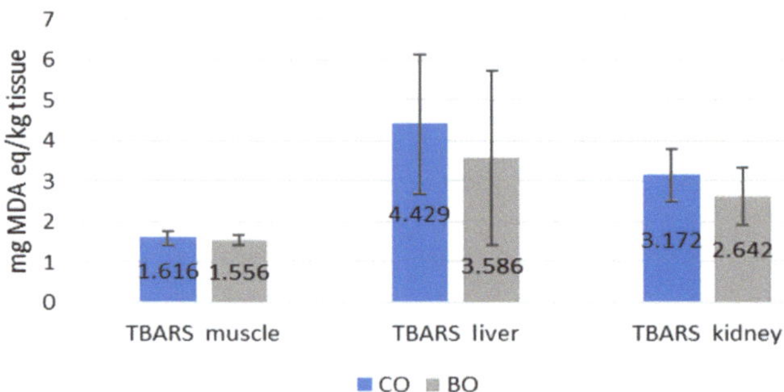

**Figure 1.** Content of TBARS (reactive thiobarbituric acid substances) in muscle (musculus semimembranosus), liver, and kidney of lambs from the CO (control oat) and BO groups (black oat).

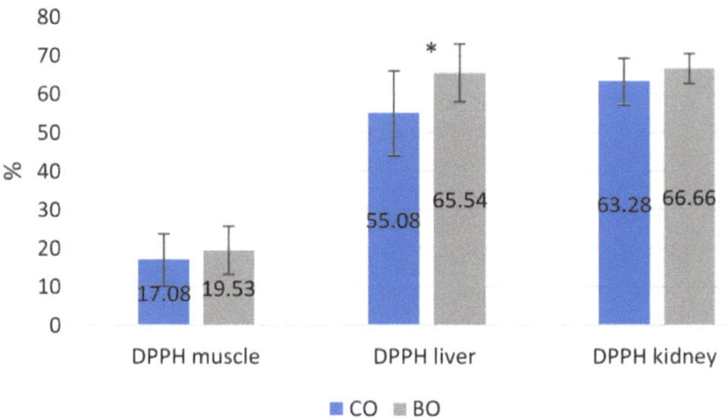

**Figure 2.** DPPH (2,2-diphenyl-1-picrylhydrazyl radical) scavenging activity in muscle (musculus semimembranosus), liver, and kidney of lambs from CO (control oat) and BO groups (black oat); * *p*-values for DPPH-liver is 0.031.

## 4. Discussion

### 4.1. Production Traits

It is well known that weaning is a very stressful period for the ruminants since it is a process in which milk is gradually replaced with concentrates or diets based on grains and forage. It has been confirmed that sheep's and goats' high metabolic demands make them more vulnerable to oxidative stress. The occurrence of oxidative stress in ruminants jeopardizes their productivity and health. One of the main dietary sources of polyphenols is cereal, particularly whole grains [51]. Whole grains contain non-nutritive chemicals called polyphenols, which are mostly flavonoids, lignans, and phenolic acids and found in all structural domains of cereal components. The influence of polyphenols in stress during weaning of young domestic animals (calves, lambs, piglets) improves digestion and absorption of nutrients, improves the function of the intestinal barrier, improves the function of the intestinal microbiota, and thus provides positive effects [52]. In the present study, no significant effect of black oats in the diet was found on the production traits of lambs (daily gain, BCS, conversion, and feed consumption), the slaughter traits of lamb carcasses (weight, individual parts of the carcass and yield, developmental measures, and physical characteristics of carcasses, with the exception of pH1). Bernardes et al. [23,24] obtained satisfactory production and slaughter indicators for lambs when using black oats together with other grains in the meals of fattening lambs. In the study by Tian et al. [53] conducted with goat kids at the age of 90 days (weight 21.38 kg) with the aim of investigating the influence of anthocyanins from purple corn on traits, meat quality and antioxidant activity of muscles, no influence on the performance of kids (daily gain, feed conversion, feed consumption) was determined. In the study by Orzuna-Orzuna et al. [54], the effects of a polyphenol-containing polyherbal mixture (PM) on growth performance, carcass characteristics, and meat quality in lambs during the fattening period were investigated. The above authors found no changes in final body weight, body condition, carcass characteristics, and meat quality between treatments ($p > 0.05$). Cimmino et al. [55] found that there were no differences in productive traits (daily gain, carcass weight, pH and percentage of dressing) in a study with additional dietary supplementation with polyphenols (3.2 mg/day for 78 days) from the residual water after olive oil extraction in Saanen goat kids. Since consumers are guided by color when choosing fresh meat, the same authors have demonstrated the importance of color for meat quality. The meat of the lambs did not change in color during the current investigation. Similar conclusions were reached in studies in which the addition of polyphenols and flavonoids to the diet had no effect on the color of the meat of lambs [56].

*4.2. Blood Parameters and Antioxidative Status*

The metabolic profile of blood can be used to diagnose and predict disease and to assess the physiology and welfare of animals [57]. The beneficial effects of polyphenols in animals depend largely on their bioavailability in target tissues [58], cellular distribution, and metabolism after absorption. The analysis of the metabolic profile of the lambs' blood showed a significant increase in the content of hematological indicators (RBC, HCT, and MCV) and a decrease in the content of EOS in the blood of the lambs in the BO group compared to the CO group. All hematological parameters in the blood of lambs showed very little variation overall, with the results being within the reference range for lambs [37,47,59]. In their study, Teng et al. [60] concluded that the positive effect of polyphenols on erythropoiesis could be beneficial both in healthy people and in patients with metabolic diseases. Youdim et al. [61] showed that polyphenols are known to protect against reactive oxygen-species-induced hemolysis by increasing red blood cell integrity in conjunction with inhibition of lipid peroxidation. In the study by Orzuna-Orzuna et al. [54], the effects of a polyphenol-containing polyherbal mixture (PM) on blood metabolite concentrations in lambs during the fattening period was studied. Authors found an increase ($p < 0.05$) in MCHC and lymphocytes and a decrease in SEG in experimental lamb's blood fed the PM diet compared to the CON treatment ($p < 0.05$), which they classify as beneficial for the health status of the lambs. The analysis of biochemical indicators and enzyme activity in the blood revealed significantly higher concentrations of TP, ALB, and GLOB as well as GPx and SOD activities but also a significant decrease in glucose concentrations and ALT and ALP enzyme activity at the end of the study in the blood of the BO groups compared to the CO group. It is also noticeable that there was a decrease in total, high-density lipoprotein and low-density lipoprotein and triglycerides in the blood of the lambs in the BO group compared to the CO group, but without significant differences ($p > 0.05$). The biochemical indicators determined in the blood were within the reference values for lambs [46,49,62]. Higher protein concentration in the lamb's blood of the BO group can be associated with increased amino acid and microbial protein flux in the duodenum of ruminants fed flavonoids [63], while increased globulin levels in serum are an indicator of an improved immune response of the lamb's body [57]. The liver plays a very important role in protein metabolism. Any damage to its cells is reflected in the total serum proteins [64]. Giannenas et al. [65] showed that the addition of a plant additive rich in polyphenolic compounds to the ration of cows reduced blood urea and increased serum globulin and total protein concentrations, improved nutritional status, increased microbial protein synthesis and minimized protein catabolism. When compared to the CO group, the glucose levels in the blood serum of the BO lambs was significantly lower at the last of the research. In a study by Antunović et al. [37], a lower blood glucose concentration was found when red corn in the ration of fattening lambs compared to the blood of lambs fed yellow corn. Even at low concentrations (1%), oat $\beta$-glucans are thought to have a high viscosity, which may contribute to lower plasma glucose and insulin levels as seen in type 2 diabetics by delaying intestinal transit, gastric emptying, and glucose and sterol absorption [66]. The activity of aspartate transferase in blood, gamma-glutamyl transferase, alkaline phosphatases, and cholesterol concentrations are used to diagnose hepatic injury in people and animals, alkaline phosphatases activity and cholesterol concentrations are used to indicate biliary blockage or mild and progressive liver damage [67]. ALT, a particular hepatic enzyme generated following hepatocellular injury, is utilized to measure liver damage instead of GGT. The current study's much lower ALT and ALP at the conclusion imply that neither liver nor muscle damage occurred because their activities were within the reference levels [48]. Antioxidant enzymes such as SOD, GPx, and catalase help to maintain a healthy level of antioxidants within cells [68].

In the present study, significantly higher GPx and SOD activities in the blood, significantly higher DPPH scavenging activity in the liver of BO lambs compared to CO lambs were found. In the study by Tian et al. [53] conducted study with goat kids at the age of 90 days with the aim to research the influence of anthocyanins from purple, determined a significantly higher DPPH scavenging activity and peroxidase levels in muscle (longissimus thoracis et lumborum). In the muscle and kidney samples, no significant changes ($p > 0.05$) were found in the TBARS content and DPPH scavenging activity, although lower TBARS content and higher DPPH scavenging activity were determined in the BO groups in comparison to the CO group, especially in the liver. The changes mentioned indicate a moderate antioxidant activity in the BO group compared to the CO group. It is to be expected that a longer duration of the study would result in a significantly more pronounced positive influence on the antioxidant status of the meat and kidney of lambs in the BO group. Similar results in lambs fed with feed mixture with added red corn were determined by Antunović et al. [37]. Changes in antioxidant status in the kidneys may be associated with the specific uptake or excretion of some tissue metabolites or intracellular metabolism [69]. In recent years, the health effects of whole-grain products have been closely associated with their phenolic compounds and antioxidant effects [51]. Polyphenols not only have strong antioxidant properties [70] but also have bacteriostatic [71], hepatoprotective [72], cholesterol-lowering [73], and immune-boosting [74] effects. In rations rich in polyphenols, an inhibitory effect on the lipid oxidation of meat (reduction of MDA levels) in lamb [75], an increase in the percentage of inhibition of lipid oxidation, an increase in superoxide dismutase activity in the blood of goats [43], and an inhibitory effect on lipid oxidation (reduction of MDA levels) in goat kids [55] were observed in small ruminants. In the current study, a lower $pH_1$ value was found in the carcasses of BO lambs compared to the CO group. The lamb's $pH_2$ was within the usual range of 5.5 to 5.8, despite the lamb's pH being similar among treatments [76]. According to Dalle's research [77], meat preservation during storage depends on maintaining an acting bacteriostatic pH below 5.8. Higher pH values, on the other hand, promote the proliferation of proteolytic microbes. The above opens up the possibility of using black oats in other types of ruminants in the future research. This is especially important during the stressful conditions during the weaning period.

## 5. Conclusions

The observed changes in glucose, protein metabolism in the of lamb's blood and tissues antioxidant status indicate that the use of polyphenol-rich black oats is justified when feeding lambs under stressful conditions, such as in the post-weaning period. In the future, it would be more beneficial to obtain the study with a longer period of feeding lambs with black oats, with its larger amount in the mixture, especially in older sheep, which may possibly improve the antioxidant activity of the meat. Therefore, a polyphenol profile in feed and meat also needs to be carried out in the future.

**Author Contributions:** Investigation, methodology, writing—original draft preparation conceptualization, visualization Z.A.; investigation, formal analysis, methodology, statistical analyses J.N.;

investigation, visualization, editing Ž.K.Š.; field experiment V.S.; formal analysis, methodology Z.S., M.Đ., V.P., L.M. and L.J.; visualization B.M. All authors have read and agreed to the published version of the manuscript.

**Funding:** This research received no external funding.

**Institutional Review Board Statement:** The study was conducted in accordance with the Declaration of Helsinki by obeying legal provisions determined by the Animal Protection Act (Republic of Croatia Official Gazette No. 133 (2006), No. 37 (2013), and No. 125 (2013)) and approved by the Committee for Animal Welfare of the Faculty of Agrobiotechnical Sciences Osijek (644-01/22-01/01; 22-03 from 17 March 2022).

**Data Availability Statement:** The data presented in this study are available on request from the corresponding author.

**Acknowledgments:** The research was carried out with the Breeding and Technological Processes in Animal Production (no. 1152) research team at the Faculty of Agrobiotechnical Sciences Osijek.

**Conflicts of Interest:** The authors declare no conflicts of interest.

# References

1. Lammer, P. *Small Grains for Livestock. A Meta-Analysis*; University of Wisconsin-Platteville: Platteville, WI, USA, 2017.
2. Klose, C.; Schehl, B.D.; Arendt, E.K. Fundamental study on protein changes taking place during malting of oats. *J. Cereal Sci.* **2009**, *49*, 83–91. [CrossRef]
3. Čech, M.; Ivanišová, E.; Hozlár, P.; Tokár, M.; Zagula, G.; Gumul, D.; Kačániová, M.; Sterczyńska, M.; Haščík, P. Nutritional composition, biological activity and technological properties of new Slovakian black oat varieties. *Microbiol. Biotech. Food Sci.* **2021**, *11*, e4238. [CrossRef]
4. Xie, F.; Lei, Y.; Han, X.; Zhao, Y.; Zhang, S. Antioxidant ability of polyphenols from black rice, buckwheat and oats: In vitro and in vivo. *Czech J. Food Sci.* **2020**, *38*, 242–247. [CrossRef]
5. Hussain, T.; Wang, J.; Murtaza, G.; Metwally, E.; Yang, H.; Kalhoro, M.S.; Kalhoro, D.H.; Rahu, B.A.; Tan, B.; Sahito, R.G.A.; et al. The role of polyphenols in regulation of heat shock proteins and gut microbiota in weaning stress. *Oxid. Med. Cell. Longev.* **2021**, *6*, 6676444. [CrossRef] [PubMed]
6. Ivanišová, E.; Čech, M.; Hozlár, P.; Zaguła, G.; Gumul, D.; Grygorieva, O.; Makowska, A.; Kowalczewski, P.Ł. Nutritional, antioxidant and sensory characteristics of bread enriched with whole meal flour from Slovakian black aat varieties. *Appl. Sci.* **2023**, *13*, 4485. [CrossRef]
7. Paudel, D.; Dhungana, B.; Caffe, M.; Krishnan, P. A Review of Health-Beneficial Properties of Oats. *Foods* **2021**, *10*, 2591. [CrossRef] [PubMed]
8. Sur, R.; Nigam, A.; Grote, D.; Liebel, F.; Southall, M.D. Avenanthramides, polyphenols from oats, exhibit anti-inflammatory and anti-itch activity. *Arch. Dermatol. Res.* **2008**, *300*, 569–574. [CrossRef]
9. Tang, Y.; Li, S.; Yan, J.; Peng, Y.; Weng, W.; Yao, X.; Gao, A.; Cheng, J.; Ruan, J.; Xu, B. Bioactive components and health functions of oat. *Food Rev. Int.* **2022**, *39*, 4545–4564. [CrossRef]
10. Singh, R.; De, S.; Belkheir, A. *Avena sativa* (oat) a potential nutraceutical and therapeutic agent: An overview. *Crit. Rev. Food Sci. Nutr.* **2013**, *53*, 126–144. [CrossRef]
11. Alemayehu, G.F.; Forsido, S.F.; Tola, Y.B.; Teshager, M.A.; Assegie, A.A.; Amare, E. Proximate, mineral and anti-nutrient compositions of oat grains (*Avena sativa*) cultivated in Ethiopia: Implications for nutrition and mineral bioavailability. *Heliyon* **2021**, *7*, e07722. [CrossRef]
12. Menon, R.; Gonzalez, T.; Ferruzzi, M.; Jackson, E.; Winderl, D.; Watson, J. Oats—from farm to fork. *Adv. Food Nutr. Res.* **2016**, *77*, 1–55. [PubMed]
13. Scott, M.B.; Styring, A.K.; McCullagh, J.S.O. Polyphenols: Bioavailability, microbiome interactions and cellular effects on health in humans and animals. *Pathogens* **2022**, *11*, 770. [CrossRef]
14. Serra, V.; Salvatori, G.; Pastorelli, G. Dietary polyphenol supplementation in food producing animals: Effects on the quality of derived products. *Animals* **2021**, *11*, 401. [CrossRef]
15. Stover, M.G.; Watson, R.R. Polyphenols in foods and dietary supplements: Role in veterinary medicine and animal health. In *Polyphenols in Human Health and Disease*; Academic Press: Cambridge, MA, USA, 2014; Chapter 1; pp. 3–7.
16. Abdel-Aal, E.S.; Young, J.C.; Rabalski, I. Anthocyanin composition in black, blue, pink, purple and red cereal grains. *J. Agric. Food Che.* **2006**, *54*, 4696–4704. [CrossRef] [PubMed]
17. Varga, M.; Jójárt, R.; Fónad, P.; Mihály, R.; Palágyi, A. Phenolic composition and antioxidant activity of colored oats. *Food Chem.* **2018**, *268*, 153–161. [CrossRef] [PubMed]
18. Fontaneli, R.S.; dos Santos, H.P.; Fontaneli, R.S.; Oliveira, J.T.; Lehmen, R.I.; Dreon, G. Winter forage grasses. In *Forrageiras Para Integração Lavoura-Pecuária-Floresta NA Região Sul-Brasileira*; Fontaneli, R.S., dos Santos, H.P., Fontaneli, R.S., Eds.; EMBRAPA: Brasilia, Brazi, 2012.

19. Derenevicz Faisca, L.; Peres, M.T.P.; Fernandes, S.R.; Bonnet, O.J.F.; Batista, R.; Deiss, L.; Monteiro, A.G. A new insight about the selection and intake of forage by ewes and lambs in different production systems on pasture. *Small Rumin. Res.* **2023**, *221*, 106949. [CrossRef]

20. Vega-García, J.I.; López-González, F.; Estrada-Flores, J.G.; Flores-Calvete, G.; Prospero-Bernal, F.; Arriaga-Jordán, C.M. Black oat (*Avena strigosa* Schreb.) grazing or silage for small-scale dairy systems in the highlands of central Mexico. Part I. Crop and dairy cow performance. *Chil. J. Agric. Res.* **2020**, *80*, 515–525. [CrossRef]

21. Vonz, D.; Menezes, L.F.G.; Paris, W.; Kuss, F.; Silveira, M.F.; Venturini, T.; Stanqueviski, F.; Boito, B. Performance of steers fed on pasture receiving different seeding rates of vetch in an integrated crop-livestock system. *Span. J. Agric. Res.* **2021**, *19*, e06SC01. [CrossRef]

22. Silva, T.B.P.; Del Valle, T.A.; Ghizzi, L.G.; Silva, G.G.; Gheller, L.S.; Marques, J.A.; Dias, M.S.S.; Nunes, A.T.; Grigoletto, N.T.S.; Takiya, C.S.; et al. Partial replacement of corn silage with whole-plant soybean and black oat silages for dairy cows. *J. Dairy Sci.* **2021**, *104*, 9842–9852. [CrossRef]

23. Bernardes, G.M.C.; Carvalho, S.; Pires, C.C.; Motta, J.H.; Teixeira, W.S.; Borges, L.I.; Fleig, M.; Pilecco, V.M.; Farinha, E.T.; Venturini, R.S. Consumption, performance and economic analysis of the feeding of lambs finished in feedlot as the use of high-grain diets. *Arq. Bras. Med. Vet. Zootec.* **2015**, *67*, 1684–1692. [CrossRef]

24. Bernardes, G.M.C.; Carvalho, S.; Venturini, R.S.; Teixeira, W.S.; Motta, J.H.; Borges, L.I.; Rosa, J.S.; Pesamosca, A.C.; Cocco, A.; Mello, V.L.; et al. Carcass characteristics and tissue composition of the meat of feedlot lambs fed high-grain diets. *Semin. Ciências Agrárias* **2018**, *39*, 2637–2646.

25. Karakuş, F. Weaning stress in lambs. *J. Int. Sci. Pub. Agric. Food* **2014**, *2*, 165–170.

26. Lynch, E.; McGee, M.; Earley, B. Weaning management of beef calves with implications 322 for animal health and welfare. *J. Appl. Anim. Res.* **2019**, *47*, 167–175. [CrossRef]

27. Freitas-de-Melo, A.; Orihuela, A.; Hötzel, M.J.; Ungerfeld, R. What do we know and need to know about weaning in sheep? An overview of weaning practises, stress and welfare. *Front. Anim. Sci.* **2022**, *3*, 823188.

28. Freitas-de-Melo, A.; Ungerfeld, R. Destete artificial en ovinos: Respuesta de estrés y bienestar animal. *Rev. Mex. Cienc. Pecu.* **2016**, *7*, 361–375. [CrossRef]

29. Damián, J.P.; Hötzel, M.J.; Banchero, G.; Ungerfeld, R. Behavioural response of grazing lambs to changes associated with feeding and separation from their mothers at weaning. *Res. Vet. Sci.* **2013**, *95*, 913–918. [CrossRef]

30. Freitas-de-Melo, A.; Ungerfeld, R.; Pérez-Clariget, R. Behavioral pattern in Texel × Corriedale terminal crossbreeding: Maternal behavior score at birth, lambs' feeding behaviors, and behavioral responses of lambs to abrupt weaning. *J. Vet. Behav.* **2019**, *30*, 9–15. [CrossRef]

31. Schichowski, C.; Moors, E.; Gauly, M. Effects of weaning lambs in two stages or by abrupt separation on their behavior and growth rate. *J. Anim. Sci.* **2008**, *86*, 220–225. [CrossRef]

32. Selaive-Villarroel, A.B.; Maciel, M.B.; de Oliveira, N.M. Effects of weaning age and weight on lamb growth rate of Morada Nova breed raised in a tropical extensive production system. *Ciência Rural. Santa Maria* **2008**, *38*, 784–788. [CrossRef]

33. Xu, Y.; Yin, F.; Wang, J.; Wu, P.; Qiu, X.; He, X.; Xiao, Y.; Gan, S. Effect of tea polyphenols on intestinal barrier and immune function in weaned lambs. *Front. Vet. Sci.* **2024**, *11*, 1361507. [CrossRef]

34. Russel, A. Body condition scoring of sheep. In *Sheep and Goat Practice*; Boden, E., Ed.; Bailliere Tindall: Philadelphia, PA, USA, 1991; p. 3.

35. National Research Council (NRC). *Nutrient Requirements of Small Ruminants: Sheep, Goats, Cervids, and New World Camelids*; The National Academy Press: Washington, DC, USA, 2007; p. 293.

36. Association of Official Analytical Chemists (AOAC). *Official Methods of Analysis*, 18th ed.; AOAC: Arlington, VA, USA, 2006.

37. Antunovic, Z.; Novoselec, J.; Klir Šalavardic, Ž.; Steiner, Z.; Šperanda, M.; Jakobek Barron, L.; Ronta, M.; Pavic, V. Influence of red corn rich in anthocyanins on productive traits, blood metabolic profile, and antioxidative status of fattening lambs. *Animals* **2022**, *12*, 612. [CrossRef]

38. INRAE-CIRAD-AFZ. Feed tables composition and nutritive values of feeds for cattle, sheep, goats, pigs, poultry, rabbits, horses and salmonids. Available online: https://www.feedtables.com (accessed on 10 July 2024).

39. Jakobek, L.; Matić, P.; Ištuk, J.; Barron, A.R. Study of interactions between individual phenolics of aronia with barley β-glucan. *Pol. J. Food Nutr. Sci.* **2021**, *71*, 187–196. [CrossRef]

40. Antunović, Z.; Domaćinović, M.; Šperanda, M.; Liker, B.; Mioč, B.; Šerić, V.; Šperanda, T. Effect of roasted cereals and soybean in feed mixtures on fattening and slaughter traits as well as blood composition in fattening lambs. *Arch. Tierz.* **2009**, *52*, 512–526. [CrossRef]

41. CIE (Commission Internationale de l'Eclairage). *Colorimetry, Official Recommendations of the International Commission on Illumination*; Publication CIE No. 15 (E-1.3.1); Bureau Central dela CIE: Paris, France, 1976.

42. Sierra, I. Contributions to the study of the Belgian White_Landrace cross: Productive characters, carcass quality and meat quality. *Rev. Inst. Econom. Prod. Ebro.* **1973**, *16*, 43–48.

43. Qwele, K.; Hugo, A.; Oyedemi, S.O.; Moyo, B.; Masika, P.J.; Muchenje, V. Chemical composition, fatty acid content and antioxidant potential of meat from goats supplemented with Moringa (*Moringa oleifera*) leaves, sunflower cake and grass hay. *Meat Sci.* **2013**, *93*, 455–462. [CrossRef] [PubMed]

44. Liu, F.; Dai, R.; Zhu, J.; Li, X. Optimizing color and lipid stability of beef patties with a mixture design incorporating with tea catechins, carnosine, and α-tocopherol. *J. Food Eng.* **2010**, *98*, 170–177. [CrossRef]
45. TIBCO. *Statistica, version 13.3.0*; TIBCO Software Inc.: Palo Alto, CA, USA, 2017.
46. Lepherd, M.L.; Canfield, P.J.; Hunt, G.B.; Bosward, K.L. Haematological, biochemical and selected acute phase protein reference intervals for weaned female Merino lambs. *Aust. Vet. J.* **2009**, *87*, 5–11. [CrossRef]
47. Latimer, K.S.; Maheffey, E.A.; Prasse, K.W. Duncan and Prasse's Veterinary Labaratory Medicine: Clinical Pathology, 4th ed. In *Plumb's Veterinary Drug Handbook*, 5th ed.; Wiley-Blackwell: Hoboken, NJ, USA, 2003; pp. 1241–1249.
48. Kaneko, J.J.; Harvey, J.W.; Bruss, M.L. *Clinical Biochemistry of Domestic Animals*, 6th ed.; Elsevier Academic Press: Amsterdam, The Netherlands, 2008; p. 963.
49. Antunović, Z.; Novoselec, J.; Šperanda, M.; Domaćinović, M.; Đidara, M. Monitoring nutritional status of lambs in organic breeding. *Krmiva* **2010**, *52*, 27–34.
50. Antunović, Z.; Klir Šalavardić, Ž.; Steiner, Z.; Đidara, M.; Drenjančević, M.; Ronta, M.; Pavić, V.; Jakobek Barron, L.; Novoselec, J. Meat quality, metabolic profile and antioxidant status of lambs fed on seedless grape pomace (*Vitis vinifera* L.). *Ann. Anim. Sci.* **2023**, *23*, 809–818. [CrossRef]
51. Tian, S.; Sun, Y.; Chen, Z.; Yang, Y.; Wang, Y. Functional properties of polyphenols in grains and effects of physicochemical processing on polyphenols. *Hindawi J. Food Qual.* **2019**, *2019*, 2793973. [CrossRef]
52. Jiao, X.; Wang, Y.; Lin, Y.; Lang, Y.; Li, E.; Zhang, X.; Zhang, Q.; Feng, Y.; Meng, X.; Li, B. Blueberry polyphenols extract as a potential prebiotic with anti-obesity effects on C57BL/6 J mice by modulating the gut microbiota. *J. Nutr. Biochem.* **2019**, *64*, 88–100. [CrossRef] [PubMed]
53. Tian, X.; Li, J.; Luo, Q.; Wang, X.; Wang, T.; Zhou, D.; Xie, L.; Ban, C.; Lu, Q. Effects of purple corn Anthocyanin on growth performance, meat quality, muscle antioxidant status, and fatty acid profiles in goats. *Foods* **2022**, *11*, 1255. [CrossRef] [PubMed]
54. Orzuna-Orzuna, J.F.; Dorantes-Iturbide, G.; Lara-Bueno, A.; Mendoza-Martínez, G.D.; Miranda-Romero, L.A.; Hernández-García, P.A. Growth performance, carcass characteristics, and blood metabolites of lambs supplemented with a polyherbal mixture. *Animals* **2021**, *11*, 955. [CrossRef] [PubMed]
55. Cimmino, R.; Barone, C.M.A.; Claps, S.; Varrivvhio, E.; Rufrano, D.; Caroprese, M.; Albenzi, M.; De Palo, P.; Neglia, G. Effect of dietary supplementation with polyphenols on meat quality in Saanen goat kids. *BMC Vet. Res.* **2018**, *14*, 181. [CrossRef] [PubMed]
56. Muela, E.; Alonso, V.; Campo, M.M.; Sañudo, C.; Beltrán, J.A. Antioxidant diet supplementation and lamb quality throughout preservation time. *Meat Sci.* **2014**, *98*, 289–295. [CrossRef]
57. Kholif, A.E.; Hassan, A.A.; El Ashry, G.M.; Bakr, M.H.; El-Zaiat, H.M.; Olafadehan, O.A.; Matloup, O.H. Phytogenic feed additives mixture enhances the lactational performance, feed utilization and ruminal fermentation of Friesian cows. *Anim. Biotechnol.* **2020**, *32*, 708–718. [CrossRef]
58. Fraga, C.G.; Oteiza, P.I. Dietary flavonoids: Dietary flavonoids: Role of (−)-epicatechin and related procyanidins in cell signaling. *Free Rad. Biol. Med.* **2011**, *51*, 813–823. [CrossRef]
59. Antunović, Z.; Novoselec, J.; Klir, Ž. Hematological parameters in ewes during lactation in organic farming. *Poljoprivreda* **2017**, *23*, 46–52. [CrossRef]
60. Teng, X.; Zhang, X.; Liu, Y.; Wang, H.; Luo, J.; Luo, Y.; An, P. Effects of dietary polyphenol supplementation on iron status and erythropoiesis: A systematic review and meta-analysis of randomized controlled trials. *Am. J. Clin. Nutr.* **2021**, *114*, 780–793.
61. Youdim, K.A.; Shukitt-Hale, B.; MacKinnon, S.; Kalt, W.; Joseph, J.A. Polyphenolics enhance red blood cell resistance to oxidative stress: In vitro and in vivo. *Biochim. Biophys. Acta* **2000**, *1523*, 117–122. [CrossRef]
62. Antunović, Z.; Mioč, B.; Klir Šalavardić, Ž.; Širić, I.; Držaić, V.; Đidara, M.; Novoselec, J. The effect of lactation stage on the hematological and serum-related biochemical parameters of the Travnik pramenka ewe. *Poljoprivreda* **2021**, *27*, 56–62. [CrossRef]
63. Balcells, J.; Aris, A.; Serrano, A.; Seradj, A.R.; Crespo, J.; Devant, M. Effects of an extract of plant flavonoids (Bioflavex) on rumen fermentation and performance in heifers fed high-concentrate diets. *J. Anim. Sci.* **2012**, *90*, 4975–4984. [CrossRef]
64. Mbuh, J.V.; Mbwaye, J. Serological changes in goats experimentally infected with Fasciola gigantica in Buea sub-division of S.W.P. Cameroon. *Vet. Parasitol.* **2005**, *131*, 255–259. [CrossRef] [PubMed]
65. Giannenas, I.; Skoufos, J.; Giannakopoulos, C.; Wiemann, M.; Gortzi, O.; Lalas, S.; Kyriazakis, I. Effects of essential oils on milk production, milk composition, and rumen microbiota in Chios dairy ewes. *J. Dairy Sci.* **2011**, *94*, 5569–5577. [CrossRef]
66. Butt, M.S.; Tahir-Nadeem, M.; Khan, M.K.I.; Shabir, R.; Butt, M.S. Oat: Unique among the cereals. *Eur. J. Nutr.* **2008**, *47*, 68–79. [CrossRef]
67. Silanikove, N.; Tiomokin, D. Toxicity induced by poultry litter consumption: Effect on parameters reflecting liver function in beef cows. *Anim. Prod.* **1992**, *54*, 203–209. [CrossRef]
68. Peng, K.; Shirley, D.C.; Xu, Z.; Huang, Q.; McAllister, T.A.; Chaves, A.V.; Acharya, S.; Liu, C.; Wang, S.; Wang, Y. Effect of purple prairie clover (*Dalea purpurea* Vent.) hay and its condensed tannins on growth performance, wool growth, nutrient digestibility, blood metabolites and ruminal fermentation in lambs fed total mixed rations. *Anim. Feed Sci. Technol.* **2016**, *222*, 100–110. [CrossRef]
69. Serra, A.; Macià, A.; Romero, M.-P.; Anglès, N.; Morelló, J.R.; Motilva, M.J. Distribution of procyanidins and their metabolites in rat plasma and tissues after an acute intake of hazelnut extract. *Food Funct.* **2011**, *2*, 562–568. [CrossRef] [PubMed]
70. Zhang, R.F.; Zeng, Q.S.; Deng, Y.Y.; Zhang, M.W.; Wei, Z.C.; Tang, X.J. Phenolic profiles and antioxidant activity of litchi pulp of different cultivars cultivated in Southern China. *Food Chem.* **2012**, *136*, 11691176. [CrossRef]

71. Ng, K.R.; Lyu, X.; Mark, R.; Chen, W.N. Antimicrobial and antioxidant activities of phenolic metabolites from flavonoid-producing yeast: Potential as natural food preservatives. *Food Chem.* **2019**, *70*, 123–129. [CrossRef]
72. Callcott, E.T.; Santhakumar, A.B.; Luo, J.; Blanchard, C.L. Terapeutic potential of rice-derived polyphenols on obesity-related oxidative stress and inflammation. *J. Appl. Biomed.* **2018**, *16*, 255–262. [CrossRef]
73. Liu, S.; You, L.; Zhao, Y.; Chang, X. Wild lonicera caerulea berry polyphenol extract reduces cholesterol accumulation and enhances antioxidant capacity in vitro and in vivo. *Food Res. Int.* **2018**, *107*, 73–83. [CrossRef] [PubMed]
74. Cuevas, A.; Saavedra, N.; Salazar, L.; Abdalla, D. Modulation of immune function by polyphenols: Possible contribution of epigenetic factors: Possible con-tribution of epigenetic factors. *Nutrients* **2013**, *5*, 2314–2332. [CrossRef]
75. Bañón, S.; Méndez, L.; Almela, E. Effects of dietary rosemary extract on lamb spoilage under retail display conditions. *Meat Sci.* **2012**, *90*, 579–583. [CrossRef]
76. Sañudo, C.; Santolaria, M.P.; Maria, G.; Osorio, M.; Sierra, I. Influence of carcass weight on instrumental and sensory lamb meat quality in intensive production systems. *Meat Sci.* **1996**, *42*, 195–202. [CrossRef] [PubMed]
77. Dalle, Z.O. Perception of rabbit meat quality and major factors influencing the rabbit carcass and meat quality. *Livest. Prod. Sci.* **2002**, *75*, 11–32. [CrossRef]

*Article*

# Impact of Feeding Systems on Performance, Blood Parameters, Carcass Traits, Meat Quality, and Gene Expressions of Lambs

Isabela J. dos Santos [1,2], Paulo C. G. Dias Junior [1,3], Tharcilla I. R. C. Alvarenga [4,*], Idalmo G. Pereira [5], Sarita B. Gallo [6], Flavio A. P. Alvarenga [4] and Iraides F. Furusho-Garcia [1]

[1] Department of Animal Science, Federal University of Lavras, Lavras 37200-000, MG, Brazil
[2] Faculty of Veterinary Medicine, University of Rio Doce Valley, Governador Valadares 35020-220, MG, Brazil
[3] Department of Veterinary Medicine, Faculty of Americana, Americana 13477-360, SP, Brazil
[4] Armidale Livestock Industries Centre, NSW Department of Primary Industries, Armidale, NSW 2351, Australia
[5] Department of Animal Science, Federal University of Minas Gerais, Belo Horizonte 31620-295, MG, Brazil
[6] Department of Nutrition and Animal Production, University of São Paulo, Pirassununga 13635-900, SP, Brazil
* Correspondence: tharcilla.alvarenga@dpi.nsw.gov.au

**Abstract:** The aim of this study was to investigate the effects of feeding systems on the growth performance of Santa Inês x Dorper lambs, meat quality, fatty acid profile, and gene expression. Thirty lambs at an initial body weight of 22.6 ± 2.59 kg were randomly assigned to one of three feed systems: a grazing system with 1.2% body weight concentrate supplementation (GS); a feedlot system with 28% forage and 72% concentrate (FFC); or feedlot with 85% whole corn grain and 15% pellets (FHG). The lambs were slaughtered after 60 days of experiment. Average daily gain, glucose, and insulin concentration were higher for lambs on FHC than lambs on a GS feeding system. The fatty acid profile in the meat of the lambs fed GS showed a higher proportion of c9t11-C18:2, C20:5, C22:5, and C22:6 compared with FFC and FHC ($p < 0.05$). Meat tenderness was lower for lambs under FFC treatment compared with GS and FHG. FHG treatment provides better performance and higher deposition of lipid content in meat compared with GS and FHG. The expression of the genes SCD-1, SREBP1-c, and EVOL6 was greater in lambs undergoing GS and FHC treatments compared with FFC. Results of this research showed a reduced performance of grazing lambs compared with the feedlot system; however, it enhanced the fatty acid profile with increased levels of polyunsaturated acids and reduced n6/n3 ratio.

**Keywords:** grain fed; grass fed; fatty acids; lipid metabolism; pasture; feedlot; whole grain

**Citation:** dos Santos, I.J.; Dias Junior, P.C.G.; Alvarenga, T.I.R.C.; Pereira, I.G.; Gallo, S.B.; Alvarenga, F.A.P.; Furusho-Garcia, I.F. Impact of Feeding Systems on Performance, Blood Parameters, Carcass Traits, Meat Quality, and Gene Expressions of Lambs. *Agriculture* **2024**, *14*, 957. https://doi.org/10.3390/agriculture14060957

Academic Editors: Petru Alexandru Vlaicu, Arabela Elena Untea and Mihaela Saracila

Received: 15 May 2024
Revised: 14 June 2024
Accepted: 14 June 2024
Published: 18 June 2024

## 1. Introduction

Nutritional strategies have been proposed to enhance the profitability and productivity of sheep systems. Pasture and feedlot are traditional management systems applied in livestock to promote greater growth rates and market-targeted fat deposition. The type of forage and grain has been shown to affect meat quality, mainly sensory characteristics, and the concentration of polyunsaturated fatty acids in the meat of lambs [1–3]. Pasture systems are richer in polyunsaturated fatty acids, which are beneficial for human health, while grain systems are known to enhance daily weight gain [4]. Overall, strategies combining feed management and nutrition aim to produce meat with better sensory acceptability and increased polyunsaturated fatty acids, which promote better human health [5].

Lamb meat from grass-fed systems contains a lower concentration of saturated fatty acids and a higher proportion of mono and polyunsaturated fatty acids, especially alphalinolenic (ALA), eicosapentaenoic (EPA), docosahexaenoic (DHA), conjugated linoleic acid (CLA), and lower n6/n3 ratio, which are beneficial to human health [2,5,6]. These fatty acids are well known to promote better nutritional value to meat, have anticarcinogenic properties, and reduce the incidence of coronary heart disease in humans [6,7].

Although meat from grass-fed lamb provides a better fatty acid profile for human consumption, from the point of view of productivity and feed efficiency, finishing lambs in feedlots is more efficient. Feedlot diets are rich in grains and high in energy; there are diets with forage and diets without forage based on whole corn grains and protein pellets [8]. These diets provide rapid growth and greater deposition of fat in the carcass [4,8].

Carcass fat deposition is a trait commercially important to the industry and target markets. Adipose tissue accumulation occurs when lipogenesis is higher than lipolysis, and these processes are controlled by some key hormones, transcription factors and enzyme-coding genes [9]. Fatty acids from meat can be controlled even more using specific factors, such as genes related to lipid metabolism [10].

Steroid-binding protein transcription factor (SREBP-1c) plays a central role in energy homeostasis, promoting lipogenesis and adipogenesis [11,12]. SREBP-1c activates the transcription of genes such as acetyl-CoA carboxylase (ACACA), fatty acid synthase (FASN), elongases (ELOVL), stearoyl-CoAdesaturase (SCD-1), and other genes that participate in bioenergetics, adipogenesis, lipolysis, and synthesis of new fatty acids and triglycerides [13–15]. Previous studies have reported gene expression related to fat metabolism in lambs [16,17]; however, there is a lack of information on gene expression due to the effects of different feeding systems.

The aim of this study was to evaluate growth performance, blood parameters, carcass traits, meat quality, and gene expression related to lipid metabolism in the *Longissimus lumborum* muscle of lambs under three different feeding systems: GS, grazing with supplement (*Cynodon* spp. cv Tifton—85 with supplement); FFC, feedlot with forage (72% concentrate; 28% *Cynodon* spp. cv Tifton—85 hay); or FHC, feedlot whole grain without forage (85% whole-grain corn; 15% pellets).

## 2. Materials and Methods

Experimental procedures used in this study were approved by the Ethics and Animal Welfare Committee of the Federal University of Lavras (UFLA), Brazil (protocol No. 063/16). Meat sensory analysis was approved by the Ethics Committee on Research with Human Beings of UFLA (process No. 2.984.593. CAAE No. 99920918.7.0000.5148), following the Resolution of the National Health Council 196/96 [18].

This study was carried out in the Sheep Farm Facility of the Department of Animal Science of the Federal University of Lavras, MG, Brazil (21°13′38″ S, wet subtropical mesothermal Cwa from dry winter). There was an average rainfall and temperature of 110.33 mm and 22.4 °C, respectively, during the experiment, from February to April 2017.

### 2.1. Animals and Feeding Systems

Thirty male lambs ($\frac{1}{2}$Santa Inês $\frac{1}{2}$Dorper) at an average body weight of 22.6 $\pm$ 2.59 kg (mean $\pm$ SD) and age of 70.0 $\pm$ 10 days (mean $\pm$ SD) were used. Previously, at 60 days of age, the lambs were weaned and were fed an initial diet (70% ground corn, 15% soybean meal, 5% vitamin mineral premix, and 10% cane molasses) and Tifton pasture. Prior to the experiment period, the lambs were prescribed 1 mL/10 kg of 5% levamisole hydrochloride (Ripercol® L, Zoetis Indústria de Produtos Veterinários Ltd.a., Campinas, SP, Brazil) for deworming. The experiment was a randomized design with three treatments and 10 animals in each treatment group. Data were collected for a period of 60 days. Experimental feeding systems consisted of GS grazing with supplementation ($n = 10$): *Cynodon* spp. cv Tifton—85 pasture plus 1.2% body weight concentrate supplementation; FFC conventional feedlot ($n = 10$): a diet containing 72% concentrate and 28% *Cynodon* spp. cv Tifton—85 hay; and FHG high grain feedlot ($n = 10$): a diet containing 85% whole-grain corn and 15% pellets (Premix Mineral Confipeso Alto Grão, Presence-Nutrição Animal, Paulínia, SP, Brazil). Data provided for FFC treatment have been used as a control treatment in Dias Junior et al. [17] in another context of analysis.

The ingredients, chemical composition and fatty acid profile of the diets are listed in Table 1. The diets for FFC and GS feeding systems were formulated to meet the nutritional requirements of lambs with 30 kg live weight and ADG of 300 g/day [19]. The supplement provided to the lambs on the GS feeding system was calculated for 1.2% body weight.

Lambs from FFC and FHC treatments were kept in individual 1.3 m$^2$ pens with free access to feed and water. The lambs were fed twice a day, at 8:00 a.m. and 4:00 p.m., ad libitum. The feeding offered, and the orts were recorded daily, calculated for a daily surplus of 15% ad libitum.

**Table 1.** Ingredients, chemical composition, and fatty acid profile (g/kg dry matter) of the experimental diets.

| Ingredients, g/kg of Dry Matter | Feeding System | | |
|---|---|---|---|
| | GS | FFC | FHG |
| Tifton—85 hay (*Cynodon* spp.) | - | 279 | - |
| Pasture Tifton 85 | 611 | - | - |
| Soybean meal | 334 | 400 | - |
| Ground corn | 41 | 297 | 85 |
| Mineral premix | 11 | 22 | - |
| Dicalcium phosphate | 2 | 2 | |
| Pellets | - | - | 15 |
| Chemical composition, g/kg of dry matter | | | |
| Dry matter | 562 | 926 | 893 |
| Crude protein | 216 | 217 | 130 |
| Ether extract | 14 | 21 | 39 |
| Neutral detergent fiber | 455 | 215 | 124 |
| Minerals | 60 | 50 | 46 |
| Non-fibrous carbohydrate | 291 | 392 | 660 |
| Metabolizable energy, Mcal/kg * | 26 | 27 | 25 |
| Fatty acids, g/kg of dry matter | | | |
| C14:0 | 11 | 18 | 1 |
| C16:0 | 50 | 29 | 22 |
| c9-C16:1 | 7 | 2 | 2 |
| C18:0 | 9 | 33 | 22 |
| c9-C18:1 | 101 | 261 | 320 |
| C18:2 n6 | 320 | 415 | 5 |
| C18:3 n3 | 213 | 2 | 2 |

GS: Grazing supplemented (*Cynodon* spp. cv Tifton—85 pasture ad libitum and supplement in the proportion of 1.2% body weight); FFC: Feedlot with forage (72% concentrate; 28% *Cynodon* spp. cv Tifton—85 hay); FHG: Feedlot whole grain without forage (85% whole-grain corn; 15% pellets). * Metabolizable energy estimated by Cannas et al. [20].

The GS lambs grazed from 7 a.m. to 6 p.m. and were kept overnight in individual 1.3 m$^2$ pens with free access to water. The supplement was individually provided during the overnight period at 1.2% body weight, with an average value of 0.365 kg of supplement throughout the experimental period. Daily in the morning, the orts of the supplements were collected, weighed, and sampled per animal to determine the individual animal intake. The GS lambs rotationally grazed a total area of 9000 m$^2$, subdivided into five paddocks of 1800 m$^2$ composed of Tifton-85 (*Cynodon* spp.). Each paddock contains a water trough and shed area with a polyethylene fabric of 20 m$^2$ in each paddock to provide 80% shade. The stocking rate was fixed, and the grazing cycle was 35 days, with 7 of occupation and 28 of rest. Pasture composition was, on average, 30.98 g/kg of dry matter, 12.57 g/kg of crude protein, 1.4 g/kg of ether extract, and 74.75 g/kg of neutral detergent fiber (NDF).

Pasture sampling was performed manually, simulating the grazing of the animals on the first, third, and seventh day of grazing of each paddock, always in the morning, until a sample of approximately 400 g of forage for each experimental group was obtained. The morphological characteristics at the time of entry and exit are shown in Table 2. Pasture samples were dried in a forced ventilation oven at 65 °C for 72 h for pre-drying. Then, the samples were ground using a 1 mm sieve for the bromatological analyses and 5 mm for the degradability test.

**Table 2.** Pasture characterization—mean of morphological constituents of pasture and height of the forage (*Cynodon* spp.) at the time of entry and exit of the animals to the paddock.

| Pasture Constituents (%) | Entry | Exit |
| --- | --- | --- |
| Leaf | 52.95 | 46.66 |
| Thatch | 23.53 | 26.86 |
| Senescent material | 11.76 | 13.03 |
| Other forages and weeds | 11.76 | 13.45 |
| Height, cm | 25.50 | 15.50 |

Pasture intake was evaluated from the 23rd to the 30th day and from the 53rd to the 60th day of the experiment using titanium dioxide ($TiO_2$) as an external indicator, according to Willians et al. [21], cited by Silva and Queiroz [22]. Pasture intake was evaluated for twelve days (seven days of adaptation followed by five days of feces and feeding sampling). In addition, 4 g of $TiO_2$ was provided via the esophagus, 2 g at 7 a.m., and 2 g at 6 p.m. Fecal samples were collected from each animal in the morning and afternoon, summing ten samples per animal during each evaluation period, which were composed per animal in each period. The composed were stored at $-18\ °C$, then dried at 65 °C, and ground using a 1 mm sieve mill for subsequent analysis. The concentration of $TiO_2$ in the feces was determined [23].

To determine indigestible neutral detergent fiber (NDFi), five bags of textile non-textile (TNT) per sample, containing 0.5 g forage and feces, were previously dried, weighed, and incubated for 264 h in the rumen of a cannulated cow [3,24]. Bags were removed, cleaned with water, dried at 65 °C and boiled for 1 h in a neutral detergent solution. Then, it was washed with hot water and acetone, dried and weighed [25]. The remaining residue was recorded as NDFi. The production of fecal dry matter was determined using the following formula: Fecal production = intake of the indicator (kg)/concentration of the indicator in feces (%), which allowed to obtain pasture intake as DMIpasture (kg/day) = (fecal production × NDFifeces)/NDFipasture.

The diets of the FFC and FHC treatments, the supplement of GS treatment, and the orts were sampled every day and combined every 15 days to obtain a fortnightly sample per experimental unit. The samples were stored at $-18\ °C$. Dry matter of feed, orts, and pasture was determined according to the method of the Association of Official Analytical Chemists [26]. All samples were ground with a Wiley mill (Marconi, Piracicaba, São Paulo, Brazil) to pass a 1 mm screen. Ash was obtained by incinerating the sample in a muffle furnace at 550 °C for 4 h [26], and ether extract (EE) was measured according to AOAC International [27]. Crude protein (CP) was determined using micro-Kjeldahl analysis [28]. The ash-free neutral detergent fiber was determined by Van Soest et al. [25]. Non-fibrous carbohydrate fraction determined as non-fibrous carbohydrate = 100 − (crude protein + ether extract + ash + neutral detergent fiber). Metabolizable energy intake was determined by Cannas et al. [20].

### 2.2. Performance and Digestibility

Body weight was recorded fortnightly after 16 h of fasting during the 60 days of the experimental period to determine average daily gain and feed efficiency. Feed efficiency was calculated as the average dry matter intake divided by the average daily gain, both in kg/day.

The nutrient intake and dry matter intake of the FFC and FHC treatments and the supplement intake of GS were measured daily (Dry matter intake = feed offered − orts). Pasture intake was measured using indigestible neutral detergent fiber (NDFi) as an internal marker [29].

The apparent digestibility coefficient was calculated for dry matter, crude protein, ether extract, neutral detergent fiber, total digestible nutrients, and non-fibrous carbohydrates according to the digestibility formula: (Feed nutrient − Fecal nutrient)/Feed nutrient. The total digestible nutrient content was determined according to the following equation: Total digestible nutrient = crude protein digestible + (ether extract digestible × 2.25) + non-fibrous carbohydrate digestible + neutral detergent fiber digestible [30].

### 2.3. Blood Biochemical Analysis

At 58 days of the experiment, blood samples were collected minutes before the first feeding of the day. Blood samples were collected by performing jugular venipuncture using 10 mL vacutainer tubes with sodium fluoride + EDTA. Immediately after collection, the samples were centrifuged (1500× $g$, room temperature for 10 min) and frozen and stored at $-20\,°C$ in 1.5 mL plastic tubes until laboratory analysis. The colorimetry technique was applied using a 96-well plate spectrophotometer reader (Multskan GO, Thermo Scientific, Waltham, MA, USA). Glucose (Bioclin Glucose Monoreagent, Belo Horizonte, Brazil), cholesterol (Bioclin Cholesterol Monoreagent, Belo Horizonte, Brazil), insulin (DRG ELISA D-35039, DRG Instruments, Marburg, Germany), and triglycerides (Bioclin Triglicerides Monoreagente, Belo Horizonte, Brazil) were quantified using commercial kits.

### 2.4. Slaughter and Carcass Sampling

At the end of the feeding trial (60 days), the animals were individually weighed in the morning, and pre-slaughter live weight was determined. The lambs were transported to a commercial abattoir located 128 km from the experimental facility. The lambs underwent 16 h of fasting for solids and ad libitum access to water. Muscle samples (5 g) were taken approximately seven minutes after bleeding, between the 12th and 13th ribs (right side), placed in cryogenic tubes, transported in liquid nitrogen, and stored at $-80\,°C$ for gene expression analysis. Hot carcass weight was recorded immediately after evisceration and skin removal. Non-carcass components and internal fat (mesenteric and visceral fat) were weighed using a scale with 0.100 kg accuracy (Filizola, Campo Grande, MS, Brazil).

pH was measured pre-rigor (right carcass side between the 12th and 13th ribs) using a pH meter TESTO-205 (Testo, Campinas, Brazil). The pH meter was calibrated at room temperature on pH 7.0 and 4.0 with standard buffers (Testo buffer, Campinas, Brazil).

Carcasses were hung at 4 °C for 24 h in the slaughterhouse. Carcasses were transferred to the university laboratory in a refrigerated truck. Cold carcass weight was recorded, and a second measure of pH was undertaken (post-rigor). Subcutaneous fat thickness was recorded by making two incisions through the fat along lines extending over the greatest depth of the muscle between the 12th and 13th ribs [31] using a digital caliper (Battery, model SR44) on the left side of the carcass. Acetate paper was placed on the LL muscle to record muscle area using ImageJ software, https://imagej.net/software/imagej/ (accessed on 13 June 2024) (National Institutes of Health, Bethesda, MD, USA). After deboning, the left and right sides of the LL muscle were wrapped in aluminum foil, vacuum-packed, and stored at $-20\,°C$ for further analysis.

### 2.5. Meat Quality

The frozen longissimus muscle (left side) was divided into six steaks of 2.5 cm thickness each, labeled, vacuum-packed, and held at $-20\,°C$. The steaks were allocated in each analysis in the anterior–posterior direction: qualitative analyses (three steaks), thiobarbituric acid reactive substances (TBARSs) (one steak), proximate composition (one steak), and fatty acid analysis (one steak). Thawing loss was determined using three steaks that were thawed for 12 h at 2 °C. Following this, color measurements were taken after opening the vacuum bag and 30 min of blooming. A Minolta CR700 Chroma Meter (Konica Minolta, Osaka, Japan) was used set on illuminant A and a 10° standard observer. The displays per steak were undertaken to record lightness ($L^*$), redness ($a^*$), and yellowness ($b^*$). Meat color parameters were averaged from nine readings per animal.

Water-holding capacity was determined using the Hamm and Deatherage [32] method and expressed as the ratio Ap/Ae [33].

Three 2.5 cm steaks were used to determine cooking loss. Meat samples were randomly placed in polyethylene bags and cooked in a water bath set at 98 ± 1 °C (aimed at 71 °C internal temperature using a digital thermometer [34]). Meat samples were weighed after 2 h of cooling at room temperature. Cooking loss was determined in percentage as the weight difference before and after cooking. For shear force analysis, each steak used for cooking loss analysis was subdivided into three subsamples (1 cm$^2$) cut in parallel to the direction of the muscle fibers (removing connective tissue and fat). A texturometer (Modle TA-TX2, Stable Micro Systems Ltd., Godalming, Surrey, UK) attached to a Warner–Bratzler slide was used and calibrated using a weight of 2 kg with an adjusted speed of 200 mm/min. The maximum positive peak values were recorded as kgf/cm$^2$.

LL muscle (100 g) was trimmed of external fat and connective tissue for proximate analysis. A multiprocessor (Philips RI7630, Itapevi, Brazil) was used to homogeneously ground the samples. Crude protein, fat, moisture, and ash were determined with a near-infrared method [35] using a FoodScanTM device (FOSS, Hillerod, Denmark).

### 2.6. Sensory Analysis

After thawing and trimming, longissimus muscles were weighed, and salt was added (1%) for sensory analysis. A grill (SFSE Croydon) was set at 250 °C, and the samples were cooked to an internal temperature of 70 °C using a digital thermometer. After cooking, the muscle was cut into cubes of 12–15 g [36]. A total of 55 panelists (25 men and 30 women ranging from 18 to 60 years old) evaluated the samples on a white plastic plate, coded with three random digits in an individual chamber. Panelist assessment was carried out in one day. Sensory evaluation was undertaken as described previously [37] using appearance, flavor, tenderness, and overall liking as sensory parameters on a hedonic scale of nine points [37].

### 2.7. Fatty Acids and Enzyme Activity

For fatty acid analysis, extraction [38], hydrolysis, and methylation [39] were performed as previously reported. A gas chromatograph fitted with a 100 m Supelco SP 2560 capillary column (0.25 mm and 0.2 µm film thickness) was used as described previously [17]. Carrier gas helium flow of 1.8 mL per minute, set at 45 mL per minute for make-up gas (N$_2$), 40 mL per minute for hydrogen, and 450 mL per minute for synthetic flame gas. The oven temperature was set at 70 °C for 4 min, increased to 170 °C at a rate of 13 °C for 1 min, subsequently increased to 250 °C at a rate of 35 °C for 1 min, and maintained at 250 °C for 5 min. The percentage of fatty acid in total lipids (fatty acids methyl ether—FAME) was obtained by individual area of fatty acid $\times$ 100/total area of fatty acid. Estimates of $\Delta$9-desaturase and elongase enzyme activity were determined as described by Malau-Aduli et al. [40] and Kelsey et al. [41]. The overall desaturase index was calculated as follows: $\Delta 9 - \text{overall} = \Delta 9 - 14 + \Delta 9 - 16 + \Delta 9 - 18$, as modified by Archibeque et al. [42].

### 2.8. Gene Expression Analysis

Reference and target primers design were carried out by the sequences registered and published in the public database of Genbank (NCBI platform—National Center for Biotechnology Information). Open Reading Frames (OFRs) of the selected sequences were obtained using the NCBI ORFinder tool and associated with the coded protein bank using the translate tool (ExPASY protein bank). Oligo PerfectTM Designer software (https://www.thermofisher.com/pl/en/home/life-science/oligonucleotides-primers-probes-genes/custom-dna-oligos/oligo-design-tools.html, accessed on 13 June 2024) was used considering the sequences obtained from Genbank. Primer's information is detailed in Dias Junior et al. [17].

Total RNA extraction from LL muscle is detailed described in Dias Junior et al. [17]. Relative gene expression was based on the corrected cycle threshold values for the amplification efficiency of each primer pair [43].

*2.9. Statistical Analysis*

Statistical analysis was performed using the GLM procedure of SAS (SAS Version 9.1, SAS Institute, Cary, NC, USA). Experimental diets were used as a fixed effect and the initial weight as a covariate. Only meat sensory analysis was considered a randomized block design, with each panel member representing one block (block effect was considered random). The MIXED procedure of SAS was used to verify the effect of the experimental diets (fixed effect) on acceptability, and the score of each panel member was considered as a repeated measure. The comparison between the mean of the different parameters evaluated was made using the $t$-test, with a difference considered significant when $p < 0.05$.

## 3. Results

*3.1. Nutrient Intake, Digestibility, and Performance*

The intake of dry matter, crude protein, ether extract, and non-fibrous carbohydrates was higher in FFC than FHG and GS ($p < 0.0001$), subsequently (Table 3). GS treatment showed the greatest neutral detergent fiber intake ($p < 0.0001$), and FFC and FHG showed superior total digestible nutrient intake compared with GS ($p < 0.0001$).

**Table 3.** Intake and digestibility of nutrients, average daily gain, and feed efficiency of lambs under three feeding systems.

| Parameter | Feeding System | | | SEM | $p$ Value |
|---|---|---|---|---|---|
| | GS | FFC | FHG | | |
| Intake, kg/day | | | | | |
| Dry matter | 0.76 [c] | 1.17 [a] | 0.90 [b] | 0.048 | 0.0001 |
| Crude protein | 0.15 [c] | 0.28 [a] | 0.22 [b] | 0.007 | 0.0001 |
| Neutral detergent fiber | 0.38 [a] | 0.23 [b] | 0.12 [c] | 0.007 | 0.0001 |
| Ether extract | 0.01 [c] | 0.02 [b] | 0.03 [a] | 0.001 | 0.0001 |
| Non-fibrous carbohydrate | 0.15 [c] | 0.44 [b] | 0.54 [a] | 0.023 | 0.0001 |
| Total digestible nutrient | 0.53 [b] | 0.80 [a] | 0.75 [a] | 0.032 | 0.0001 |
| Digestibility coefficient | | | | | |
| Dry matter | 0.63 [b] | 0.75 [a] | 0.77 [a] | 0.013 | 0.0478 |
| Crude protein | 0.73 [b] | 0.80 [a] | 0.79 [a] | 0.010 | 0.0001 |
| Neutral detergent fiber | 0.69 [a] | 0.46 [b] | 0.39 [c] | 0.017 | 0.0001 |
| Ether extract | 0.62 [b] | 0.70 [a] | 0.72 [a] | 0.014 | 0.0001 |
| Non-fibrous carbohydrate | 0.68 [c] | 0.77 [b] | 0.84 [a] | 0.013 | 0.0001 |
| Total digestible nutrient | 0.71 [b] | 0.82 [a] | 0.84 [a] | 0.012 | 0.0001 |
| Performance | | | | | |
| Average daily gain, kg/day | 0.182 [b] | 0.222 [b] | 0.311 [a] | 0.015 | 0.0001 |
| Feed efficiency | 0.342 [b] | 0.279 [b] | 0.426 [a] | 0.023 | 0.0006 |

GS: Grazing supplemented (*Cynodon* spp. cv Tifton—85 pasture ad libitum and supplement in the proportion of 1.2% body weight); FFC: Feedlot with forage (72% concentrate; 28% *Cynodon* spp. cv Tifton—85 hay); FHG: Feedlot whole grain without forage (85% whole-grain corn; 15% pellets). SEM: Standard error of the mean. Mean in the same row with different superscripts differ significantly ($p < 0.05$).

The apparent digestibility coefficients for dry matter, crude protein, ether extract, and total digestible nutrients were lower in GS compared with FFC and FHG ($p < 0.05$). GS showed the greatest digestibility of neutral detergent fiber and the lowest for non-fibrous carbohydrates ($p < 0.0001$).

Average daily gain and feed efficiency were greater for lambs raised under FHG compared with GS and FFC ($p < 0.0001$).

*3.2. Blood Biochemical Parameters*

The concentration of cholesterol and triglycerides was not affected by the feeding system ($p > 0.05$); however, glucose and insulin were lower in lambs raised under GS compared with FFC and FHG ($p < 0.001$) (Table 4).

**Table 4.** Blood biochemical parameters (mg/dL) of lambs under three feeding systems.

| Parameter | Feeding System | | | SEM | *p* Value |
|---|---|---|---|---|---|
| | **GS** | **FFC** | **FHG** | | |
| Cholesterol | 133.28 | 137.83 | 162.34 | 12.340 | 0.2202 |
| Glucose | 59.29 [c] | 73.18 [b] | 89.49 [a] | 3.641 | 0.0001 |
| Triglycerides | 71.46 | 65.99 | 66.95 | 4.611 | 0.6736 |
| Insulin | 55.33 [b] | 90.54 [a] | 104.82 [a] | 8.722 | 0.0019 |

GS: Grazing supplemented (*Cynodon* spp. cv Tifton—85 pasture ad libitum and supplement in the proportion of 1.2% body weight); FFC: Feedlot with forage (72% concentrate; 28% *Cynodon* spp. cv Tifton—85 hay); FHG: Feedlot whole grain without forage (85% whole-grain corn; 15% pellets). SEM: Standard error of the mean. Mean in the same row with different superscripts differ significantly (*p* < 0.05).

### 3.3. Carcass and Meat Quality Parameters

The pre-slaughter live weight of lambs raised under FFC was greater than FHG but lower than GS (*p* < 0.01) (Table 5). Lambs raised under FHG had lower weights of non-carcass components, but in proportion, they were greater (*p* < 0.01). Hot carcass weight was not affected by the feeding system (*p* > 0.05), but cold carcass weight was lower under GS compared with FFC and FHG (*p* < 0.05). Loin muscle area and subcutaneous fat thickness were inferior for GS lambs compared with FFC and FHG (*p* < 0.01). Internal fat (mesenteric + visceral fat) was greater in lambs raised under FFC compared with GS but lower than FHG (*p* = 0.0016). Carcass pH (hot and cold) was not affected by the feeding system (*p* > 0.05).

**Table 5.** Carcass characteristics of lambs under three feeding systems.

| Parameter | Feeding System | | | SEM | *p* Value |
|---|---|---|---|---|---|
| | **GS** | **FFC** | **FHG** | | |
| Pre-slaughter live weight, kg | 33.46 [c] | 38.53 [b] | 43.17 [a] | 0.860 | 0.0001 |
| Non-carcass components, kg | 15.40 [b] | 16.13 [ab] | 17.75 [a] | 0.480 | 0.0061 |
| Non-carcass components, % | 46.64 [a] | 41.97 [b] | 41.53 [b] | 1.219 | 0.011 |
| Hot carcass weight, kg | 15.79 [b] | 19.40 [a] | 20.81 [a] | 0.443 | 0.0001 |
| Hot carcass yield, % | 47.25 | 50.28 | 48.24 | 1.035 | 0.1308 |
| Cold carcass weight, kg | 15.35 [b] | 18.91 [a] | 20.30 [a] | 0.464 | 0.0001 |
| Cold carcass yield, % | 45.98 | 48.97 | 47.04 | 1.089 | 0.1703 |
| Loin muscle area, cm$^2$ | 13.75 [b] | 17.53 [a] | 18.01 [a] | 0.697 | 0.0004 |
| Subcutaneous fat thickness, mm | 1.51 [b] | 3.32 [a] | 3.34 [a] | 0.346 | 0.0068 |
| pH hot carcass | 6.94 | 6.75 | 6.78 | 0.085 | 0.2598 |
| pH cold carcass | 5.74 | 5.90 | 5.93 | 0.106 | 0.3948 |
| Internal fat | 0.73 [c] | 0.97 [b] | 1.38 [a] | 0.110 | 0.0016 |

GS: Grazing supplemented (*Cynodon* spp. cv Tifton—85 pasture ad libitum and supplement in the proportion of 1.2% body weight); FFC: Feedlot with forage (72% concentrate; 28% *Cynodon* spp. cv Tifton—85 hay); FHG: Feedlot whole grain without forage (85% whole-grain corn; 15% pellets). SEM: Standard error of the mean. Mean in the same row with different superscripts differ significantly (*p* < 0.05).

*L** was lower on the longissimus muscle of GS lambs compared with FFC and FHG (*p* = 0.0043), and no effect was observed on *a** and *b** values (*p* > 0.05) (Table 6). The feeding system did not affect thawing loss, cooking loss, shear force, and water-holding capacity (*p* > 0.05).

The proximate composition of the meat did not differ in crude protein and ash content (*p* > 0.05), whereas the lipid content in the meat was higher in FHC lambs compared with GS (*p* = 0.0269), and the moisture content was higher in FHG lambs compared with GS (*p* = 0.0464).

**Table 6.** Parameters of meat quality and chemical composition of lambs under three feeding systems.

| Parameter | Feeding System | | | SEM | *p* Value |
|---|---|---|---|---|---|
| | GS | FFC | FHG | | |
| *L** | 38.07 [b] | 41.28 [a] | 41.99 [a] | 0.802 | 0.0043 |
| *a** | 14.78 | 15.28 | 14.78 | 0.647 | 0.8188 |
| *b** | 10.68 | 9.83 | 10.45 | 0.617 | 0.6111 |
| Thawing loss, % | 5.37 | 4.39 | 5.09 | 0.343 | 0.1388 |
| Cooking loss, % | 14.19 | 12.74 | 14.18 | 1.388 | 0.7015 |
| Shear force, kgf/cm$^2$ | 3.44 | 3.59 | 3.27 | 0.182 | 0.469 |
| Water-holding capacity | 0.15 | 0.18 | 0.17 | 0.011 | 0.283 |
| Moisture, % | 74.57 [a] | 73.88 [ab] | 73.29 [b] | 0.343 | 0.0464 |
| Crude protein, % | 21.61 | 21.40 | 21.70 | 0.256 | 0.7057 |
| Fat, % | 2.19 [b] | 2.52 [ab] | 3.44 [a] | 0.316 | 0.0269 |
| Ash, % | 1.62 | 2.18 | 1.88 | 0.200 | 0.1581 |

GS: Grazing supplemented (*Cynodon* spp. cv Tifton—85 pasture ad libitum and supplement in the proportion of 1.2% body weight); FFC: Feedlot with forage (72% concentrate; 28% *Cynodon* spp. cv Tifton—85 hay); FHG: Feedlot whole grain without forage (85% whole-grain corn; 15% pellets). SEM: Standard error of the mean. Mean in the same row with different superscripts differ significantly ($p < 0.05$).

### 3.4. Sensory Analysis

Appearance and overall liking of the meat were not affected by the feeding system ($p > 0.05$) (Table 7). Consumers scored higher for the flavor of meat from lambs raised under GS compared with FFC and FHG ($p = 0.0023$), but scored meat tenderness lower in lambs raised under FFC compared with GS and FHG ($p = 0.0246$).

**Table 7.** Sensory analysis of muscle Longissimus lumborum muscle of lambs under three feeding systems.

| Parameter | Feeding System | | | SEM | *p* Value |
|---|---|---|---|---|---|
| | GS | FFC | FHG | | |
| Appearance | 5.87 | 6.02 | 6.20 | 0.231 | 0.6699 |
| Flavor | 7.20 [a] | 6.47 [b] | 6.18 [b] | 0.239 | 0.0023 |
| Tenderness | 7.05 [a] | 6.70 [b] | 7.46 [a] | 0.19 | 0.0246 |
| Overall liking | 6.73 | 6.18 | 6.67 | 0.229 | 0.1111 |

GS: Grazing supplemented (*Cynodon* spp. cv Tifton—85 pasture ad libitum and supplement in the proportion of 1.2% body weight); FFC: Feedlot with forage (72% concentrate; 28% *Cynodon* spp. cv Tifton—85 hay); FHG: Feedlot whole grain without forage (85% whole-grain corn; 15% pellets). SEM: Standard error of the mean. Mean in the same row with different superscripts differ significantly ($p < 0.05$).

### 3.5. Fatty Acids

Fatty acids were affected by feeding systems; the main and most beneficial enhancement was observed in the meat of lambs raised under GS. C16:0 and c9-C18:1 were inferior to GS compared with FFC and FHG ($p < 0.01$), and the sum of SFA tended to be inferior as well ($p = 0.0570$) (Table 8). The PUFAs C20:4 (arachidonic), C20:5 (EPA), C22:5 (DPA), and C22:6 (DPA), as well as CLA c9 t11-C18:2 (CLA) and the sum of n3 and n6 and the sum of PUFAs, were higher for GS compared with FFC and FHG ($p < 0.01$). The same pattern was observed in the elongase activity estimate ($p = 0.0131$). Lambs raised under FFC had greater levels of MUFA and Δ 9–18 compared with GS but lower levels than FHG ($p < 0.01$).

**Table 8.** Fatty acid composition (% of total fatty acid) of the Longissimus lumborum muscle of lambs under three feeding systems.

| Fatty Acid, % FAME | Feeding System | | | SEM | *p* Value |
|---|---|---|---|---|---|
| | GS | FFC | FHG | | |
| C10:0 | 0.15 | 0.14 | 0.14 | 0.020 | 0.9416 |
| C14:0 | 2.63 | 2.44 | 2.41 | 0.232 | 0.6418 |

**Table 8.** *Cont.*

| Fatty Acid, % FAME | Feeding System | | | SEM | *p* Value |
|---|---|---|---|---|---|
| | **GS** | **FFC** | **FHG** | | |
| C16:0 | 20.60 [b] | 23.88 [a] | 23.14 [a] | 0.547 | 0.0006 |
| C18:0 | 16.20 [a] | 13.92 [b] | 10.62 [c] | 0.549 | 0.0001 |
| c9-C16:1 n7 | 1.79 | 1.89 | 2.49 | 0.217 | 0.0676 |
| t11 C18:1 | 3.61 [b] | 2.47 [b] | 5.36 [a] | 0.356 | 0.0001 |
| c11-C18:1 | 1.67 | 1.77 | 1.92 | 0.138 | 0.468 |
| c9-c18:1 | 34.98 [b] | 41.47 [a] | 40.52 [a] | 0.836 | 0.0001 |
| C18:2 n6 | 0.05 | 0.04 | 0.04 | 0.005 | 0.8816 |
| C20:4 n6 | 3.00 [a] | 1.70 [b] | 1.12 [b] | 0.374 | 0.0046 |
| C18:3 n3 | 0.12 | 0.09 | 0.10 | 0.008 | 0.0735 |
| C20:5 (EPA) | 0.31 [a] | 0.09 [b] | 0.04 [b] | 0.037 | 0.0001 |
| C22:5 (DPA) | 0.65 [a] | 0.28 [b] | 0.14 [b] | 0.071 | 0.0001 |
| C22:6 (DHA) | 0.17 [a] | 0.07 [b] | 0.03 [b] | 0.021 | 0.0002 |
| c9 t11-C18:2 (CLA) | 0.82 [a] | 0.49 [b] | 0.33 [c] | 0.043 | 0.0001 |
| Σ n3 | 0.60 [a] | 0.25 [b] | 0.18 [b] | 0.056 | 0.0001 |
| Σ n6 | 3.279 [a] | 2.00 [b] | 1.26 [b] | 0.404 | 0.0051 |
| n6/n3 | 5.09 [b] | 7.59 [a] | 7.18 [a] | 0.606 | 0.0159 |
| Σ SFA | 42.82 | 43.03 | 39.82 | 1.001 | 0.0570 |
| Σ MUFA | 44.30 [c] | 49.33 [b] | 52.58 [a] | 0.878 | 0.0001 |
| Σ PUFA | 12.01 [a] | 7.25 [b] | 7.11 [b] | 1.127 | 0.0057 |
| Δ 9–14 | 6.80 | 4.33 | 4.70 | 1.137 | 0.2693 |
| Δ 9–16 | 7.94 | 7.25 | 9.68 | 0.818 | 0.1152 |
| Δ 9–18 | 68.33 [c] | 74.84 [b] | 79.25 [a] | 0.875 | 0.0001 |
| Δ 9 Overall | 48.33 [b] | 51.91 [b] | 54.37 [a] | 0.776 | 0.0001 |
| Elongase | 69.60 [a] | 68.25 [ab] | 66.57 [b] | 0.669 | 0.0131 |

GS: Grazing supplemented (*Cynodon* spp. cv Tifton—85 pasture ad libitum and supplement in the proportion of 1.2% body weight); FFC: Feedlot with forage (72% concentrate; 28% *Cynodon* spp. cv Tifton—85 hay); FHG: Feedlot whole grain without forage (85% whole-grain corn; 15% pellets). SEM: Standard error of the mean. Σ n3: sum omega 3 fatty acids. Σ n6: sum omega 6 fatty acids. n6/n3: omega6/omega3. Σ SFA: sum saturated fatty acids. Σ MUFA: sum monounsaturated fatty acids. Σ PUFA: sum of polyunsaturated fatty acids. Mean in the same row with different superscripts differ significantly ($p < 0.05$).

### 3.6. Gene Expression

The expression of the SCD-1 gene was lower in lambs under FFC treatment compared with GS and FHC ($p = 0.0375$) (Figure 1). The feeding system did not affect the expression of the PPARα gene ($p > 0.05$). The SREBP-1c gene was more expressed in GS compared with FFC but less than FHC ($p < 0.0001$). ELOV6 gene expression was higher in lambs under GS compared with FFC and FHC ($p = 0.0114$).

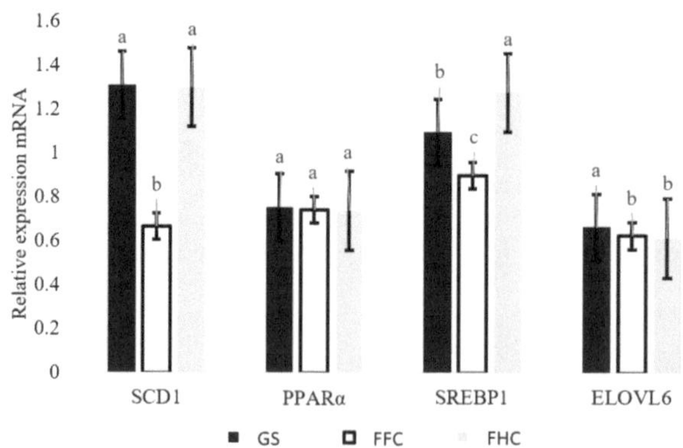


**Figure 1.** Relative expression of the Sterol regulatory element-binding proteins (SREBP-1), peroxisome proliferator-activated receptor alpha (PPAR-α), Stearoyl-CoA desaturase (SCD1) and Elongase 6 (ELOVL6)

genes in the Longissimus lumborum muscle of lambs under three feeding systems. GS: Grazing supplemented (*Cynodon* spp. cv Tifton—85 pasture ad libitum and supplement in the proportion of 1.2% body weight); FFC: Feedlot with forage (72% concentrate; 28% *Cynodon* spp. cv Tifton—85 hay); FHG: Feedlot whole grain without forage (85% whole-grain corn; 15% pellets). SEM: Standard error of the mean. Mean in the same row with different superscripts differ significantly ($p < 0.05$).

## 4. Discussion

Dry matter intake was lower in lambs under GS, as ruminants often reduce dry matter intake when dietary energy density decreases [44,45]. Also, the neutral detergent fiber in the GS diet contributes to the reduction in feed intake [46,47]. Lambs under FHC showed intermediary dry matter intake, which is associated with physiological, physical, or psychogenic mechanisms [48]. The physiological mechanism could be observed when high-energy diets, such as FHC, were provided in this study. Consequently, lambs from the FHC treatment had the highest intake of non-fibrous carbohydrates and total digestible nutrients, which are inherent to the diet.

A lower digestibility of dry matter for GS treatment was expected due to its higher fiber content. De Paula Carlis et al. [47] observed similar results when studying increased levels of neutral detergent fiber in feedlot lambs. This result is due to the replacement of corn with hay, whereas corn has greater digestion potential compared with hay [19]. The total digestible nutrients of corn (820 g/kg) are 1.4 times higher than the total digestible nutrients of coast cross hay (500 g/kg) [49,50], and it explains the higher digestibility of total digestible nutrients for lambs raised under FFC and FHC.

Neutral detergent fiber digestibility was higher in lambs raised under GS due to a higher proportion of this nutrient in the pasture and its higher intake. Pereira et al. [51] observed a similar result in lambs under feedlot-fed diets differing in fiber levels. The intake of neutral detergent fiber from forage is probably favored for maintaining rumen pH in the optimal range for the growth of fibrolytic bacteria. The favorable environment may have stimulated the increase in the population of these bacteria, which resulted in the greater digestibility of neutral detergent fiber [52].

The higher intake and digestibility of non-fibrous carbohydrates and total digestible nutrients in the FHC treatment may explain a greater weight gain in those lambs. As suggested by De Paula Carlis et al. [47], these are conditions that promote a greater production of short-chain fatty acids in the rumen and, consequently, a better performance. An increase in short-chain fatty acids reflects an increase in the amount of metabolizable energy for weight gain [53], which culminated in a greater average daily gain for lambs raised under FHC. Feed intake is associated with a higher weight gain and feed efficiency improvement to FHC treatment. Similar results of increased feed efficiency for sheep consuming high-grain diets are reported in the literature [8,54].

The highest glucose concentration in lambs raised under FHC and the lowest under GS is associated with the type of fermented carbohydrate in the diet. Diets with higher proportions of non-fibrous carbohydrates, especially starch, provide greater production and absorption of propionate, which is the main precursor of gluconeogenesis in ruminants [55] and may lead to an increased concentration of glucose in the plasma. [56] reported that an increase in glucose concentration is directly related to an increase in propionate production. In addition, the results of insulin concentration can be associated with the results of glucose concentration. Insulin is an anabolic hormone essential in maintaining glucose homeostasis. It is secreted by pancreatic islet β-cells in response to increased concentrations of circulating glucose and amino acids [57]. The higher glucose input from gluconeogenesis in lambs raised under FHC and FFC stimulated the production and release of insulin to maintain glucose homeostasis. A higher concentration of insulin and glucose in ruminants is associated with a higher fat deposit, as insulin stimulates the uptake and use of glucose by cells [58]. In ruminants, blood is not a site of synthesis or storage of excess energy, as these processes occur in the adipose tissues.

The lower hot and cold carcass weight in lambs raised under GS is explained by a lower intake and digestibility of dry matter, resulting in lower average daily gain and live weight in GS lambs. Lambs raised under FFC showed lower live weight compared with FHC, but their hot and cold carcass weight were similar. This is associated with a higher amount of internal fat in FHC lambs.

Subcutaneous fat thickness was lower for lambs raised under GS, which is related to a lower insulin concentration, an anabolic hormone that stimulates lipogenesis [58]. Lambs raised under FHC and FFC showed higher concentrations of insulin, which stimulated the lipogenesis process, reflecting higher saturated fatty acids. Loin muscle area was superior in lambs raised under FHC and FFC, which is related to carcass weight, whereas loin muscle area reflects the degree of muscularity of the carcasses [59]

$L^*$ was higher in meat from lambs raised under FHC and FFC, which is associated with a greater deposition of intramuscular fat. According to Realini et al. [60], fat is the chemical component of meat that has the highest luminosity. These effects on luminosity were also reported by Brito et al. [61] and Holman et al. [62].

Water loss from thawing or cooking is directly related to the water-holding capacity, which may influence shear force values. In the present study, water-holding capacity and shear force were similar between feeding systems and had characteristic values of meat considered tender by consumers. Protein and mineral contents in the meat were similar in the feeding systems. However, the fat proportion of the meat was higher in lambs raised under FHC. These results are related to the increased intake and digestibility of total digestible nutrients and a higher concentration of glucose and insulin, resulting in greater lipogenesis and hypertrophy of adipocytes. Moisture content was negatively correlated with the fat results, and it agrees with data published by D'Alessandro et al. [63], as the lipid meat content is inversely associated with the moisture content.

Meat quality is an important factor for product acceptance. It is closely related to several aspects, including sensory characteristics. Meat from lambs raised under GS showed better flavor than the other treatments. This result is related to a lower fat content in the meat of these animals. Bravo-Lamas et al. [64] found that meat from animals consuming high-grain diets had a lower intensity of species-specific flavor due to a lower proportion of C18:3 n3, which is associated with a strong flavor of sheep meat. However, the contribution of C18:3 n3 in the amount of total fatty acids is low, and in the present study, there was no greater deposition of C18:3 in the meat of lambs raised under GS.

Meat texture is one of the main attributes of meat acceptance [65]. This attribute showed a significant difference between feeding systems, and more tender meat was observed in lambs under FFC. This result agrees with Muela et al. [66] and Sañudo et al. [1], who reported that grazing lambs tended to have tougher meat than lambs finished in confinement. In the present study, lamb meat from the GS e FHC treatments showed a similar texture. This may be associated with the age at which the animals were slaughtered since the age of slaughter has a strong impact on the quality of lamb meat [67].

Overall liking was not affected by the feeding system, and sensory analysis indicated a moderate to high acceptance of the lamb meat of all three groups. Commercially, moderate acceptance demonstrates consumption and recommendation of the product [68].

The highest concentration of C18:0 in lambs raised under GS is explained by the greater extent of biohydrogenation in the rumen and the greater relative expression of the ELOVL6 gene in the muscle LL of GS lambs (Figure 1). A longer feed retention time in the rumen allows complete biohydrogenation to form the final product C18:0 [69], favoring its deposition in the meat. The ELOVL6 is the main enzyme responsible for the elongation process of fatty acids in ruminants [70], and it showed higher expression in the LL of GS lambs, which is consistent with the higher proportion of C18:0. A higher C18:0 content accompanied by the reduction in the C16:0 present in muscle of animals from GS is beneficial; C16:0 is associated with cardiovascular diseases, while C18:0 may have a neutral or protective effect against cardiovascular diseases [71].

The highest concentration of c9-C18:1 in FHC agrees with the expression of the SCD-1 gene. The SCD-1 has desaturation activity between carbons 9 and 10, converting C18:0 into c9-C18:1 [72]. According to Campbell et al. [73] and Smith et al. [74], more energetic diets stimulate lipogenesis, promoting a greater activity of the enzyme stearoyl-CoA desaturase (SCD-1). The gene expression results corroborate with a higher desaturase activity on the C18:0 fatty acid. These results are also associated with the desaturation index ($\Delta$9 total), and the FHC treatment showed the highest value of this index.

Although FHC showed higher amounts of t11-C18:1, this did not result in a higher proportion of c9t11-C18:2 (conjugated linoleic acid, CLA) as expected since t11-C18:1 is a precursor of c9t11-C18:2. During endogenous synthesis, the desaturation of t11-C18:1 occurs through the action of the enzyme $\Delta$9 desaturase, converting it into c9t11-C18:2. On the other hand, a higher proportion of c9t11-C18:2 was observed under GS due to a higher proportion of C18:3 n3 and C18:2 n6 in lambs raised under GS. The increase in c9t11-C18:2 is associated with positive effects on human health due to anticarcinogenic, antioxidant, antidiabetic, and immunostimulatory actions [75], and it was associated with an increase in C18:0 and a reduction in C16:0 in the LL of GS lambs, increasing the quality of the final product and making it healthier for human consumption.

There was no effect of the feeding system on PUFA C18:3 n3 or C18:2 n6, though the n6/n3 ratio was the lowest in GS lambs. Even so, at 5.093, it was still higher than the recommended 4:1 ratio for promoting a human health benefit while being far lower than the 10:1 ratio that causes damage to health (n6/n3) [76].

C20:4 (arachidonic acid) is an intermediate in the metabolism of C18:2 n6 and C18:3 n3. Lambs raised under GS showed a higher sum of these precursors of C20:4, which resulted in a greater deposition in the meat (Table 8).

Fatty acids of the n3 series, C20:5 n3 (eicosapentaenoic acid, EPA), C22:5 n3 (docosapentaenoic acid, DPA), and C22:6 n3 (docosahexanoic, DHA), which have anticarcinogenic and anti-inflammatory actions and act in the development and protection of the nervous system [77], showed the highest values in lambs raised under GS. Lambs managed in the pasture have higher proportions of the fatty acids EPA, DHA, and DPA due to the action of elongases and desaturases, which act in the biosynthesis of long-chain fatty acids from C18:3 n3 present in pastures. This result is associated with the gene expression of ELOVL6 (Figure 1), which was higher in lambs raised under GS.

The expression of PPAR$\alpha$ transcription factor mRNA was similar among feeding systems. The activation of the PPAR$\alpha$ starts in response to the need for energy, resulting in the catabolism of fatty acids [9,78]. In the present study, the absence of differences between the different feeding systems indicates that none of the treatments required the mobilization of fat for energy, corroborating the results found for triglycerides.

A higher SCD-1 gene expression in lambs raised under GS and FHC corroborates the higher proportion of c9-C18:1 under FHC and c9-t11-C18:2 under GS. SCD-1 gene expression is related to the activity of the stearoyl-CoA desaturase enzyme; this enzyme's main products are c9-C18:1 and c9-t11-C18:2 [72], corroborating the results for these fatty acids.

The results obtained for SREBP-1c and SCD-1 for lambs raised under FHC were associated with a higher insulin concentration, thereby stimulating the SREBP-1c transcription factor. According to Ricoult et al. [12], the greatest stimulus to produce the transcription factor SREBP-1c is the concentration of insulin in the bloodstream. The intermediate values for SREBP-1c transcription factors in lambs raised under GS may be related to a higher proportion of acetate, the main short-chain fatty acids produced from the fermentation of neutral detergent fiber [25], which was present in high concentrations in GS treatment. After absorption, acetate is distributed to peripheral tissues, where it is converted into acetyl-CoA through the action of the enzyme acetyl-CoA synthetase, which is encoded by SREBP-1c [79].

Lambs raised under GS showed a higher expression of the ELOVL6 gene, which acts in the process of the elongation of fatty acid chains [80]. The ELOVL6 gene encodes the enzyme that catalyzes the elongation of palmitic acid (C16:0) to stearic acid (C18:0) [81]. In the present study, C18:0 was higher in lambs raised under GS, corroborating the higher ELOVL6 expression and higher elongase activity index.

## 5. Conclusions

The results indicated no major differences in carcass characteristics and parameters of meat quality between lambs raised under feedlot and high-grain systems; however, the high-grain system showed superiority in feed efficiency and glucose concentration. Nonetheless, the results of this study confirm that grazing lambs under supplementation increases the expression of the gene ELOVL6 and enhances meat flavor and the concentration of PUFA, which are important to human health. Differences between the feeding systems can be used by producers to determine target markets. More research is needed to overcome the challenges and limitations of experimental grazing research, e.g., replication groups and a larger number of animals, and assess the economic profitability among the different systems.

**Author Contributions:** Conceptualization, I.J.d.S. and I.F.F.-G.; methodology, I.J.d.S. and I.F.F.-G.; validation, I.J.d.S. and I.F.F.-G.; formal analysis, I.J.d.S., I.G.P. and I.F.F.-G.; investigation, I.J.d.S., P.C.G.D.J. and I.F.F.-G.; resources, I.F.F.-G.; data curation, I.J.d.S.; writing—original draft preparation, I.J.d.S.; writing—review and editing, I.J.d.S., P.C.G.D.J., T.I.R.C.A., I.G.P., F.A.P.A. and I.F.F.-G.; visualization, S.B.G.; supervision, I.F.F.-G.; project administration, I.F.F.-G.; funding acquisition, I.F.F.-G. All authors have read and agreed to the published version of the manuscript.

**Funding:** This research was funded by FAPEMIG-Brazil (Minas Gerais State Research Support Foundation) for student scholarship and CNPq-Brazil (National Council for Scientific and Technological Development, grant number 459789/2014-7) for project financial support.

**Institutional Review Board Statement:** The animal study protocol was approved by the Animal Care and Use Committee from the Federal University of Lavras (protocol code 063/16). The sensory study protocol was approved by the Ethics Committee in Research with Human Beings of UFLA (protocol code 2.984.593. CAAE No. 99920918.7.0000.5148).

**Data Availability Statement:** Data will be made available at a reasonable request to the last author, I.F.F.G.

**Acknowledgments:** Thanks to a research team from GAO/UFLA-Brazil (Grupo de Apoio à Ovinocaprinocultura, Federal University of Lavras) for their support in conducting the field and laboratory research.

**Conflicts of Interest:** The authors declare no conflicts of interest.

## References

1. Sañudo, C.; Muela, E.; del Mar Campo, M. Key factors involved in lamb quality from farm to fork in Europe. *J. Integr. Agric.* **2013**, *12*, 1919–1930. [CrossRef]
2. Alvarenga, T.I.R.C.; Chen, Y.; Furusho-Garcia, I.F.; Perez, J.R.O.; Hopkins, D.L. Manipulation of Omega-3 PUFAs in lamb: Phenotypic and genotypic views. *Compr. Rev. Food Sci. Food Saf.* **2015**, *14*, 189–204. [CrossRef] [PubMed]
3. Alvarenga, F.A.P.; Furusho-Garcia, I.F.; Alvarenga, T.I.R.C.; Dias Junior, P.C.G.; Alves, F.A.N.; dos Santos, E.P.; Casagrande, D.R.; Teofilo, T.S.; Sales, L.A.; Almeida, A.K.; et al. Performance, fecal egg count and feeding behavior of lambs grazing elephant grass (*Pennisetum purpureum* Schum.) with increased levels of protein supplementation. *Small Rumin. Res.* **2022**, *216*, 106826. [CrossRef]
4. Ates, S.; Keles, G.; Demirci, U.; Dogan, S.; Kirbas, M.; Filley, S.J.; Parker, N.B. The effects of feeding system and breed on the performance and meat quality of weaned lambs. *Small Rumin. Res.* **2020**, *192*, 106225. [CrossRef]
5. Margetín, M.; Oravcová, M.; Margetínová, J.; Kubinec, R. Fatty acids in intramuscular fat of Ile de France lambs in two different production systems. *Arch. Anim. Breed.* **2018**, *61*, 395–403. [CrossRef] [PubMed]
6. Chikwanha, O.C.; Vahmani, P.; Muchenje, V.; Dugan, M.E.; Mapiye, C. Nutritional enhancement of sheep meat fatty acid profile for human health and wellbeing. *Food Res. Int.* **2018**, *104*, 25–38. [CrossRef] [PubMed]
7. Gebauer, S.K.; Chardigny, J.M.; Jakobsen, M.U.; Lamarche, B.; Lock, A.L.; Proctor, S.D.; Baer, D.J. Effects of ruminant trans fatty acids on cardiovascular disease and cancer: A comprehensive review of epidemiological, clinical, and mechanistic studies. *Adv. Nutr.* **2011**, *2*, 332–354. [CrossRef]

8. Gallo, S.B.; de Almeida Merlin, F.; de Macedo, C.M.; de Oliveira Silveira, R.D. Whole grain diet for Feedlot Lambs. *Small Rumin. Res.* **2014**, *120*, 185–188. [CrossRef]

9. Ladeira, M.M.; Schoonmaker, J.P.; Swanson, K.C.; Duckett, S.K.; Gionbelli, M.P.; Rodrigues, L.M.; Teixeira, P.D. Review: Nutrigenomics of marbling and fatty acid profile in ruminant meat. *Animal* **2018**, *12*, S282–S294. [CrossRef]

10. Hiller, B.; Herdmann, A.; Nuernberg, K. Dietary n-3 fatty acids significantly suppress lipogenesis in bovine muscle and adipose tissue: A functional genomics approach. *Lipids* **2011**, *46*, 557–567. [CrossRef] [PubMed]

11. Goldstein, J.L.; DeBose-Boyd, R.A.; Brown, M.S. Protein sensors for membrane sterols. *Cell* **2006**, *124*, 35–46. [CrossRef] [PubMed]

12. Ricoult, S.J.H.; Dibble, C.C.; Asara, J.M.; Manning, B.D. Sterol regulatory element binding protein regulates the expression and metabolic functions of wild-type and oncogenic IDH. *Mol. Cell. Biol.* **2016**, *36*, 2384–2395. [CrossRef] [PubMed]

13. Mead, J.R.; Irvine, S.A.; Ramji, D.P. Lipoprotein lipase: Structure, function, regulation, and role in disease. *J. Mol. Med.* **2002**, *80*, 753–769. [CrossRef] [PubMed]

14. Jakobsson, A.; Westerberg, R.; Jacobsson, A. Fatty acid elongases in mammals: Their regulation and roles in metabolism. *Prog. Lipid Res.* **2006**, *45*, 237–249. [CrossRef] [PubMed]

15. Matsumoto, C.; Hanson, N.Q.; Tsai, M.Y.; Glynn, R.J.; Gaziano, J.M.; Djoussé, L. Plasma phospholipid saturated fatty acids and heart failure risk in the physicians' health study. *Clin. Nutr.* **2013**, *32*, 819–823. [CrossRef] [PubMed]

16. Alvarenga, T.I.R.C.; Chen, Y.; Lewandowski, P.; Ponnampalam, E.N.; Sadiq, S.; Clayton, E.H.; van de Ven, R.J.; Perez, J.R.O.; Hopkins, D.L. The expression of genes encoding enzymes regulating fat metabolism is affected by maternal nutrition when lambs are fed algae high in omega-3. *Livest. Sci.* **2016**, *187*, 53–60. [CrossRef]

17. Dias Junior, P.G.; Santos, I.J.; do Nascimento, F.L.; Paternina, E.A.S.; Alves, B.A.; Pereira, I.G.; Ramos, A.L.S.; Alvarenga, T.I.R.C.; Furusho-Garcia, I.F. Macadamia oil and vitamin E for lambs: Performance, blood parameters, meat quality, fatty acid profile and gene expression. *Anim. Feed Sci. Technol.* **2022**, *293*, 115475. [CrossRef]

18. National Health Council. Resolution No. 196/96. 1996. Available online: http://bvsms.saude.gov.br/bvs/saudelegis/cns/1996 /res0196_10_10_1996.html (accessed on 20 November 2020).

19. NRC. *Nutrient Requirements of Dairy Cattle*; National Research Council, 519; National Research Council: Washington, DC, USA, 2001.

20. Cannas, A.; Tedeschi, L.O.; Fox, D.G.; Pell, A.N.; Van Soest, P.J. A mechanistic model for predicting the nutrient requirements and feed biological values for sheep. *J. Anim. Sci.* **2004**, *82*, 149–169. [CrossRef]

21. Williams, C.H.; David, D.J.; Iismaa, O. The determination of chromic oxide in faeces samples by atomic absorption spectrophotometry. *J. Agric. Sci.* **1962**, *59*, 381–385. [CrossRef]

22. Silva, D.J.; Queiroz, A.C. *Analise de Alimentos: Metodos Químicos e Biologicos*; Universidade Federal de Viçosa: Viçosa, Brazil, 2002; 235p.

23. Myers, W.D.; Ludden, P.A.; Nayigihugu, V.; Hess, B.W. Technical Note: A procedure for the preparation and quantitative analysis of samples for titanium dioxide 1. *J. Anim. Sci.* **2004**, *82*, 179–183. [CrossRef]

24. Casali, A.O.; Detmann, E.; Valadares Filho, S.C.; Pereira, J.C.; Henriques, L.T.; Freitas, S.G.; Paulino, M.F. Influencia do tempo de incubaçao e do tamanho de partículas sobre os teores de compostos indigestíveis em alimentos e fezes bovinas obtidos por procedimentos in situ. *R. Bras. Zootec.* **2008**, *37*, 335–342. [CrossRef]

25. Van Soest, P.J.; Robertson, J.B.; Lewis, B.A. Methods for Dietary Fiber, Neutral Detergent Fiber, and Nonstarch Polysaccharides in Relation to Animal Nutrition. *J. Dairy Sci.* **1991**, *74*, 3583–3597. [CrossRef] [PubMed]

26. AOAC. *Association of Official Analytical Chemists*, 15th ed.; Official Methods of Analysis; AOAC: Arlington, VA, USA, 1990.

27. AOAC. *Association of Official Analytical Chemists*, 17th ed.; Official Methods of Analysis; AOAC: Arlington, VA, USA, 2000.

28. AOAC. *Association of Official Analytical Chemists*, 19th ed.; Official Methods of Analysis; AOAC: Gaithersburg, MD, USA, 2012.

29. Valente, T.N.P.; Detmann, E.; Queiroz, A.C.D.; Valadares Filho, S.D.C.; Gomes, D.I.; Figueiras, J.F. Evaluation of ruminal degradation profiles of forages using bags made from different textiles. *Rev. Bras. Zootec.* **2011**, *40*, 2565–2573. [CrossRef]

30. Weiss, W.P.; Conrad, H.R.; St. Pierre, N.R. A theoretically based model for predicting total digestible nutrient values of forages and concentrates. *Anim. Feed Sci. Technol.* **1992**, *39*, 95–110. [CrossRef]

31. Fisher, A.V.; Boer, H. The EAAP standard method of sheep carcass assessment. Carcass measurements and dissection procedures report of the EAAP working group on carcass evaluation, in cooperation with the CIHEAM instituto agronomico Mediterraneo of Zaragoza and the CEC directora. *Livest. Prod. Sci.* **1994**, *38*, 149–159. [CrossRef]

32. Hamm, R.; Deatherage, F.E. Changes in hydration, solubility and charges of muscle 662 proteins during heating of meat. *J. Food Sci.* **1960**, *25*, 587–610. [CrossRef]

33. Sales, L.A.; Rodrigues, L.M.; Silva, D.R.G.; Fontes, P.R.; Torres Filho, R.A.; Ramos, A.L.S.; Ramos, E.M. Effect of freezing/irradiation/thawing processes and subsequent aging on tenderness, color, and oxidative properties of beef. *Meat Sci.* **2020**, *163*, 108078. [CrossRef] [PubMed]

34. Fabre, R.; Dalzotto, G.; Perlo, F.; Bonato, P.; Teira, G.; Tisocco, O. Cooking method effect on Warner-Bratzlershear force of different beef muscles. *Meat Sci.* **2018**, *138*, 10–14. [CrossRef] [PubMed]

35. AOAC. *Association of Official Analytical Chemists*, 18th ed.; Official Methods of Analysis; AOAC: Gaithersburg, MD, USA, 2007.

36. Abreu, K.S.F.; Vérasa, A.S.C.; Ferreira, M.A.; Madruga, M.S.; Maciel, M.I.S.; Félix, S.C.R.; Urbano, S.A. Quality of meat from sheep fed diets containing spineless cactus (*Nopalea cochenillifera* Salm Dyck). *Meat Sci.* **2018**, *148*, 229–235. [CrossRef]

37. Massingue, A.A.; Torres Filho, R.A.; Fontes, P.R.; Ramos, A.L.S.; Fontes, E.A.F.; Perez, J.R.O.; Ramos, E.M. Effect of mechanically deboned poultry meat content on technological properties and sensory characteristics of lamb and mutton sausages. *Asian-Australas. J. Anim. Sci.* **2018**, *31*, 576–584. [CrossRef] [PubMed]
38. Hara, A.; Radin, N.S. Lipid extraction of tissues with a low-toxicity solvent. *Anal. Biochem.* **1978**, *90*, 420–426. [CrossRef] [PubMed]
39. Christie, W.W. *Lipid Analysis*, 2nd ed.; Pergamon Press: Elmsford, NY, USA; University of Wisconsin: Madison, WI, USA, 1982; 207p.
40. Malau-Aduli, A.E.O.; Siebert, B.D.; Bottema, C.D.K.; Pitchford, W.S. A comparison of the fatty acid composition of triacylglycerols in adipose tissue from Limousin and Jersey cattle. *Aust. J. Agric. Res.* **1997**, *48*, 715–722. [CrossRef]
41. Kelsey, J.A.; Corl, B.A.; Collier, R.J.; Bauman, D.E. The effect of breed, parity, and stage of lactation on conjugated linoleic acid (CLA) in milk fat from dairy cows. *J. Dairy Sci.* **2003**, *86*, 2588–2597. [CrossRef] [PubMed]
42. Archibeque, S.L.; Lunt, D.K.; Gilbert, C.D.; Tume, R.K.; Smith, S.B. Fatty acid indices of stearoyl-CoA desaturase do not reflect actual stearoyl-CoA desaturase enzyme activities in adipose tissues of beef steers finished with corn-, flaxseed-, or sorghum-based diets. *J. Anim. Sci.* **2005**, *83*, 1153–1166. [CrossRef] [PubMed]
43. Pfaffl, M.W. A new mathematical model for relative quantification in real-time RT-PCR. *Nucleic Acids Res.* **2001**, *29*, 2002–2007. [CrossRef] [PubMed]
44. Galyean, M.L.; Defoor, P.J. Effects of roughage source and level on intake by feedlot cattle. *J. Anim. Sci.* **2003**, *81*, E8–E16.
45. Goulart, R.S.; Vieira, R.A.; Daniel, J.L.; Amaral, R.C.; Santos, V.P.; Toledo Filho, G.; Cabezas-Garcia, E.H.; Tedeschi, L.O.; Nussio, L.G. Effects of source and concentration of neutral detergent fiber from roughage in beef cattle diets on feed intake, ingestive behavior, and ruminal kinetics. *J. Anim. Sci.* **2020**, *98*, skaa107. [CrossRef]
46. Riaz, M.Q.; Südekum, K.H.; Clauss, M.A. Voluntary feed intake and digestibility of four domestic ruminant species as influenced by dietary constituents: A meta-analysis. *Livest. Sci.* **2014**, *162*, 76–85. [CrossRef]
47. De Paula Carlis, M.S.; Sturion, T.U.; da Silva, A.L.A.; Eckermann, N.R.; Polizel, D.M.; de Assis, R.G.; Souza, T.T.; Dias Junior, P.C.G.; Vicente, A.C.S.; Santos, I.J.; et al. Whole corn grain-based diet and levels of physically effective neutral detergent fiber from forage (pefNDF) for feedlot lambs: Digestibility, ruminal fermentation, nitrogen balance and ruminal pH. *Small Rumin. Res.* **2021**, *205*, 106567. [CrossRef]
48. Van Soest, P.J. Symposium on factors influencing the voluntary intake of herbage by ruminants: Voluntary intake in relation to chemical composition and digestibility. *J. Anim. Sci.* **1965**, *24*, 834–843. [CrossRef]
49. Ítavo, L.C.V.; Valadares Filho, S.C.; Silva, F.F.; Valadares, R.F.D.; Cecon, P.R.; Itavo, C.C.B.; Moraes, E.H.B.K.; Paulino, P.V.R. Nutritional value of Cynodon grass hay. Intake, degradability and apparent digestibility by means of internal markers. *Rev. Bras. Zootec.* **2002**, *31*, 1024–1032. [CrossRef]
50. NRC. *Nutrient Requirements of Small Ruminants: Sheep, Goats, Cervids and New World Camelids*; National Research Council: Washington, DC, USA, 2007.
51. Pereira, D.M.; Oliveira, S.J.; Santos, E.M.; Carvalho, G.G.P.; Azevedo, J.A.V.; Corrêa, Y.R.; Perazzo, A.F.; Assis, D.Y.C.; Leite, G.M.; Gomes, P.G.B.; et al. Productive performance and ingestive behaviour of feedlot finished Santa Ines lambs on diets containing differing fibre levels. *N. Z. J. Agric. Res.* **2022**, *65*, 213–226. [CrossRef]
52. Hoover, W.H. Chemical factors involved in ruminal fiber digestion. *J. Dairy Sci.* **1986**, *69*, 2755–2766. [CrossRef]
53. Penner, G.B.; Steele, M.A.; Aschenbach, J.R.; McBride, B.W. Ruminant Nutrition Symposium: Molecular adaptation of ruminal epithelia to highly fermentable diets. *J. Anim. Sci.* **2011**, *89*, 1108–1119. [CrossRef] [PubMed]
54. Archimède, H.; Pellonde, P.; Despois, P.; Etienne, T.; Alexandre, G. Growth performances and carcass traits of OvinMartinik lambs fed various ratios of tropical forage to concentrate under intensive conditions. *Small Rumin. Res.* **2008**, *75*, 162–170. [CrossRef]
55. Theurer, C.B.; Huber, J.T.; Delgado-Elorduy, A.; Wanderley, R. Invited review: Summary of steam-flaking corn or sorghum grain for lactating dairy cows. *J. Dairy Sci.* **1999**, *82*, 1950–1959. [CrossRef] [PubMed]
56. Bines, J.A.; Hart, I.C. The response of plasma insulin and other hormones to intraruminal infusion of vfa mixtures in cattle. *Can. J. Anim. Sci.* **1984**, *64*, 304–305. [CrossRef]
57. Lehninger, T.M.; Nelson, D.L.; Cox, M.M. *Principles of Biochemistry*, 6th ed.; Artmed: New York, NY, USA, 2018.
58. Koster, J.D.; Opsomer, G. Insulin resistance in dairy cows. *Vet. Clin. N. Am. Food Anim. Pract.* **2013**, *29*, 299–322. [CrossRef]
59. Sañudo, C.; Alfonso, M.; SaâNchez, A.; Delfa, R.; Teixeira, A. Carcass and meat quality in light lambs from different fat classes in the EU carcass classification system. *Meat Sci.* **2000**, *56*, 89–94. [CrossRef] [PubMed]
60. Realini, C.E.; Duckett, S.K.; Brito, G.W.; DallaRizza, M.; de Mattos, D. Effect of pasture vs. concentrate feeding with or without antioxidants on carcass characteristics, fatty acid composition, and quality of Uruguayan beef. *Meat Sci.* **2004**, *66*, 567–577. [CrossRef]
61. Brito, G.F.; Ponnampalam, E.N.; Hopkins, D.L. The effect of extensive feeding systems on growth rate, carcass traits, and meat quality of finishing lambs. *Compr. Rev. Food Sci. Food Saf.* **2017**, *16*, 23–38. [CrossRef] [PubMed]
62. Holman, B.W.B.; van de Ven, R.J.; Mao, Y.; Coombs, C.E.O.; Hopkins, D.L. Using instrumental (CIE and reflectance) measures to predict consumers' acceptance of beef colour. *Meat Sci.* **2017**, *127*, 57–62. [CrossRef] [PubMed]
63. D'Alessandro, A.G.; Palazzo, M.; Petrotos, K.; Goulas, P.; Martemucci, G. Fatty acid composition of light lamb meat from Leccese and Comisana dairy breeds as affected by slaughter age. *Small Rumin. Res.* **2015**, *127*, 36–43. [CrossRef]
64. Bravo-Lamas, L.; Barron, L.J.R.; Farmer, L.; Aldai, N. Fatty acid composition of intramuscular fat and odour-active compounds of lamb commercialized in northern Spain. *Meat Sci.* **2018**, *139*, 231–238. [CrossRef] [PubMed]

65. Sañudo, C.; Sanchez, A.; Alfonso, M. Small ruminant production systems and factors affecting lamb meat quality. *Meat Sci.* **1998**, *49*, S29–S64. [CrossRef]

66. Muela, E.; Monge, P.; Sañudo, C.; Campo, M.M.; Beltrán, J.A. Sensory quality of lamb following long-term frozen storage. *Meat Sci.* **2016**, *114*, 32–37. [CrossRef] [PubMed]

67. Arsenos, G.; Banos, G.; Fortomaris, P.; Katsaounis, N.; Stamataris, C.; Tsaras, L.; Zygoyiannis, D. Eating quality of lamb meat: Effects of breed, sex, degree of maturity and nutritional management. *Meat Sci.* **2002**, *60*, 379–387. [CrossRef] [PubMed]

68. Costa, J.B.; Oliveira, R.L.; Silva, T.M.; Barbosa, A.M.; Borja, M.S.; de Pellegrini, C.B.; da Silva Oliveira, V.; Xavier Ribeiro, R.D.; Bezerra, L.R. Fatty acid, physicochemical composition and sensory attributes of meat from lambs fed diets containing licuri cake. *PLoS ONE* **2018**, *13*, e0206863. [CrossRef] [PubMed]

69. Enjalbert, F.; Combes, S.; Zened, A.; Meynadier, A. Rumen microbiota and dietary fat: A mutual shaping. *J. Appl. Microbiol.* **2017**, *123*, 782–797. [CrossRef]

70. Zhu, J.; Shi, X.E.; Lu, H.; Xia, B.; Li, Y.; Li, X.; Zhang, Q.; Yang, G. RNA-seqtranscriptome analysis of extensor digitorumlongus and soleus muscles in large white pigs. *Mol. Genet. Genom.* **2016**, *291*, 687–701. [CrossRef]

71. Hunter, J.E.; Zhang, J.; Kris-Etherton, P.M.; Childs, L. Cardiovascular disease risk of dietary stearic acid compared with trans, other saturated, and unsaturated fatty acids: A systematic review. *Am. J. Clin. Nutr.* **2010**, *91*, 46–63. [CrossRef] [PubMed]

72. Bessa, R.J.B.; Alves, S.P.; Santos-Silva, J. Constraints and potentials for the nutritional modulation of the fatty acid composition of ruminant meat. *Eur. J. Lipid Sci. Technol.* **2015**, *117*, 1325–1344. [CrossRef]

73. Campbell, E.M.G.; Sanders, J.O.; Lunt, D.K.; Gill, C.A.; Taylor, J.F.; Davis, S.K.; Riley, D.G.; Smith, S.B. Adiposity, lipogenesis, and fatty acid composition of subcutaneous and intramuscular adipose tissues of Brahman and Angus crossbred cattle. *J. Anim. Sci* **2016**, *94*, 1415–1425. [CrossRef] [PubMed]

74. Smith, S.B.; Gill, C.A.; Lunt, D.K.; Brooks, M.A. Regulation of fat and fatty acid composition in beef cattle. *Asian-Australas. J. Anim. Sci.* **2009**, *22*, 1225–1233. [CrossRef]

75. Manso, T.; Gallardo, B.; Guerra-Rivas, C. Modifying milk and meat fat quality through feed changes. *Small Rumin. Res.* **2016**, *142*, 31–37. [CrossRef]

76. Manni, A.; El-Bayoumy, K.; Skibinski, C.G.; Thompson, H.J.; Santucci-Pereira, J.; Bidinotto, L.T.; Russo, J. The role of omega-3 fatty acids in breast cancer prevention. In *Trends in Breast Cancer Prevention*; Springer: Berlin/Heidelberg, Germany, 2016; pp. 51–81. [CrossRef]

77. Sawada, N.; Inoue, M.; Iwasaki, M.; Sasazuki, S.; Shimazu, T.; Yamaji, T.; Takachi, R.; Tanaka, Y.; Mizokami, M.; Tsugane, S. Consumption of n-3 fatty acids and fish reduces risk of hepatocellular carcinoma. *Gastroenterology* **2012**, *142*, 1468–1475. [CrossRef] [PubMed]

78. Bionaz, M.; Thering, B.J.; Loor, J.J. Fine metabolic regulation in ruminants via nutrient-gene interactions: Saturated long-chain fatty acids increase expression of genes involved in lipid metabolism and immune response partly through PPAR-α activation. *Br. J. Nutr.* **2012**, *107*, 179–191. [CrossRef] [PubMed]

79. Fujino, T.; Kondo, J.; Ishikawa, M.; Morikawa, K.; Yamamoto, T.T. Acetyl-CoA Synthetase 2, a Mitochondrial Matrix Enzyme Involved in the Oxidation of Acetate. *J. Biol. Chem.* **2001**, *276*, 11420–11426. [CrossRef]

80. Scollan, N.D.; Price, E.M.; Morgan, S.A.; Huws, S.A.; Shingfield, K.J. Can we improve the nutritional quality of meat? *Proc. Nutr. Soc.* **2017**, *76*, 603–618. [CrossRef]

81. Green, C.D.; Ozguden-Akkoc, C.G.; Wang, Y.; Jump, D.B.; Olson, L.K. Role of fatty acid elongases in determination of de novo sysnthesized monounsaturated fatty acid species. *J. Lipid Res.* **2010**, *51*, 1871–1877. [CrossRef]

# Production Traits, Blood Metabolic Profile, and Antioxidative Status of Dairy Goats Fed a Red Corn Supplemented Feed Mixture

Zvonko Antunović [1], Željka Klir Šalavardić [1], Josip Novoselec [1,*], Zvonimir Steiner [1], Mislav Đidara [1], Valentina Pavić [2], Lidija Jakobek Barron [3], Mario Ronta [1] and Boro Mioč [4]

[1] Faculty of Agrobiotechnical Sciences Osijek, Josip Juraj Strossmayer University of Osijek, Trg Sv. Trojstva 3, 31000 Osijek, Croatia; zantunovic@fazos.hr (Z.A.); zklir@fazos.hr (Ž.K.Š.); zsteiner@fazos.hr (Z.S.); mdidara@fazos.hr (M.Đ.); mronta@fazos.hr (M.R.)

[2] Department of Biology, Josip Juraj Strossmayer University of Osijek, Cara Hadrijana 8, 31000 Osijek, Croatia; vpavic@biologija.unios.hr

[3] Faculty of Food and Technology Osijek, Josip Juraj Strossmayer University of Osijek, Trg Sv. Trojstva 3, 31000 Osijek, Croatia; lidija.jakobek@ptfos.hr

[4] Department of Animal Science and Technology, Faculty of Agriculture, University of Zagreb, Svetošimunska cesta 25, 10000 Zagreb, Croatia; bmioc@agr.hr

* Correspondence: jnovoselec@fazos.hr; Tel.: +385-31-554906

**Abstract:** This study investigated the effect of red corn in the feed mixture of dairy goats on production traits, blood metabolic profile, and antioxidative status. The study was conducted on 30 French Alpine dairy goats. The feed mixture for the goats in the control group (CC) contained yellow corn (100%). In the first experimental group (RC50), yellow corn was partially (50%) replaced by red corn (RC), and in the second experimental group (RC100), yellow corn was completely (100%) replaced by red corn. No significance variations ($p > 0.05$) were determined in production traits of dairy goats between dietary treatments. A significant increase in hemoglobin (84.43, 100.00 and 106.55 g/L), mean corpuscular hemoglobin (7.98, 9.70 and 12.54 pg), and mean corpuscular hemoglobin concentration (293.57, 357.50 and 462.78 g/L) was found in the RC groups, and a decrease in erythrocytes in the RC100 compared with the RC50 group of goats (from 8.71 to 10.45 × $10^{12}$ L). A significant increase in blood superoxide dismutase (SOD) activity in the RC groups was found (0.29, 0.53, and 0.44 U/mL). The results indicate maintaining production traits and a moderate effect on blood metabolic profile (most hematologic parameters) as well as a positive antioxidative effect RC.

**Keywords:** lactating goats; corn with increased anthocyanins; milk performance; blood metabolic profile

Citation: Antunović, Z.; Klir Šalavardić, Ž.; Novoselec, J.; Steiner, Z.; Đidara, M.; Pavić, V.; Jakobek Barron, L.; Ronta, M.; Mioč, B. Production Traits, Blood Metabolic Profile, and Antioxidative Status of Dairy Goats Fed a Red Corn Supplemented Feed Mixture. *Agriculture* **2024**, *14*, 82. https://doi.org/10.3390/agriculture14010082

Academic Editors: Petru Alexandru Vlaicu, Arabela Elena Untea and Mihaela Saracila

Received: 5 December 2023
Revised: 28 December 2023
Accepted: 29 December 2023
Published: 30 December 2023



## 1. Introduction

Domestic animals are pressured to increase their production, i.e., financial profit of their farmers. This is why an increase in cellular respiration and production of free radicals and oxidative stress often occurs, afterward [1]. This is the primary cause of various metabolic disorders in animals, especially in the transition period, which leads to impairment of welfare, production properties, and quality of obtained products [2]. However, various antioxidants can be used for prevention [3,4]. The prohibition on the use of antibiotics and the interest for the use of various phyto-additives as food supplements such as flavonoids, including anthocyanins, is growing [5]. Flavonoids can promote the growth and development of animals and improve the health and quality of animal products [3]. Feeding anthocyanin- and antioxidant-rich forages to sheep and dairy cows can improve performance and product quality [6]. Anthocyanins are water-soluble glycosides of polyhydroxy and polymethoxy derivatives of 2-phenyl-benzopyrylium or flavylium salts responsible for the colors like red, purple, and blue presented in many fruits, vegetables, and cereal kernels as well as included in a class of flavonoids [7]. Further efforts by scientists are being made toward the use of anthocyanins in natural sources, through the consumption of certain feeds rich in their content. Corn is the most commonly concentrated fodder used in animal feeding. However, the different varieties of corn contain significantly different concentrations of anthocyanidins, as blue, red, and purple corn have higher concentration of anthocyanidins (up to 325 mg/100 g dry matter of corn), while yellow corn is rich in carotenoids (up to 823 µg/100 g dry matter of corn) [8]. Anthocyanins, besides giving various striking colors, also have a positive health influence [9]. However, the regulatory mechanisms in the metabolic pathway for anthocyanins are not yet fully elucidated [10]. It is known that anthocyanins as one of the flavonoids are absorbed in the gut poorly and thus concentrations in tissues are too low to obtain an effective antioxidant defense [4]. Changxing et al. [11] reported that anthocyanins have excellent antioxidant, anti-inflammatory, antimicrobial, and anticancer effects as well are proven to be safe and potent feed additives. In the available literature, we found only one paper that investigated the application of red corn in weaned lamb feeding [12], while there is no research on goats. Therefore, for the discussion, we used research conducted with purple corn in ruminant nutrition while monitoring the content of anthocyanins and polyphenols in them. Tian et al. [13] propose the use of purple corn in ruminant nutrition due to improved antioxidant resistance, better transfer of anthocyanins to goat's milk, and the lack of impact on milk composition. The present study investigates the effect of different levels of red corn in the feed mixture of dairy goats on productive traits (especially milk yield and composition), blood metabolic profile, and antioxidant status.

## 2. Materials and Methods

### 2.1. Experimental Design

The study was conducted in accordance with the Declaration of Helsinki, by obeying legal provisions determined by the Animal Protection Act (Republic of Croatia Official Gazette No. 133 (2006), No. 37 (2013), and No. 125 (2013)), and approved by the Committee for Animal Welfare of the Faculty of Agrobiotechnical Sciences Osijek (644-01/23-01/23 from 14 March 2023).

The study was conducted with 30 French Alpine dairy goats during lactation on a family farm in Osijek-Baranja County in Croatia. The goats were approximately 4 years old and in their third lactation. All the goats included in the study were healthy and in good physical condition. The research goats were selected from a flock of 60 animals. The goats were divided into three groups of 10 animals each, depending on their nutrition. In each experimental group, there was the same number of goats with singles and twins. The goats were housed in the barn throughout the study, and milking took place in the milking parlor in a separate room. All selected goats had kidded within seven days. The experiment started 60 days after kidding. Eight days before the experiment started goats had an adaptation period and were fed with experimental diet. In total, the experiment lasted 30 days. The index of body condition score (BCS) was determined using a scale of 1 to 5 points according to Santucci and Maestrini [14]. This scale uses points from 1 (thin) to 5 (obese), with intervals of 0.25. All personnel dealing with live goats have been trained and educated.

### 2.2. Feedstuff Analysis and Nutrition

The goats were given individual feed mixtures (1.2 kg per day) according to their requirements according to NRC [15]. Feeding with prepared feed mixtures was conducted in separate feeding troughs in the milking parlor during twice-daily milking by a machine. Furthermore, the goats had hay (*ad libitum*). During the research, 24 h before each goat's milk sampling (1st, 15th, and 30th days), the kids were separated from their mothers. Milk samples for the research were taken only during morning hand milking. A measuring cylinder was used to determine the morning milk yield of each goat. During the rest of the time, that is, except for the days of sampling, kids stayed with their mothers and suckled milk. Experimental diets differed in the type of corn. Yellow corn (100%) was included in the control group diet. In the first experimental group (RC50), yellow corn was partially replaced by 50% red corn in the feed mixture. In the feed mixture of the second experimental group (RC100), the yellow corn was completely replaced by 100% red corn. Red corn is an old native Croatian variety. The feed samples (feed mixture, hay, and corn) were dried and ground to a powder using an ultra-centrifugal mill (heavy metal-free, Retsch ZM 200, Haan, Germany). The standard methods of AOAC [16] were used to determine the feed composition of goats. The feed mixture (g/kg feed mixture) contained: corn 600, barley 120, wheat flour 23, soybean meal (46% crude proteins) 100, extruded soybean 120, salt 4, calcium carbonate 3, mineral vitamin premix 30. Goats mineral vitamin premix contained: 21% Ca; 5% P; 6% Na; 5% Mg; 1,200,000 IU/kg vitamin A; 140,000 IU/kg vitamin $D_3$; 3500 mg/kg vitamin E; 600 mg/kg Fe (iron sulphate monohydrate); 490 mg/kg Cu (copper sulphate pentahydrate); 500 mg/kg Cu (in the form of chelates); 5 mg/kg Mn (manganese sulphate pentahydrate); 6500 mg/kg Zn (zinc oxide); 1500 mg/kg Zn (in the form of chelates); 60 mg/kg I (anhydrous calcium iodate); 40 mg/kg Co (cobalt carbonate monohydrate); 50 mg/kg Se (disodium selenite). Chemical compositions of the food are presented in Table 1. The methods used for the analysis of the nutritional values of feed are described in [12]. Extraction and determination of total polyphenols and anthocyanins from feed were performed according to the procedures and equipment described by [17]. Concentrations of total polyphenols and anthocyanins in the RC50 group were 513.06 and 217.32 mg/kg, while in the RC100 group they were 1276.70 and 485.40 mg/kg, respectively.

**Table 1.** Chemical composition of feed mixture, hay, yellow corn, and red corn used in investigation with dairy goats.

| Chemical Content (g/kg DM) | Feed Mixture | Yellow Corn | Red Corn | Hay |
|---|---|---|---|---|
| DM | 912 | 905 | 907 | 932 |
| Crude protein | 157 | 100 | 105 | 91 |
| Crude fiber (g/kg DM) | 36 | 25 | 23 | 328 |
| Crude ash (g/kg DM) | 30 | 13 | 14 | 67 |
| EE (g/kg DM) | 51 | 38 | 37 | 8 |
| ME (MJ/kg DM) | 12 | - | - | 7 |
| Polyphenols * (total), mg/kg | 144.77 | 179.87 | 298.69 | - |
| Anthocyanins ** (total), mg/kg | 0 | 125.37 | 253.04 | - |

DM—dry matter; * The concentration of polyphenols in RC50 and RC100 was 513.06 and 1276.70 mg/kg, respectively; ** The concentration of anthocyanins in RC50 and RC100 was 217.32 and 485.40 mg/kg, respectively.

## 2.3. Milk Sampling and Analysis

At the 1st, 15th, and 30th day of the research, two milk samples (one for chemical analyses and other for biochemical parameters) were taken from each goat at morning (7:00 a.m.) during routine hand milking. Total morning milk yield of each goat was determined using a measuring cylinder. Immediately after the collection of these two milk samples, a milk sample (30 mL) from each goat was placed in plastic bottles containing 0.3 mL of azidiol, cooled to 4 °C, and prepared for analysis. The chemical composition of the goat's milk was determined using the MilkoScan FT 6000 analyzer (Foss Electric, Hillerød, Denmark) according to the principle of infrared spectroscopy in accordance with HRN ISO 9622:2017 [18]. The somatic cell count (SCC) was determined according to the Fluor-electronic method HRN ISO 13366-2/Ispr.1:2007 [19] using a Fossomatic 5000 analyzer (Foss Electric, Hillerød, Denmark). To determine the biochemical parameters, the milk was centrifuged at $5000 \times g$ (30 min) after fat separation and the milk plasma was frozen at $-80$ °C until analysis. In blood serum, biochemical parameters (aspartate aminotransferase—AST; alanine aminotransferase—ALT; gamma-glutamyl transferase—GGT; iron—Fe; calcium—Ca; and phosphorus—P) were determined using the Olympus AU 400 biochemical analyzer (Olympus, Tokyo, Japan), while the activity of SOD was determined using RANSOD (Randox Laboratories, Crumlin, UK) on the same Olympus AU 400 biochemical analyzer (Olympus, Tokyo, Japan).

### 2.3.1. Antioxidant Indicators

Fresh goat's milk samples were stored at $-80\ °C$. The extraction of the antioxidant compounds followed the method described in the paper of Alyaqoubi et al. [20] with some modifications. Fresh milk (1 g) was added to 10 mL extraction solution (1 N HC1/95% ethanol (*v/v*, 15/85)) in brown 50 mL bottles and shaken for 1 h at 30 °C in a rotary shaker (Brunswick™ Innova® 43/43R—Console Incubator Shaker, Eppendorf AG, Hamburg, Germany) at 300 rpm. The mixture of solvent and samples was then centrifuged at $4400\times g$ (Hermle Z 326 K, Hermle Labortechnik GmbH, Wehingen, Germany) at 5 °C for 40 min. The supernatant liquids were stored at $-20\ °C$ in the dark until further analysis for DPPH radical scavenging activity and $TBARS_{450}$. The extraction was performed three times.

### 2,2-Diphenyl-1-Picrylhydrazyl (DPPH) Radical Scavenging Activity

In a modification of the methods of [21], the DPPH radical scavenging test was carried out to determine the total antioxidant activity of milk extracts. Briefly, an aliquot of 200 µL of milk extracts was added to 1.8 mL of a 0.2 mM methanol DPPH radical solution, so the final concentration was 0.18 mM DPPH. The mixture was shaken vigorously using a vortex mixer and then left to stand for 30 min in the dark at room temperature. Negative control was carried under the same conditions, which contained 200 µL of milk extraction solution and 1800 µL of 0.2 mM DPPH in methanol. All measurements were performed three times. Absorbance was measured at 517 nm using a spectrophotometer (Lambda 25, Perkin Elmer, Waltham, MA, USA). The percentage of DPPH scavenging activity was determined using Equation (1):

$$\text{DPPH scavenging activity (\%)} = ((Ab + As) - Am)/Ab \times 100 \tag{1}$$

where Ab is the absorbance of the negative control, 0.18 mM DPPH radical methanolic solution with the milk extraction solution, As is the absorbance of the milk sample control, and Am is the absorbance of samples of tested milk extracts (final concentration 10 µL/mL) and DPPH radical.

### Thiobarbituric Acid Reactive Substances ($TBARS_{450}$)

Slightly modified methods from Oancea et al. [22] based on the absorbance values (at 450 nm, 532 nm and 600 nm) from Sun et al. [23] with a spectrophotometer (Lambda 25, Perkin Elmer, Waltham, MA, USA) were used to measure $TBARS_{450}$. Determinations were performed after defrosting of milk samples overnight in the refrigerator. Each sample used for analytical determination consisted of 1.5 mL of defrosted milk, and 0.5 mL of 0.1% TCA was used for the protein-removal step. After centrifugation at $3000\times g$ for 5 min at 4 °C (Hermle Z 326 K, Hermle Labortechnik GmbH, Wehingen, Germany), the supernatants were incubated with 1 mL of TBA/TCA reagent (0.5% TBA in a 20% TCA) for 90 min at 80 °C.

After incubation and cooling of samples on ice, the absorbance values were read at different wavelengths, specific for milk degradation products (450 nm—saturated aldehydes, 532 nm—MDA and 600 nm—nonspecific absorption). The results were expressed as the absorbance values at 450 nm after subtraction of non-specific absorption at 600 nm.

### 2.4. Blood Sampling and Analysis of Blood Metabolic Profile

Blood samples were taken from the jugular vein of each goat (10 mL) into a sterile vacuum tube Venoject® (Sterile Terumo Europe, Leuven, Belgium) containing ethylenediaminetetraacetic acid (EDTA) as an anticoagulant for hematologic analysis. After blood collection, samples were transferred on ice at 0–6 °C to the laboratory of the Department of Animal Production and Biotechnology, Faculty of Agrobiotechnical Sciences, Josip Juraj Strossmayer University of Osijek, Osijek Croatia. The EDTA tubes containing the anticoagulant were inverted several times to ensure adequate mixing of the blood. Hematological parameters in goat whole blood, such as leukocyte (WBC, $10^9$ L) and erythrocytes count (RBC, $10^{12}$ L) as well as the content of hemoglobin (HGB, g/L), hematocrit (HCT,

L/L), mean corpuscular volume (MCV, fL), the mean hemoglobin content in erythrocytes (MCH, pg), and the mean hemoglobin concentration in the erythrocytes (MCHC, g/L) were determined using the Sysmex PocH-100iV automated three-differential hematology analyzer (Sysmex Europe GmbH, Hamburg, Germany). Blood samples collected in sterile vacuum tubes (Venoject®, Leuven, Belgium) were centrifuged within 10 min at $1609.92\times g$ by centrifuge ROTOFIX 32A (Hettich GmbH & Co. KG, Tuttlingen, Germany) and the serum samples obtained were analyzed in the Olympus AU400. In the blood serum, the concentration of biochemical parameters was determined and showed in the reagent code, such as calcium (Ca, mmol/L; OSR 60117), phosphorus-inorganic (P-inorg, mmol/L; OSR6122), magnesium (Mg, mmol/L; OSR6189), iron (Fe, µmol/L; OSR6186), urea (mmol/L; OSR6134), glucose (GUK, mmol/L; OSR6140) total proteins (TPROT, g/L; OSR6132), albumin (ALB, g/L; OSR6102), cholesterol (CHOL, mmol/L; OSR6116), LDL—low-density lipoprotein (LDL, mmol/L; OSR6183), HDL—high-*density* lipoprotein (HDL, mmol/L; OSR6187), triglycerides (TGC, mmol/L; OSR60118); β-hydroxybutyrate (BHB, mmol/L; RB1007), and non-esterified fatty acids (NEFA, mmol/L; FA115). Also, in the blood serum, activities of enzymes were determined and showed in the reagent code, such as alanine aminotransferase (ALT, U/L; OSR 6007), aspartate aminotransferase (AST, U/L; OSR6009), γ-glutamyl transferase (GGT; U/L; OSR6020), and glutathione reductase (GR; U/L; GR2368) and were measured by using Olympus System reagents (Lismeehan, Ireland). The globulin content (GLOB) was calculated as the difference between the total protein and albumin content. In blood serum, the activity of glutathione peroxidase (GPx, U/L; RA505) was determined using a Ransel® kit (RANSEL, RANDOX, London, UK), and the activity of total superoxide dismutase (SOD, U/mL; SD125) was determined using a Ransod® kit (RANDOX, San Diego, CA, USA) on an automatic Olympus AU 400 (Olympus, Tokyo, Japan) analyzer.

*2.5. Statistical Analyses*

Data for milk yield, milk composition, blood metabolic profile parameters, and antioxidant status for each goat and sampling period were analyzed by repeated measure using the PROC MIXED procedure of SAS 9.4 [24] according to the following model:

$$Yijk = \mu + di + hij + wk + dwik + eijk,$$

where:

$\mu$ = overall mean,
di = fixed effect of diet (i = CC, RC50, RC100),
hij = animal within diet as subject (j = CC, RC540, RC100),
wk = fixed effect of period during lactation (k = 1–3),
dwik = interaction between diet and period (diet × period), and
eijk = random error variation (residual error).

Means were compared using Tukey's honestly significant difference test [24], and $p \leq 0.05$ was indicated as significant. The values for SCC were logarithmically transformed to obtain a linear value that approximates the normal distribution.

## 3. Results

Table 2 shows the body weight, body condition score, and milk yield and composition of milk of dairy goats during lactation when fed feed mixtures containing different levels of red corn.

No significant ($p > 0.05$) variations were determined in milk yield and composition between dietary treatments. The timing of sampling and the interaction between feed and timing of sampling did not affect the parameters studied, except for urea concentration depending on the lactation stage (Table 2). It is evident that the activity of enzymes (AST, ALT and GGT) in milk showed an insignificant decrease ($p > 0.05$) when feeding with a higher content of red corn in the feed mixtures (RC50 and RC100) compared to the CC group

**Table 2.** Production traits and milk composition of dairy goat fed feed mixtures with different levels of red corn.

| Traits | Diets | | | SEM | p-Value | | |
|---|---|---|---|---|---|---|---|
| | CC | RC50 | RC100 | | D | T | D*T |
| Morning milk yield (g) | 1461.47 | 1323.54 | 1367.18 | 54.176 | 0.555 | 0.443 | 0.779 |
| Body weight (kg) | 46.10 | 48.22 | 47.83 | 0.712 | 0.465 | 0.991 | 0.993 |
| BCS (point) | 2.76 | 2.90 | 2.80 | 0.051 | 0.577 | 0.687 | 0.856 |
| Milk composition | | | | | | | |
| Fat (g/100 g) | 3.14 | 3.22 | 3.09 | 0.083 | 0.840 | 0.939 | 0.980 |
| Protein (g/100 g) | 2.84 | 2.89 | 3.00 | 0.029 | 0.086 | 0.569 | 0.995 |
| Lactose (g/100 g) | 4.45 | 4.48 | 4.53 | 0.022 | 0.800 | 0.285 | 0.335 |
| NFDM (g/100 g) | 8.34 | 8.43 | 8.58 | 0.042 | 0.101 | 0.650 | 0.688 |
| AST (U/L) | 171.29 | 130.43 | 145.66 | 11.812 | 0.415 | 0.734 | 0.614 |
| ALT (U/L) | 386.68 | 346.95 | 352.00 | 46.341 | 0.904 | 0.460 | 0.784 |
| GGT (U/L) | 364.92 | 360.19 | 327.88 | 15.362 | 0.584 | 0.766 | 0.446 |
| Fe ($\mu$mol/L) | 17.21 | 15.79 | 13.76 | 1.261 | 0.603 | 0.774 | 0.769 |
| Ca (mmol/L) | 41.03 | 45.30 | 45.84 | 1.673 | 0.417 | 0.081 | 0.484 |
| P-inorg (mmol/L) | 28.47 | 32.63 | 33.96 | 1.200 | 0.154 | 0.172 | 0.435 |
| Urea (mg/dL) | 19.88 | 22.51 | 23.06 | 0.257 | 0.572 | 0.032 | 0.911 |
| SSC (log) | 5.43 | 5.51 | 5.37 | 0.049 | 0.645 | 0.940 | 0.562 |

CC—control corn; RC50—red corn 50%, RC100—red corn 100%; D—diet, T—time of sampling, D*T—interaction (diet × time of sampling), SEM—standard error of mean; BCS—body condition score; NFDM—non-fat dry matter; AST—aspartate aminotransferase, ALT—alanine aminotransferase, GGT—$\gamma$-glutamyl transferase; SSC—somatic cell count.

From the results of the analysis (Table 3), the increase in HGB, MCH, and MCHC was found in feeding goats with an increased content of RC in both groups (RC50 and RC100) compared with CC, and a decrease in RBC in RC100 compared with RC50. These parameters were affected by dietary treatment and time of milk sampling and their interaction.

**Table 3.** Hematologic parameters in dairy goat fed feed mixtures containing red corn.

| Parameter | Diets | | | SEM | p-Value | | |
|---|---|---|---|---|---|---|---|
| | CC | RC50 | RC100 | | D | T | D*T |
| RBC ($\times 10^{12}$ L) | 9.74 [ab] | 10.45 [a] | 8.71 [b] | 0.228 | 0.015 | 0.180 | 0.843 |
| WBC ($\times 10^{9}$ L) | 9.74 | 8.99 | 9.74 | 0.350 | 0.313 | 0.382 | 0.901 |
| HGB (g/L) | 84.43 [b] | 100.00 [a] | 106.55 [a] | 5.212 | <0.001 | <0.001 | <0.001 |
| HCT (L/L) | 0.27 | 0.28 | 0.26 | 0.006 | 0.169 | 0.015 | 0.728 |
| MCH (pg) | 7.98 [c] | 9.70 [b] | 12.54 [a] | 0.661 | <0.001 | <0.001 | <0.001 |
| MCV (fL) | 27.24 | 27.25 | 30.54 | 0.992 | 0.300 | 0.207 | 0.474 |
| MCHC (g/L) | 293.57 [c] | 357.50 [b] | 462.78 [a] | 18.755 | <0.001 | <0.001 | <0.001 |

CC—control corn; RC50—red corn 50%; RC100—red corn 100%; D—diet; T—time of sampling; D*T—interaction (diet × time of sampling); SEM—standard error of mean; RBC—erythrocytes, WBC—number of leukocytes; HGB—hemoglobin; HCT—hematocrit; MCV—mean corpuscular volume; MCH—mean hemoglobin content in erythrocytes; MCHC—mean hemoglobin concentration in erythrocytes; [a–c] Values within a row with different superscripts differ significantly at $p < 0.05$.

Analysis of blood metabolic profile (biochemical parameters and enzyme activities) in dairy goats concerning the consumption of feed mixtures with red corn did not reveal statistically significant ($p > 0.05$) differences, except for the concentrations of urea, glucose, and Mg influence of milk sampling time (Table 4).

In the RC100 group, a significant decrease in TBARS$_{450}$ values in milk compared with CC was found, as well as a smaller increase in the DPPH scavenging, but without significant differences in RC50 and RC100 compared with CC. A sampling time affected TBARS$_{450}$ concentration significantly as well (Table 5). The TBARS450 results were expressed as the absorbance values at 450 nm after subtraction of non-specific absorption at 600 nm, while the absorbance values at 532 nm were insignificant.

**Table 4.** Blood biochemical parameters and enzyme activities of dairy goat fed with feed mixtures containing red corn.

| Parameter | Diets | | | SEM | *p*-Value | | |
|---|---|---|---|---|---|---|---|
| | CC | RC50 | RC100 | | D | T | D*T |
| Urea (mmol/L) | 7.58 | 7.92 | 7.44 | 0.220 | 0.626 | <0.001 | 0.488 |
| TPROT (g/L) | 74.74 | 77.13 | 73.71 | 0.867 | 0.295 | 0.788 | 0.505 |
| ALB (g/L) | 27.56 | 27.83 | 26.46 | 0.439 | 0.442 | 0.760 | 0.541 |
| GLOB (g/L) | 47.18 | 49.30 | 47.25 | 0.616 | 0.314 | 0.454 | 0.741 |
| Ca (mmol/L) | 2.31 | 2.29 | 2.24 | 0.024 | 0.463 | 0.097 | 0.720 |
| P-inorganic (mmol/L) | 2.25 | 2.11 | 2.20 | 0.065 | 0.696 | 0.975 | 0.375 |
| Mg (mmol/L) | 1.30 | 1.30 | 1.29 | 0.011 | 0.960 | 0.009 | 0.273 |
| Fe (μmol/L) | 25.57 | 24.90 | 24.33 | 0.950 | 0.869 | 0.489 | 0.314 |
| GUK (mmol/L) | 3.55 | 3.49 | 3.60 | 0.045 | 0.636 | 0.031 | 0.803 |
| CHOL (mmol/L) | 2.74 | 3.09 | 2.67 | 0.080 | 0.099 | 0.656 | 0.801 |
| HDL (mmol/L) | 1.64 | 1.83 | 1.63 | 0.044 | 0.143 | 0.907 | 0.891 |
| LDL (mmol/L) | 1.03 | 1.18 | 0.95 | 0.041 | 0.098 | 0.316 | 0.736 |
| TGC (mmol/L) | 0.17 | 0.18 | 0.19 | 0.011 | 0.734 | 0.235 | 0.921 |
| NEFA (mmol/L) | 0.11 | 0.14 | 0.10 | 0.009 | 0.311 | 0.157 | 0.863 |
| BHB (mmol/L) | 0.71 | 0.67 | 0.57 | 0.024 | 0.071 | 0.562 | 0.766 |
| AST (U/L) | 130.53 | 121.43 | 137.93 | 3.423 | 0.187 | 0.512 | 0.914 |
| ALT (U/L) | 27.71 | 24.59 | 27.54 | 0.679 | 0.136 | 0.326 | 0.878 |
| GGT (U/L) | 45.29 | 43.92 | 43.44 | 1.027 | 0.753 | 0.681 | 0.506 |

CC—control corn; RC50—red corn 50%; RC100—red corn 100%; D—diet, T—time of sampling; D*T—interaction (diet × time of sampling); SEM—standard error of mean; TPROT—total proteins; ALB—albumins; GLOB—globulins; GUK—glucose; CHOL—cholesterol, HDL—HDL-cholesterol; LDL—LDL-cholesterol; TGC—triglycerides; NEFA—non-esterified fatty acids, BHB—β-hydroxybutyrate; AST—aspartate aminotransferase; ALT—alanine aminotransferase, GGT—γ-glutamyl transferase.

**Table 5.** Antioxidative status of milk and blood of dairy goats fed with feed mixtures containing red corn.

| Parameter | Diets | | | SEM | *p*-Value | | |
|---|---|---|---|---|---|---|---|
| | CC | RC50 | RC100 | | D | T | D*T |
| **Milk** | | | | | | | |
| TBARS$_{450}$ | 0.025 [a] | 0.022 [ab] | 0.019 [b] | 0.017 | 0.022 | <0.001 | 0.186 |
| DPPH scavenging (%) | 87.78 | 91.17 | 92.54 | 1.107 | 0.187 | 0.099 | 0.324 |
| SODm (U/mL) | 10.37 | 11.45 | 11.63 | 0.668 | 0.735 | 0.532 | 0.616 |
| **Blood serum** | | | | | | | |
| GPx (U/L) | 1633.48 | 1669.75 | 1531.92 | 99.589 | 0.726 | <0.001 | 0.043 |
| SODb (U/mL) | 0.29 [b] | 0.53 [a] | 0.44 [a] | 0.025 | 0.001 | 0.356 | 0.916 |
| GR (U/L) | 96.16 | 95.36 | 95.06 | 2.561 | 0.990 | 0.003 | 0.512 |

CC—control corn; RC50—red corn 50%; RC100—red corn 100%; D—diet; T—time of milk sampling; D*T—interaction (diet × time of sampling); SEM—standard error of mean; TBARS$_{450}$—Thiobarbituric acid reactive substances; DPPH—radical scavenging activity at final milk concentration 10 uL/mL; SODm—milk superoxide dismutase; GPx—blood glutathione peroxidase; SODb—blood superoxide dismutase; GR—blood glutathione reductase; [a,b] Values within a row with different superscripts differ significantly at *p* < 0.05.

It can be seen that SODm activity in milk increased when fed with a higher content of red corn in the feed mixtures (RC50 and RC100) compared to the CC group but without significant differences. A significant ($p \leq 0.05$) effect was found on the activity of antioxidant enzymes in the blood of dairy goats fed with different proportions of RC, namely an increase in SODb activity in the RC50 and RC100 groups and a significant effect of sampling time on GPx and GR activity in the blood (Table 5).

## 4. Discussion

Lactation is a very challenging physiological process for an animal. Due to their high metabolic demands, goats and other small ruminants are prone to suffering from oxidative stress [13]. Both the productivity and health of ruminants are negatively influenced by the oxidative stress. Therefore, research on the addition of anthocyanins to the feed of small ruminants is very timely. However, the high content of polyphenols in food, especially astringent polyphenols, can bind nutrients and reduce their absorption in the digestive system causing reduced food consumption and impaired animal production properties [25]. Plants rich in anthocyanins influence the immune response associated with the inhibition of inflammatory processes by promoting the normalization of microflora in the digestive tract and reducing the permeability of the gastrointestinal barrier [26]. In this research, no significant changes were observed in the production traits of goats. The lactose content of milk was slightly increased in RC50 and RC100 goats compared to the CC group (from 4.45 to 4.53% DM). In a study by Tian et al. [27], no effects of feeding goats with anthocyanins from purple corn on dry matter intake, average daily gain of goats, glucose blood concentrations, urea nitrogen, total proteins, albumins, and DPPH activity were observed. In a study [13] conducted with lactating dairy goats, no change in milk quantity and composition was observed when feeding purple corn silage, except for a significant increase in lactose content in the goat milk compared to the group fed ordinary corn (from 4.50 to 4.55%). This may be related to fermentation in the rumen, in particular by inhibiting acetic acid and increasing the proportion of propionic acid in the rumen. The authors indicated that the anthocyanin sugars can be broken down in the digestive system and are therefore involved in lactose synthesis. The absence of changes in the activity of other enzymes such as GPx and GR could be explained by the poor absorption of anthocyanins by small ruminants compared with monogastric animals. SOD is the first line of defense among the intracellular enzymes that scavenge reactive oxygen species [28]. Consequently, anthocyanins are powerful antioxidants that regulate peroxidation reactions and control the production of free radicals in the goat's body [29]. Quadros et al. [6] suggested that anthocyanins can suppress oxidation resistance and enhance plasma SOD enzyme, resulting in reduced saturated fatty acids (SFA) and increased polyunsaturated fatty acid (PUFA) levels. Antunović et al. [12] found similar deviations in blood GPx and GR activities, and an insignificant increase in blood SOD activity in lambs fed mixtures with the addition of red corn. Hosoda et al. [30] pointed out that the cause might be a sufficient number of non-enzymatic antioxidants, as no serious oxidative stress was found in sheep. Tian et al. [31] found an increase in plasma SOD activity when goats were fed silage from anthocyanin-rich purple corn, while the activity of other antioxidant enzymes (GPx and catalase) in the goats' plasma did not change. These authors pointed out that the reason could be that the increased activity of SOD in the plasma of goats reduces the load on the antioxidant defense system so that other enzymes in the plasma do not change [32]. Matsuba et al. [33] found that when feeding dairy cows with purple corn silage, the activity of SOD in the blood also increased. Corn rich in anthocyanins is suitable to provide an antioxidant effect in dairy cattle due to the stability of anthocyanins in rumen fluid [34]. Hosoda et al. [28] found an increase in plasma activity SOD in lactating sheep fed purple corn and a significant increase in plasma activity SOD in cows [35] fed anthocyanin-rich corn silage. The antioxidant enzymes, such as SOD, have the ability to eliminate reactive oxygen and prevent cell damage, which can be associated with the use of purple corn extract that can penetrate cell membranes and stimulate the production of antioxidant enzyme [36]. Tian et al. [29], in a study on the effects of anthocyanin-rich purple corn pot silage in goat feed, found a potential to improve antioxidant status by improving SOD activity in blood plasma through modulation of antioxidant genes in the mammary gland. In the present study, a significant ($p \leq 0.05$) increase in blood SOD activity was found in the RC50 and RC100 groups compared to the control group, which corresponds with previously mentioned studies. Hematologic and biochemical parameters and enzyme activities in goat blood in the present investigation were within the reference values established for goats under similar rearing conditions, except for the MCH content in RC50 and RC100, MCV in all groups, and MCHC (only in RC100), which were above the reference values [37–39] with a slight effect of period and interaction between diet and period, which was expected. A significant increase in HGB, MCH, and MCHC content with increasing RC content in both groups (RC50 and RC 100) compared with CC might be related to anthocyanins stabilizing erythrocyte membranes and inhibiting hemoglobin polymerization [40]. According to Youdim et al. [41], polyphenols protect against reactive oxygen species-induced hemolysis by increasing the integrity of red blood cells in conjunction with the inhibition of lipid peroxidation. In the research by Antunović et al. [12], which was carried out on the feeding of lambs after weaning, an increase in the HGB content was also found with an increase in the proportion of red corn in the feed. The effect of anthocyanins from the feed on the antioxidant response of the animal organism is well known. The antioxidant properties of phenols are attributed to their redox property, which allows phenols to act as reducing agents, hydrogen donors, and oxygen scavengers. In the present study, a significant decrease in $TBARS_{450}$ in milk was observed in the RC100 group compared with CC, as well as a smaller increase in the DPPH scavenging, but without significant differences in RC50 and RC100 compared with CC. These changes also indicate the antioxidative effect of nutritional treatments. Indeed, the anthocyanins in plasma can supply hydrogen atoms to DPPH [42] and thereby increase DPPH scavenging activity. In addition, antioxidants in the diet can reduce the content of superoxide anions, hydroxyl radicals, and peroxynitrite [43]. The aforementioned authors point out that it can inhibit the activities of enzymes that generate reactive oxygen species and increase the expression of antioxidant enzymes such as SOD. Tsuda et al. [44] performed a study with rats fed C3G anthocyanins (2 g/kg) for 14 days. A significant decrease in $TBARS_{450}$ formation during serum formation was observed, but no significant difference in serum lipid concentrations (CHOL, NEFA, TGC). Tian et al. [13] also found no changes in DPPH scavenging activity and GPx activity when feeding goats with purple corn silage due to the possible high toxicity of $O_2$ that was converted into low toxicity of $H_2O_2$ by SOD, thus facilitating oxidative stress. In the present research, we also did not determine significant ($p > 0.05$) differences in DPPH and GPx activity when feeding the goats with red corn.

## 5. Conclusions

From the present results, it can be concluded that the replacement of yellow corn with red corn had no effect on the production traits and blood metabolic profile of the goats, except for most of the blood hematologic parameters. Higher levels of HGB, MCH, and MCHC were observed with an increase in the content of CR in both groups (RC50 and RC100) compared to CC, as well as a decrease in RBC in RC100 compared to RC50. A significant increase in SOD activity in the blood of RC50 and RC100 groups was found. A significant decrease in $TBARS_{450}$ in milk was found compared to the CC group, as well as an increase in DPPH scavenging, but without significant differences compared to the CC group. These changes indicate an adequate antioxidative effect of red corn used in goat diet, since the SOD activity in blood, as well as antioxidant indicators ($TBARS_{450}$ and DPPH scavenging), were increased in goat milk, while the blood metabolic profile and production traits were maintained. Further studies should be undertaken to develop indicators for antioxidant status in the blood ($TBARS_{450}$, DPPH scavenging) and anthocyanin profiles in red corn and milk from lactating dairy goats.

**Author Contributions:** Investigation, writing—original draft preparation conceptualization, visualization Z.A.; investigation, formal analysis, and methodology, visualization, supervision J.N.; investigation, visualization Z.S.; formal analysis and methodology, visualization M.Đ., V.P. and L.J.B.; field investigation M.R.; investigation, visualization B.M.; investigation, visualization, Ž.K.Š. All authors have read and agreed to the published version of the manuscript.

**Funding:** This research received no external funding.

**Institutional Review Board Statement:** The study was conducted in accordance with the Declaration of Helsinki, by obeying legal provisions determined by the Animal Protection Act (Republic of Croatia Official Gazette No. 133 (2006), No. 37 (2013), and No. 125 (2013)), and approved by the Committee for Animal Welfare of the Faculty of Agrobiotechnical Sciences Osijek (644-01/23-01/23 from 14 March 2023).

**Data Availability Statement:** The data presented in this study are available on request from the corresponding author.

**Acknowledgments:** The research was carried out with the Innovative Breeding and Technological Processes in Animal Production (no. 1126) research team at the Faculty of Agrobiotechnical Sciences Osijek.

**Conflicts of Interest:** The authors declare no conflicts of interest.

# References


1. Mutinati, M.; Piccinno, M.; Roncetti, M.; Campanile, D.; Rizzo, A.; Sciorsci, R.L. Oxidative stress during pregnancy in the sheep. *Reprod. Domest. Anim.* **2013**, *48*, 353–357. [CrossRef] [PubMed]
2. Chauhan, S.S.; Celi, P.; Ponnampalam, E.; Dunshea, F.R. Antioxidant dynamics in the live animal and implications for ruminant health and product (meat/milk) quality: Role of vitamin E and selenium. *Anim. Prod. Sci.* **2014**, *54*, 1525–1536. [CrossRef]
3. Castillo, C.; Pereira, V.; Abuelo, A.; Hernández, J. Effect of supplementation with antioxidants on the quality of bovine milk and meat production. *Sci. World J.* **2013**, *2013*, 616098. [CrossRef] [PubMed]
4. Surai, P.F. Polyphenol compounds in the chicken/animal diet: From the past to the future. *J. Anim. Physiol. Anim. Nutr.* **2014**, *98*, 19–31. [CrossRef] [PubMed]
5. Giller, K.; Sinza, S.; Messadene-Chelalib, J.; Marquardt, S. Maternal and direct dietary polyphenol supplementation affect growth, carcass and meat quality of sheep and goats. *Animal* **2021**, *15*, 100333. [CrossRef] [PubMed]
6. Quadros, D.G.; Kerth, C.R.; Miller, R.; Tolleson, D.R.; Redden, R.R.; Xu, W. Intake, growth performance, carcass traits, and meat quality of feedlot lambs fed novel anthocyanin-rich corn cobs. *Transl. Anim. Sci.* **2023**, *7*, txac171. [CrossRef] [PubMed]
7. Ávila, M.; Hidalgo, M.; Sánchez-Moreno, C.; Pelaez, C.; Requena, T.; Pascual-Teresa, S. Bioconversion of anthocyanin glycosides by Bifidobacteria and Lactobacillus. *Food Res. Int.* **2009**, *42*, 1453–1461. [CrossRef]
8. Sheng, S.; Li, T.; Liu, R.H. Corn phytochemicals and their health benefits. *Food Sci. Hum. Wellness* **2018**, *7*, 185–195.
9. Kumari, P.; Raju, D.V.S.; Prasad, K.V.; Saha, S.; Panwar, S.; Paul, S.; Banyal, N.; Bains, A.; Chawla, P.; Fogarasi, M. Characterization of Anthocyanins and Their Antioxidant Activities in Indian Rose Varieties (*Rosa* × *hybrida*) Using HPLC. *Antioxidants* **2022**, *11*, 2032. [CrossRef]
10. Logo, C.; Cassani, E.; Petroni, K.; Calvenzani, V.; Tonelli, C.; Pilu, R. Study of maize genotypes rich in antocyanins for human and animal nutrition. In Proceeding of the Joint Meeting AGI-SIBV-SIGA, Assisi, Italy, 19–22 September 2011; p. 42.
11. Changxing, L.; Chenling, M.; Alagawany, M.; Jinhua, L.; Dongfand, D.; Gaichao, W.; Wenyin, Z.; Syed, S.F.; Arain, M.A.; Saeed, M.; et al. Health benefits and potential applications of anthocyanins in poultry feed industry. *World's Poult. Sci. J.* **2018**, *74*, 251–263. [CrossRef]
12. Antunović, Z.; Novoselec, J.; Klir Šalavardić, Ž.; Steiner, Z.; Šperanda, M.; Jakobek Barron, L.; Ronta, M.; Pavić, V. Influence of red corn rich in anthocyanins on productive traits, blood metabolic profile, and antioxidative status of fattening lambs. *Animals* **2022**, *12*, 612. [CrossRef] [PubMed]
13. Tian, X.Z.; Paengkoum, P.; Paengkoum, S.; Chumpawadee, S.; Ban, C.; Sorasak, T. Purple corn (*Zea mays* L.) stover silage with abundant anthocyanins transferring anthocyanin composition to the milk and increasing antioxidant status of lactating dairy goats. *J. Dairy Sci.* **2019**, *102*, 413–418. [CrossRef] [PubMed]
14. Santucci, P.M.; Maestrini, O. Body condition of dairy goats in extensive systems of production: Method of estimation. *J. Dairy Sci.* **1985**, *34*, 471–490. [CrossRef]
15. National Research Council (NRC). *Nutrient Requirements of Small Ruminants: Sheep, Goats, Cervids, and New World Camelids*; The National Academy Press: Washington, DC, USA, 2007; p. 293.
16. Association of Official Analytical Chemists (AOAC). *Official Methods of Analysis*, 18th ed.; AOAC: Arlington, VA, USA, 2006.
17. Jakobek, L.; Matić, P.; Ištuk, J.; Barron, A.R. Study of Interactions Between Individual Phenolics of Aronia with Barley β-Glucan. *Pol. J. Food Nutr. Sci.* **2021**, *71*, 187–196. [CrossRef]
18. *HRN ISO 9622:2017*; Milk and Liquid Milk Products—Guidelines for the Application of Mid-Infrared Spectrometry. Croatian Standards Institute: Zagreb, Croatia, 2017.
19. *HRN ISO 13366-2/Ispr.1:2007 (ISO 13366-2:2006)*; Milk-Enumeration of Somatic Cells-Part 2: Guidance on the Operation of Fluoro-Opto-Electronic Counters. Croatian Standards Institute: Zagreb, Croatia, 2007.
20. Alyaqoubi, S.; Abdullah, A.; Addai, Z.R. Antioxidant activity of goat's milk from three different locations in Malaysia. In Proceedings of the Universiti Kebangsaan Malaysia, Faculty of Science and Technology 2014 Postgraduate Colloquium, Selangor, Malaysia, 9–11 April 2014; Volume 1614, pp. 198–201.

21. Qwele, K.; Hugo, A.; Oyedemi, S.O.; Moyo, B.; Masika, P.J.; Muchenje, V. Chemical composition, fatty acid content and antioxidant potential of meat from goats supplemented with Moringa (*Moringa oleifera*) leaves, sunflower cake and grass hay. *Meat Sci.* **2013**, *93*, 455–462. [CrossRef] [PubMed]

22. Oancea, A.-G.; Untea, A.E.; Dragomir, C.; Radu, G.L. Determination of Optimum TBARS Conditions for Evaluation of Cow and Sheep Milk Oxidative Stability. *Appl. Sci.* **2022**, *12*, 6508. [CrossRef]

23. Sun, Q.; Faustman, C.; Senecal, A.; Wilkinson, A.L.; Furr, H. Aldehyde reactivity with 2-thiobarbituric acid and TBARS in freeze-dried beef during accelerated storage. *Meat Sci.* **2001**, *57*, 55–60. [CrossRef]

24. *SAS 9.4 Copyright (c) 2002–2012*; SAS Institute Inc.: Cary, NC, USA, 2013.

25. Frutos, P.; Hervás, G.; Giráldez, F.J.; Mantecón, A.R. Review. Tannins and ruminant nutrition. *Span. J. Agric. Res.* **2004**, *2*, 191–202. [CrossRef]

26. Pieszka, M.; Gogol, P.; Pietras, M.; Pieszka, M. Valuable components of dried pomaces of chokeberry, black currant, strawberry, apple and carrot as a source of natural antioxidants and nutraceuticals in the animal diet. *Ann. Anim. Sci.* **2015**, *15*, 475–491. [CrossRef]

27. Tian, X.Z.; Li, J.X.; Luo, Q.Y.; Zhou, D.; Long, Q.M.; Wang, X.; Lu, Q.; Wen, G.L. Effects of Purple Corn Anthocyanin on Blood Biochemical Indexes, Ruminal Fluid Fermentation, and Rumen Microbiota in Goats. *Front. Vet. Sci.* **2021**, *8*, 715710. [CrossRef]

28. Suong, N.T.M.; Paengkoum, S.; Schonewille, J.T.; Purba, R.A.P.; Paengkoum, P. Growth Performance, Blood Biochemical Indices, Rumen Bacterial Community, and Carcass Characteristics in Goats Fed Anthocyanin-Rich Black Cane Silage. *Front. Vet. Sci.* **2022**, *9*, 880838. [CrossRef] [PubMed]

29. Tian, X.; Lu, Q.; Zhao, S.; Li, J.; Luo, Q.; Wang, X.; Zhang, J.; Zheng, N. Purple corn anthocyanin affects lipid mechanism, flavor compound profiles, and related gene expression of *Longissimus Thoracis et Lumborum* muscle in goats. *Animals* **2021**, *11*, 2407. [CrossRef] [PubMed]

30. Hosoda, K.; Miyaji, M.; Matsuyama, H.; Haga, S.; Ishizaki, H.; Nonaka, K. Effect of supplementation of purple pigment from anthocyanin-rich corn (*Zea mays* L.) on blood antioxidant activity and oxidation resistance in sheep. *Livest. Sci.* **2012**, *145*, 266–270. [CrossRef]

31. Tian, X.Z.; Xin, H.; Paengkoum, P.; Paengkoum, S.; Ban, C.; Sorasak, T. Effects of anthocyanin-rich purple corn (*Zea mays* L.) stover silage on nutrient utilization, rumen fermentation, plasma antioxidant capacity, and mammary gland gene expression in dairy goats. *J. Anim. Sci.* **2019**, *97*, 1384–1397. [CrossRef] [PubMed]

32. Mittler, R. Oxidative stress, antioxidants and stress tolerance. *Trends Plant Sci.* **2002**, *7*, 405–410. [CrossRef] [PubMed]

33. Matsuba, T.; Kubozono, H.; Saegusa, A.; Obata, K.; Gotoh, K.; Miki, K.; Akiyama, T.; Oba, M. Effect of feeding purple corn (*Zea mays* L.) silage on productivity and blood superoxide dismutase concentration in lactating cows. *J. Dairy Sci.* **2019**, *102*, 7179–7182. [CrossRef]

34. Hosoda, K.; Eruden, B.; Matsuyama, H.; Shioya, S. Silage fermentative quality and characteristics of anthocyanin stability in anthocyanin-rich corn (*Zea mays* L.). *Asian-Australas. J. Anim. Sci.* **2009**, *22*, 528–533. [CrossRef]

35. Hosoda, K.; Eruden, B.; Matsuyama, H.; Shioya, S. Effect of anthocyanin-rich corn silage on digestibility, milk production and plasma enzyme activities in lactating dairy cows. *Anim. Sci. J.* **2011**, *83*, 453–459. [CrossRef]

36. Ramos-Escudero, F.; Muñoz, A.M.; Alvarado-Ortíz, C.; Alvarado, Á.; Yáñez, J.A. Purple corn (*Zea mays* L.) phenolic compounds profile and its assessment as an agent against oxidative stress in isolated mouse organs. *J. Med. Food* **2012**, *15*, 704–713. [CrossRef]

37. Kaneko, J.J.; Harvey, J.W.; Bruss, M.L. *Clinical Biochemistry of Domestic Animals*, 6th ed.; Elsevier Academic Press: Amsterdam, The Netherlands, 2008; p. 963.

38. Antunović, Z.; Novoselec, J.; Klir, Ž.; Đidara, M. Hematological parameters in the Alpine goats during lactation. *Poljoprivreda* **2013**, *19*, 40–43.

39. Antunović, Z.; Marić, I.; Klir, Ž.; Šerić, V.; Mioč, B.; Novoselec, J. Haemato-biochemical profile and acid-base status of Croatian spotted goats of different ages. *Arch. Anim. Breed.* **2019**, *62*, 455–463. [CrossRef] [PubMed]

40. Sivamaruthi, B.S.; Kesika, P.; Chaiyasut, C. The Influence of Supplementation of Anthocyanins on Obesity-Associated Comorbidities: A Concise Review. *Foods* **2020**, *9*, 687. [CrossRef] [PubMed]

41. Youdim, K.A.; Shukitt-Hale, B.; MacKinnon, S.; Kalt, W.; Joseph, J.A. Polyphenolics enhance red blood cell resistance to oxidative stress: In vitro and in vivo. *Biochim. Biophys. Acta* **2000**, *1523*, 117–122. [CrossRef] [PubMed]

42. Mattioli, R.; Francioso, A.; Mosca, L.; Silva, P. Anthocyanins: A Comprehensive Review of Their Chemical Properties and Health Effects on Cardiovascular and Neurodegenerative Diseases. *Molecules* **2020**, *25*, 3809. [CrossRef]

43. Calderón-Montaño, J.M.; Burgos-Morón, E.; Pérez-Guerrero, C.; Lopez-Lazaro, M.A. A Review on the Dietary Flavonoid Kaempferol. *Mini-Rev. Med. Chem.* **2011**, *11*, 298–344. [CrossRef]

44. Tsuda, T.; Horio, F.; Osawa, T. Dietary cyanidin 3-O-D-glucoside increases ex vivo oxidation resistance of serum in rats. *Lipids* **1998**, *33*, 583–588. [CrossRef]

*Article*

# Milk Yield, Composition, and Fatty Acid Profile in Milk of Dairy Cows Supplemented with Microalgae *Schizochytrium* sp.: A Meta-Analysis

José Felipe Orzuna-Orzuna [1], Juan Eduardo Godina-Rodríguez [2], Jonathan Raúl Garay-Martínez [3], Guillermo Reséndiz-González [4], Santiago Joaquín-Cancino [5] and Alejandro Lara-Bueno [1,*]

[1] Departamento de Zootecnia, Universidad Autónoma Chapingo, Chapingo C.P. 56230, State of Mexico, Mexico; jforzuna@gmail.com
[2] Campo Experimental Uruapan, Instituto Nacional de Investigaciones Forestales, Agrícolas y Pecuarias, Av. Latinoamérica 1001, Uruapan C.P. 60150, Michoacán, Mexico; godina.juan@inifap.gob.mx
[3] Campo Experimental Las Huastecas, Instituto Nacional de Investigaciones Forestales, Agrícolas y Pecuarias, Altamira C.P. 89610, Tamaulipas, Mexico; garay.jonathan@inifap.gob.mx
[4] Facultad de Estudios Superiores Cuautitlán, Universidad Nacional Autónoma de México, Cuautitlán C.P. 54740, State of Mexico, Mexico; guillermo.resendiz@cuautitlan.unam.mx
[5] Facultad de Ingeniería y Ciencias, Universidad Autónoma de Tamaulipas, Centro Universitario. Cd., Victoria C.P. 87000, Tamaulipas, Mexico; sjoaquin@docentes.uat.edu.mx
* Correspondence: alarab@chapingo.mx

**Abstract:** This study aimed to evaluate the effects of the microalgae (MIAs) *Schizochytrium* sp. as a dietary supplement for dairy cows with respect to the yield, composition, and fatty acid profile of milk using a meta-analytical method. The data used in the statistical analyses were obtained from 11 peer-reviewed scientific publications. The effect size was assessed using the weighted mean differences (WMDs) between MIA-supplemented and control treatments. Dry matter intake, milk fat yield, and milk fat content decreased ($p < 0.001$) in response to the dietary inclusion of *Schizochytrium* sp. MIAs. However, *Schizochytrium* sp. MIAs supplementation increased ($p = 0.029$) milk yield. The dietary inclusion of *Schizochytrium* sp. MIAs decreased ($p < 0.05$) the content of the fatty acids (FAs) butyric, caproic, caprylic, capric, undecanoic, lauric, pentadecanoic, palmitic, heptadecanoic, stearic, arachidic, and total saturated FAs, and it resulted in a ω-6/ω-3 ratio in milk. In contrast, *Schizochytrium* sp. MIAs supplementation increased ($p < 0.05$) the content of linoleic, conjugated linoleic, eicosapentaenoic, behenic, docosahexaenoic, total monounsaturated FAs, total polyunsaturated FAs, and total omega-3 FAs in milk. The results showed that *Schizochytrium* sp. MIAs could be used as a dietary supplement to improve the milk yield and fatty acid profile of milk obtained from dairy cows.

**Keywords:** *Schizochytrium* sp.; *Aurantiochytrium* sp.; meta-regression; milk quality

**Citation:** Orzuna-Orzuna, J.F.; Godina-Rodríguez, J.E.; Garay-Martínez, J.R.; Reséndiz-González, G.; Joaquín-Cancino, S.; Lara-Bueno, A. Milk Yield, Composition, and Fatty Acid Profile in Milk of Dairy Cows Supplemented with Microalgae *Schizochytrium* sp.: A Meta-Analysis. *Agriculture* **2024**, *14*, 1119. https://doi.org/10.3390/agriculture14071119

Academic Editors: Petru Alexandru Vlaicu, Arabela Elena Untea and Mihaela Saracila

Received: 25 June 2024
Revised: 9 July 2024
Accepted: 9 July 2024
Published: 11 July 2024

## 1. Introduction

In recent years, the demand for high-quality milk has increased due to the growth in the world population and improvement of citizens' quality of life [1]. According to Pereira [2], cow's milk is the most consumed milk worldwide and contains several important nutrients for human nutrition and health, such as vitamins, minerals, proteins, fats, and sugars. However, Plata-Pérez et al. [3] and Samková and Kaláč [4] mention that cow's milk has a high content of total saturated fatty acids (SFAs), which have been reported to increase the risk of metabolic and cardiovascular diseases (CVDs) in humans [2]. On the other hand, previous studies [4,5] indicate that cow's milk fat only has a low proportion of total polyunsaturated fatty acids (PUFAs), which decreases triglycerides and cholesterol in blood [6] have anti-inflammatory and antithrombotic properties [7], and benefit brain development and function in humans [8]. Consequently, in recent years, several studies have focused on the search for nutritional strategies that help improve the nutritional and nutraceutical quality of cow's milk through an increase in the total PUFA content and a reduction in the proportion of total SFAs [4,9]. Among the currently available strategies, the inclusion of PUFA-rich products (e.g., protected fats and oilseeds) in cows' diets has been successfully used to increase milk PUFA content [3,4].

Recent studies [1,10] indicate that the inclusion of microalgae (MIAs) in diets for dairy cows can also be used as a nutritional strategy to improve PUFA content and, at the same time, reduce the amount of SFAs in milk. MIAs are photosynthetic microorganisms [11], which contain a wide variety of vitamins, minerals, pigments, proteins (between 39 and 71%), carbohydrates (from 10 to 57%), and between 6 and 86% lipids, mainly long-chain PUFAs [9,11,12]. This variation in the chemical composition of MIAs depends on the cultivation and growth conditions, as well as their physical and chemical processing before being included in diets [11]. In dairy cows, large amounts (>100 g/kg DM) of MIAs have been successfully used as a protein ingredient in the diet [13]. Likewise, other studies have used MIAs to partially replace some ingredients, such as ground corn [14], soybean meal [15], and rapeseed and faba bean meals in the diets of dairy cows [13,16]. However, some recent reviews [17–19] have recommended not including high doses (>20 g/kg DM) of MIAs in dairy ruminant diets to avoid negative effects on fiber digestibility, feed intake, and milk fat content.

Particularly in dairy cows, few studies have evaluated the effects of the use of *Schizochytrium* sp. MIAs as a dietary supplement on milk yield [1,20], milk composition [6,21], and the profile of fatty acids in milk [22,23]. In addition, the results obtained in some of these studies are contradictory, which makes it difficult to draw reliable conclusions. For example, in some studies [10,21], *Schizochytrium* sp. MIAs have negatively affected milk yield and milk fat content. However, other authors have detected positive [24] or neutral [23] effects on milk yield, milk composition, and fatty acid profile in milk from dairy cows supplemented with *Schizochytrium* sp. MIAs. According to recent studies [17,19], the breed and lactation stage of the animals, the doses and species of MIAs, the duration of the experimental period, and the amount of forage in the diet are factors that influence the variability of the results observed between studies.

To date, some narrative reviews have been published [9,11,17] that suggest the use of MIAs as a dietary supplement to improve the productive performance and fatty acid profile of ruminant milk. However, none of these reviews focused only on dairy cows or *Schizochytrium* sp. MIAs, nor did they use a meta-analytic approach. According to Littell et al. [25], a meta-analysis is a set of statistical methods that allows for the quantitative results of multiple studies to be combined and statistically analyzed to obtain objective evidence on a given topic. The hypothesis of the present study states that the use of *Schizochytrium* sp. MIAs as a dietary supplement for dairy cows will positively impact the milk yield and profile of fatty acids in milk. Consequently, the present study aimed to evaluate, using a meta-analytical approach, the effects of the use of *Schizochytrium* sp. MIAs as a dietary supplement for dairy cows on the milk yield, milk composition, and milk fatty acid profile.

## 2. Materials and Methods

### 2.1. Literature Search

The research question was formulated using the PICO strategy proposed by Nishikawa-Pacher [26], in which P is the population, I is the intervention, C is the comparison, and O is the outcome. Therefore, in the current study, the population was dairy cows, the intervention was the dietary supplementation with *Schizochytrium* sp. MIAs, the comparison was between diets with and without the addition of *Schizochytrium* sp. MIAs, and the outcomes were the means of treatments obtained in the milk yield, milk composition, and milk fatty acid profile. Subsequently, the scientific documents that evaluated the effects of dietary supplementation with *Schizochytrium* sp. MIAs in dairy cows were identified, selected, chosen, and included in the database by following the guidelines of the PRISMA protocol [27], as shown in Figure 1. The identification of the literature was carried out through systematic searches using the search engines PubMed, ScienceDirect, Scopus, and Web of Science. The keywords used in all databases were: microalgae, dairy cows, dairy cattle, milk yield, milk composition, milk fatty acid profile, *Schizochitrium* sp., and *Aurantiochytrium* sp. *Aurantiochytrium* sp. MIAs were included in the keywords of the searches carried out because *Schizochytrium* sp. and *Aurantiochytrium* sp. are synonyms [28]. To obtain updated information, literature searches were restricted to studies published in the recent decade (January 2014 to May 2024).

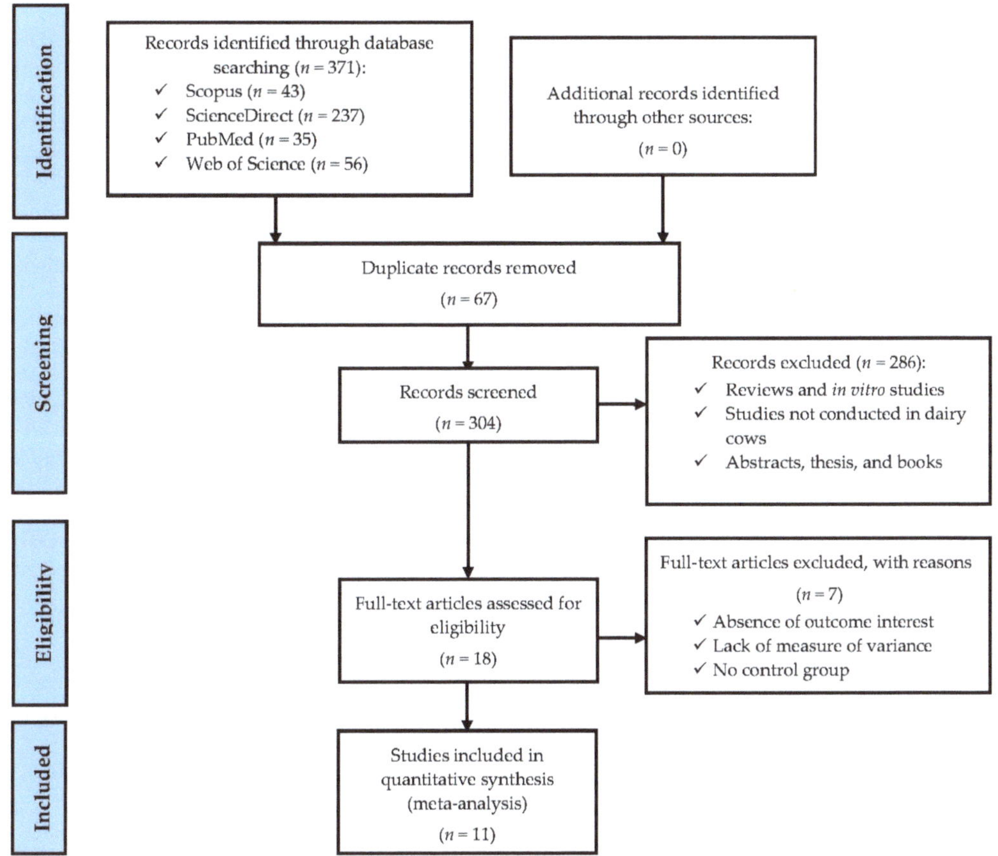


**Figure 1.** A PRISMA flow diagram detailing the literature search strategy and study selection for the meta-analysis.

### 2.2. Inclusion and Exclusion Criteria

Through the searches, 371 scientific documents were identified which, after eliminating duplicate documents, were reduced to 304. After this, documents that had any of the following traits were eliminated: (1) conference proceedings, books, theses, review articles, and simulations; (2) studies that did not use dairy cows or used experimentally infected, or cannulated dairy cows; and (3) studies that did not use MIAs *Schizochytrium* sp. or combined *Schizochytrium* sp. with another genus of microalgae. The remaining documents were evaluated, and only those that met all of the following inclusion criteria were used to form the meta-analysis database [19,29]: (1) full-text scientific articles published in English in peer-reviewed journals; (2) studies using dairy cows as experimental animals and randomized controlled study design; (3) studies that measured and reported data on milk yield, milk composition, or milk fatty acid profile; (4) studies that evaluated the impact of dietary supplementation with *Schizochytrium* sp. MIAs compared with the control using the same basal diet in all treatments; (5) studies that reported the amount (g/kg DM) of *Schizochytrium* sp. MIAs included in the diet or provided the information necessary to estimate it; and (6) studies that included data on the treatment means, number of experimental units (n), and standard error of means (SEM) from the control (diets without *Schizochytrium* sp. MIAs supplementation) and experimental (diets supplemented with *Schizochytrium* sp. MIAs) treatments.

### 2.3. Data Extraction

Table 1 shows the 11 scientific articles used to build the meta-analysis database. The following characteristics of each selected scientific article were extracted into an Excel spreadsheet: (1) name of the author; (2) year of publication; (3) breed of dairy cows; (4) doses (g/kg DM) of *Schizochytrium* sp. MIAs included in the experimental diets; (5) duration of the period of dietary supplementation with *Schizochytrium* sp. MIAs; (6) days in milk from dairy cows; and (7) amount of forage (g/kg DM) included in the diet. The final database only included response variables reported in at least three scientific articles, as recommended by other authors [25,29]. The final database included the following milk yield and milk composition variables: dry matter intake (DMI), milk yield (MY), milk fat yield (MFY), milk protein yield (MPY), milk lactose yield (MLY), milk fat content (MFC), milk protein content (MPC), milk lactose content (MLC), and somatic cell count (SCC). In addition, the database included the following milk fatty acids: butyric (C4:0), caproic (C6:0), caprylic (C8:0), capric (C10:0), undecanoic (C11:0), lauric (C12:0), myristic (C14:0), myristoleic (C14:1), pentadecanoic (C15:0), palmitic (C16:0), palmitoleic (C16:1), heptadecanoic (C17:0), margoleic (C17:1), stearic (C18:0), oleic (C18:1 n-9 cis), linoleic (C18:2 n-6 cis), conjugated linoleic (C18:2 cis-9, trans-11), α-linolenic (C18: 3 n-3), γ-linolenic (C18:3 n 6), arachidic (C20:0), eicosapentaenoic (C20:5 n-3), behenic (C22:0), docosahexaenoic (C22:6 n-3), total saturated fatty acids (SFAs), total monounsaturated fatty acids (MUFAs), total polyunsaturated fatty acids (PUFAs), total omega-3 (ω-3), total omega 6 (ω-6), and ω-6/ω-3 ratio. From the 11 scientific articles selected, the means of treatments, SEM, and n for each response variable were extracted.

**Table 1.** Description of the studies included in the meta-analysis database.

| Reference | Breed | Days in Milk | Supplementation Period, d | Dose (g/kg DM) | Forage, g/kg DM |
|---|---|---|---|---|---|
| Glover et al. [20] | Holstein | NR | 112 | 9.8 | 557 |
| Liu et al. [1] | Holstein | NR | 60 | 6.8, 10.6 | 523 |
| Marques et al. [21] | Holstein | 130 | 84 | 2, 4, 6 | 480 |
| Moate et al. [30] | Holstein | 163 | 80 | 6, 12, 18 | 700, 710, 730 |
| Moran et al. [31] | Holstein | 133 | 84 | 4.1 | 710 |
| Moran et al. [22] | Holstein | 164 | 84 | 6.6, 6.6 | 707 |
| Till et al. [6] | Holstein | 77 | 112 | 2.1, 4.3, 6.4 | 554 |
| Till et al. [23] | Holstein | 25 | 98 | 4.5 | 543 |
| Vahmani et al. [24] | Holstein | 0 | 195 | 9.4 | 51 |
| Vanbergue et al. [10] | Holstein | 100 | 70 | 7.3, 18 | 75.5, 76.1 |
| Vanbergue et al. [32] | Holstein | 100 | 70 | 7.3, 18 | 75.5, 76.1 |

NR = not reported.

### 2.4. Calculations, Statistical Analysis, Heterogeneity, and Publication Bias

The "meta" [33] and "metafor" [34] packages available in R statistical software (version 4.1.2) were used to analyze all data in the current study. The effect size (ES) of *Schizochytrium* sp. MIAs supplementation in dairy cow diets was assessed by examining the weighted mean differences (WMDs) between the experimental (diets with *Schizochytrium* sp. MIAs) and control (diets without *Schizochytrium* sp. MIAs) treatments. According to Takeshima et al. [35], WMDs have greater statistical power and are easier to interpret than other

ES measures, which makes their use advisable. The methods and procedures previously proposed by Der-Simonian and Laird [36] for random effects models were used to weigh the treatment means by inverse variance.

Heterogeneity between studies was rigorously assessed using $I^2$ and Cochran's Q statistics, and significant heterogeneity was declared when Q had $p \leq 0.05$ and $I^2 > 50\%$ [25,37,38]. Furthermore, the statistical tests of Egger [39] and Begg [40] were used to detect the possible presence of publication bias considering a significance threshold of $p \leq 0.05$.

### 2.5. Meta-Regression and Subgroup Analysis

Univariate meta-regression analyses were used to test the effects of *Schizochytrium* sp. MIAs doses, *Schizochytrium* sp. MIAs supplementation periods, days in milk from dairy cows, and amount of forage included in the diets in the heterogeneity detected in the response variables. Meta-regression analyses were performed using Der-Simonian and Laird's [37] method of moments only on response variables that had the following traits: (1) reported in at least ten scientific articles [25]; (2) have $p \leq 0.05$ for the Q test and $I^2 > 50\%$ [38]; and (3) have no publication bias (i.e., $p > 0.05$ in Egger and Begg tests) [39,40]. The doses of *Schizochytrium* sp. MIAs added to the diets ($\leq$10 and 11–18 g/kg DM), the supplementation period (experimental period) with *Schizochytrium* sp. MIAs ($\leq$80 or >80), the days in milk ($\leq$100 or >100), and the level of forage included in the diets ($\leq$500 g/kg DM and >500 g/kg DM) were used as categorical covariates. The breed of dairy cows was not used as a covariate because all studies used Holstein cows (Table 1). The significant covariates ($p \leq 0.05$) were evaluated through subgroup analysis, including only subgroups with at least three comparisons, as recommended by other authors [19,29,41].

## 3. Results

### 3.1. Milk Yield and Composition

Table 2 shows that DMI, MFY, and MFC decreased ($p < 0.001$) in response to *Schizochytrium* sp. MIAs supplementation. In contrast, MY increased in response to *Schizochytrium* sp. MIAs supplementation ($p = 0.029$). On the other hand, MPY, MLY, MPC, MLC, and SCC were not affected ($p > 0.05$) by *Schizochytrium* sp. MIAs supplementation.

**Table 2.** Milk yield and milk composition of dairy cows supplemented with *Schizochytrium* sp. microalgae.

| Item | N (NC) | Control Means (SD) | WMD (95% CI) | *p*-Value | Heterogeneity *p*-Value | $I^2$ (%) | Egger Test [1] *p*-Value | Begg Test [2] *p*-Value |
|---|---|---|---|---|---|---|---|---|
| DMI, kg/d | 10 (19) | 22.845 (1.341) | −0.341 (−0.541; −0.141) | <0.001 | <0.001 | 68.64 | 0.209 | 0.859 |
| MY, kg/d | 10 (19) | 32.06 (5.37) | 0.458 (0.048; 0.868) | 0.029 | 0.967 | 0.00 | 0.455 | 0.646 |
| MFY, kg/d | 9 (18) | 1.251 (0.167) | −0.128 (−0.182; −0.075) | <0.001 | <0.001 | 73.09 | 0.282 | 0.467 |
| MPY, kg/d | 9 (18) | 1.048 (0.173) | 0.005 (−0.01; 1 0.021) | 0.523 | 0.948 | 0.00 | 0.341 | 0.502 |
| MLY, kg/d | 7 (15) | 1.537 (0.266) | 0.015 (−0.006; 0.036) | 0.163 | 0.953 | 0.00 | 0.082 | 0.130 |
| MFC, g/kg | 10 (19) | 35.21 (5.86) | −5.095 (−6.730; −3.461) | <0.001 | <0.001 | 72.03 | 0.398 | 0.242 |
| MPC, g/kg | 10 (19) | 32.68 (1.65) | −0.248 (−0.929; 0.432) | 0.474 | <0.001 | 91.49 | 0.325 | 0.413 |
| MLC, g/kg | 10 (19) | 49.12 (1.99) | −0.054 (−0.359; 0.251) | 0.727 | <0.001 | 68.39 | 0.210 | 0.136 |
| SCC, $\times 10^3$ cell/mL | 7 (12) | 1.907 (0.709) | 0.208 (−0.102; 0.519) | 0.189 | <0.001 | 95.46 | 0.948 | 0.329 |

N = number of studies; NC = number of comparisons between *Schizochytrium* sp. microalgae treatment and control treatment; SD = standard deviation; WMD = weighted mean difference between control and treatments with *Schizochytrium* sp. microalgae; CI = confidence interval of WMD; *p*-value of the $\chi^2$ (Q) test of heterogeneity; $I^2$ = proportion of total variation of size effect estimates that is due to heterogeneity; [1] = Egger's regression asymmetry test; [2] = Begg's adjusted rank correlation; DMI = dry matter intake; MY = milk yield; MFY = milk fat yield; MPY = milk protein yield; MLY = milk lactose yield; MFC = milk fat content; MPC = milk protein content; MLC = milk lactose content; SCC = somatic cell count.

### 3.2. Milk Fatty Acid Profile

Table 3 shows that the dietary inclusion of *Schizochytrium* sp. MIAs did not affect ($p > 0.05$) the content of C14:0, C14:1, C16:1, C17:1, C18:3 n-3, C18:3 n-6, and total ω-6 in milk. However, *Schizochytrium* sp. MIAs supplementation decreased ($p < 0.05$) the content of C4:0, C6:0, C8:0, C10:0, C11:0, C12:0, C15:0, C16:0, C17:0, C18:0, C18:1 n-9 cis, C20:0, and total SFAs and the ω-6/ω-3 ratio in milk. In contrast, dietary supplementation with *Schizochytrium* sp. MIAs increased ($p < 0.05$) the content of C18:2 n-6 cis, C18:2 cis-9, trans-11, C20:5 n-3, C22:0, C22:6 n-3, total MUFAs, total PUFAs, and total ω-3 in milk (Table 3).

**Table 3.** Fatty acids (FAs) (% of total FAs) in the milk of dairy cows supplemented with *Schizochytrium* sp. microalgae.

| Item | N (NC) | Control Means (SD) | WMD (95% CI) | p-Value | Heterogeneity p-Value | Heterogeneity I² (%) | Egger Test [1] p-Value | Begg Test [2] p-Value |
|---|---|---|---|---|---|---|---|---|
| Butyric (C4:0) | 7 (13) | 2.703 (1.545) | −0.025 (−0.047; −0.002) | 0.029 | 0.772 | 0.00 | 0.139 | 0.217 |
| Caproic (C6:0) | 7 (14) | 2.454 (1.518) | −0.060 (−0.097; −0.023) | 0.002 | 0.063 | 43.09 | 0.209 | 0.749 |
| Caprylic (C8:0) | 7 (14) | 1.224 (0.326) | −0.059 (−0.090; −0.029) | <0.001 | 0.086 | 36.25 | 0.636 | 0.855 |
| Capric (C10:0) | 8 (15) | 2.846 (0.522) | −0.217 (−0.293; −0.140) | <0.001 | 0.061 | 39.95 | 0.367 | 0.588 |
| Undecanoic (C11:0) | 5 (11) | 0.665 (0.166) | −0.183 (−0.228; −0.137) | <0.001 | <0.001 | 99.45 | 0.757 | 0.465 |
| Lauric (C12:0) | 7 (14) | 3.537 (0.352) | −0.229 (−0.316; −0.142) | <0.001 | 0.077 | 44.45 | 0.270 | 0.966 |
| Myristic (C14:0) | 8 (15) | 11.640 (0.910) | −0.135 (−0.279; 0.010) | 0.068 | 0.222 | 21.36 | 0.344 | 0.429 |
| Myristoleic (C14:1) | 8 (15) | 1.189 (0.471) | 0.018 (−0.033; 0.069) | 0.494 | 0.088 | 42.77 | 0.489 | 0.478 |
| Pentadecanoic (C15:0) | 7 (13) | 1.291 (0.470) | −0.038 (−0.056; −0.019) | <0.001 | 0.219 | 22.23 | 0.141 | 0.189 |
| Palmitic (C16:0) | 8 (16) | 34.119 (3.957) | −0.728 (−1.276; −0.181) | 0.009 | 0.063 | 41.31 | 0.144 | 0.646 |
| Palmitoleic (C16:1) | 5 (11) | 1.979 (0.538) | 0.031 (−0.084; 0.146) | 0.602 | 0.076 | 40.91 | 0.230 | 0.384 |
| Heptadecanoic (C17:0) | 8 (16) | 0.601 (0.185) | −0.009 (−0.014; −0.003) | 0.002 | 0.312 | 12.33 | 0.610 | 0.615 |
| Margoleic (C17:1) | 5 (10) | 0.439 (0.342) | 0.002 (−0.003; 0.007) | 0.501 | 0.128 | 34.95 | 0.313 | 0.117 |
| Stearic (C18:0) | 8 (16) | 9.083 (2.046) | −2.026 (−2.911; −1.142) | <0.001 | <0.001 | 96.05 | 0.974 | 0.646 |
| Oleic (C18:1 n-9 cis) | 7 (13) | 19.424 (2.830) | −0.725 (−1.461; 0.012) | 0.049 | <0.001 | 96.90 | 0.830 | 0.228 |
| Linoleic (C18:2 n-6 cis) | 6 (10) | 2.057 (0.718) | 0.173 (0.107; 0.238) | <0.001 | 0.501 | 0.00 | 0.362 | 0.428 |
| Conjugated linoleic (C18:2 cis-9, trans-11) | 8 (16) | 0.489 (0.244) | 0.503 (0.350; 0.657) | <0.001 | <0.001 | 95.92 | 0.099 | 0.128 |
| α-linolenic (C18:3 n-3) | 9 (14) | 0.432 (0.201) | −0.005 (−0.030; 0.020) | 0.722 | 0.064 | 39.47 | 0.153 | 0.086 |
| γ-linolenic (C18:3 n-6) | 4 (8) | 0.038 (0.019) | 0.028 (−0.020; 0.076) | 0.252 | <0.001 | 99.55 | 0.692 | 0.406 |
| Arachidic (C20:0) | 7 (15) | 0.165 (0.042) | −0.018 (−0.031; −0.005) | 0.006 | <0.001 | 98.18 | 0.174 | 0.129 |
| Eicosapentaenoic (C20:5 n-3) | 9 (16) | 0.061 (0.032) | 0.022 (0.010; 0.034) | <0.001 | <0.001 | 98.65 | 0.275 | 0.867 |
| Behenic (C22:0) | 6 (12) | 0.079 (0.043) | 0.015 (0.003; 0.026) | 0.014 | <0.001 | 92.25 | 0.196 | 0.087 |
| Docosahexaenoic (C22:6 n-3) | 19 (16) | 0.042 (0.031) | 0.226 (0.187; 0.264) | <0.001 | <0.001 | 95.44 | 0.062 | 0.321 |
| Total SFA | 11 (21) | 69.345 (4.216) | −2.419 (−3.225; −1.613) | <0.001 | <0.001 | 74.67 | 0.082 | 0.305 |
| Total MUFA | 11 (21) | 24.452 (3.395) | 0.842 (0.260; 1.424) | 0.005 | <0.001 | 65.25 | 0.125 | 0.087 |
| Total PUFA | 11 (21) | 6.490 (9.110) | 2.001 (1.412; 2.591) | <0.001 | <0.001 | 98.83 | 0.386 | 0.333 |
| Total omega-3 (ω-3) | 8 (16) | 0.614 (0.274) | 0.183 (0.133; 0.233) | <0.001 | <0.001 | 97.47 | 0.173 | 0.478 |
| Total omega 6 (ω-6) | 8 (16) | 3.123 (1.270) | 0.253 (0.074; 0.431) | 0.006 | <0.001 | 95.78 | 0.068 | 0.227 |
| ω-6/ω-3 ratio | 6 (12) | 6.35 (4.42) | −0.611 (−0.820; −0.402) | <0.001 | <0.001 | 86.36 | 0.993 | 0.998 |

N = number of studies; NC = number of comparisons between *Schizochytrium* sp. microalgae treatment and control treatment; SD = standard deviation; WMD = weighted mean difference between control and treatments with *Schizochytrium* sp. microalgae; CI = confidence interval of WMD; p-value of the $\chi^2$ (Q) test of heterogeneity; I² = proportion of total variation of size effect estimates that is due to heterogeneity; [1] = Egger's regression asymmetry test; [2] = Begg's adjusted rank correlation; SFA = saturated fatty acid; MUFA = monounsaturated fatty acid; PUFA = polyunsaturated fatty acid.

### 3.3. Publication Bias and Meta-Regression

Tables 2 and 3 show that Egger's regression asymmetry test and Begg's adjusted rank correlation were not significant ($p > 0.05$) for any of the response variables tested, indicating that publication bias was not detected.

Table 2 shows that there was significant heterogeneity (Q) ($p \leq 0.05$) in DMI, MFY, MFC, MPC, MLC, and SCC. Likewise, Table 3 shows that there was significant Q ($p \leq 0.05$) in the content of FAs C11:0, C18:0, C18:1n-9 cis, C18:2 cis-9, trans-11, C18:3 n-6, C20:0, C20:5 n-3, C22:0, C22:6 n-3, total SFAs, total MUFAs, total PUFAs, total ω-3, and total ω-6 and the ω-6/ω-3 ratio. In the current study, meta-regression analyses were only performed on

DMI, MFC, MPC, MLC, total SFAs, total MUFAs, and total PUFAs. This is justified because the other response variables that had heterogeneity were reported in less than ten studies, and under these conditions, the power of the test is low [25].

Table 4 shows that the covariates' days in milk and supplementation period did not have a significant relationship ($p > 0.05$) with any of the response variables evaluated. None of the covariates used had a significant relationship ($p < 0.05$) with MFC, MPC, total MUFAs, and total PUFAs. The *Schizochytrium* sp. MIAs dose covariate explained ($p < 0.05$) 68.72 and 18.32% of the observed heterogeneity for MLC and total SFAs, respectively. The level of forage in the diet explained ($p < 0.05$) 43.62 and 12.39% of the observed heterogeneity for DMI and MLC, respectively.

**Table 4.** Meta-regression of the effects of dietary *Schizochytrium* sp. microalgae supplementation on the milk production and quality of dairy cows.

| Parameter | | Days in Milk | Supplementation Period | Microalgae Dose | Forage Level |
|---|---|---|---|---|---|
| Dry matter intake (DMI) | QM | 0.778 | 0.150 | 0.085 | 12.656 |
| | df | 1 | 1 | 1 | 1 |
| | *p*-Value | 0.378 | 0.699 | 0.770 | <0.001 |
| | $R^2$ (%) | 0.00 | 0.00 | 0.00 | 43.62 |
| Milk fat content (MFC) | QM | 0.220 | 0.221 | 2.996 | 0.444 |
| | df | 1 | 1 | 1 | 1 |
| | *p*-Value | 0.136 | 0.638 | 0.083 | 0.505 |
| | $R^2$ (%) | 0.27 | 0.00 | 9.22 | 0.00 |
| Milk protein content (MPC) | QM | 0.497 | 0.069 | 0.372 | 2.240 |
| | df | 1 | 1 | 1 | 1 |
| | *p*-Value | 0.481 | 0.792 | 0.542 | 0.134 |
| | $R^2$ (%) | 4.44 | 0.00 | 0.00 | 3.89 |
| Milk lactose content (MLC) | QM | 1.382 | 0.397 | 24.533 | 3.387 |
| | df | 1 | 1 | 1 | 1 |
| | *p*-Value | 0.240 | 0.529 | <0.001 | 0.046 |
| | $R^2$ (%) | 0.00 | 1.02 | 68.72 | 12.39 |
| Total SFAs | QM | 0.054 | 2.349 | 3.425 | 0.880 |
| | df | 1 | 1 | 1 | 1 |
| | *p*-Value | 0.816 | 0.125 | 0.044 | 0.127 |
| | $R^2$ (%) | 0.00 | 0.00 | 18.32 | 0.00 |
| Total MUFAs | QM | 1.620 | 1.303 | 0.292 | 2.546 |
| | df | 1 | 1 | 1 | 1 |
| | *p*-Value | 0.203 | 0.254 | 0.589 | 0.111 |
| | $R^2$ (%) | 1.70 | 7.96 | 0.00 | 0.00 |
| Total PUFAs | QM | 0.684 | 0.079 | 3.020 | 1.352 |
| | df | 1 | 1 | 1 | 1 |
| | *p*-Value | 0.408 | 0.779 | 0.082 | 0.301 |
| | $R^2$ (%) | 3.97 | 3.10 | 2.74 | 0.00 |

QM = coefficient of moderators; QM is considered significant at $p \leq 0.05$; df: degree of freedom; $R^2$ = the amount of heterogeneity accounted for; SFAs = saturated fatty acids; MUFAs = monounsaturated fatty acids; PUFAs = polyunsaturated fatty acids.

### 3.4. Subgroup Analysis

Figure 2a shows that MLC was decreased ($p < 0.05$) when low doses ($\leq$10 g/kg DM) of *Schizochytrium* sp. MIAs were used. However, the inclusion of *Schizochytrium* sp. MIAs at doses ranging between 11 and 18 g/kg DM did not affect MLC ($p > 0.05$). On the other hand, the total SFAs content decreased ($p < 0.001$) regardless of the dose of *Schizochytrium* sp. MIAs used (Figure 2b).

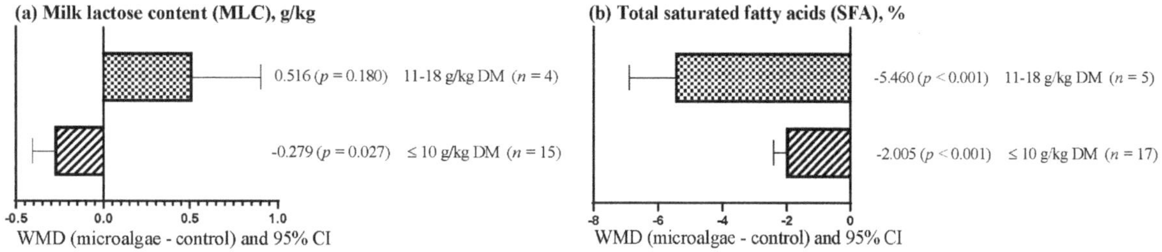



**Figure 2.** Subgroup analysis (subgroup = doses of microalgae (g/kg DM)) of the effect of including *Schizochytrium* sp. microalgae in dairy cows' diets, WMD = weighted mean differences between microalgae treatments and the control.

Figure 3a shows that DMI decreased ($p < 0.05$) regardless of the level of forage included in the diets. On the other hand, MLC decreased ($p < 0.05$) in dairy cows consuming diets with a forage level $\leq 500$ g/kg DM. However, MLC was not affected when the forage level in the diet of dairy cows was greater than 500 g/kg DM (Figure 3b).

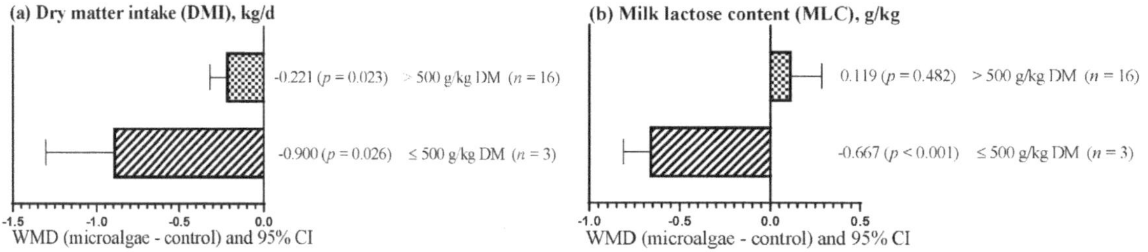



**Figure 3.** Subgroup analysis (subgroup = forage in diet (g/kg DM)) of the effect of *Schizochytrium* sp. microalgae supplementation in dairy cows' diets; WMD = weighted mean differences between microalgae treatments and the control.

## 4. Discussion

### 4.1. Milk Yield and Composition

In the present meta-analysis, supplementation with *Schizochytrium* sp. MIAs decreased DMI. This effect could be related to the unpleasant taste and smell that MIAs generally have [17]. Likewise, it has been hypothesized that the lower DMI observed in ruminants supplemented with *Schizochytrium* sp. MIAs may be related to possible toxic effects of the PUFAs contained in MIAs on ruminal microorganisms, which decreases fiber digestibility. This hypothesis is supported by recent studies [42,43], which indicate that the inclusion of *Schizochytrium* sp. MIAs as a supplement in ruminant diets decreases the relative abundance of fiber-degrading ruminal bacteria (e.g., *Fibrobacter succinogenes* and *Ruminococcus albus*). Likewise, the lower DMI observed in dairy cows supplemented with *Schizochytrium* sp. MIAs could be related to the PUFAs contained in the MIAs [21] as PUFAs stimulate the release of intestinal hormones in dairy cows, such as glucagon-like peptide-1 (GLP-1) and cholecystokinin (CKK), and DMI decreases when plasma levels of CKK and GLP-1 increase [44].

In the present meta-analysis, the dietary inclusion of *Schizochytrium* sp. MIAs increased MY; however, MFY and MFC decreased in response to *Schizochytrium* sp. MIAs supplementation. In a previous meta-analysis, Orzuna-Orzuna et al. [19] also detected higher MY and lower MFC in lactating small ruminants supplemented with MIAs as a dietary supplement. The higher MY and lower DMI observed suggest that *Schizochytrium* sp. MIAs improved feed efficiency (FE) in dairy cows; however, there was no necessary information in the database to assess FE directly. Some previous reviews [11,12,17] report that most *Schizochytrium* sp. MIAs contain folic acid, vitamin E, B–complex vitamins ($B_1$,

$B_6$, and $B_{12}$), and essential amino acids (e.g., methionine and lysine). The presence of these nutrients in the *Schizochytrium* sp. MIAs could explain the higher MY observed in the present meta-analysis as recent studies [45–48] have shown that these nutrients improve nutritional status and energy metabolism and lead to higher MY in dairy cows.

The lower MFC observed in the present meta-analysis could be negative as a positive correlation between MFC and cheese yield has been reported [49]. The exact mechanism of the lower MFC in dairy cows supplemented with *Schizochytrium* sp. MIAs has yet to be studied [6]. However, the lower MFC observed could be related to the high content of PUFAs in the *Schizochytrium* sp. MIAs [50], as Hussein et al. [51] suggest that dietary supplementation with products rich in PUFAs may inhibit circulating fatty acid uptake and de novo synthesis by the mammary gland. This hypothesis is supported by the results of Vahmani et al. [52] in dairy cows supplemented with *Schizochytrium* sp. MIAs. In that study, the authors detected lower hepatic expression of some genes (lipoprotein lipase, fatty acid synthase, stearoyl-CoA desaturase) involved in the uptake, de novo synthesis, and desaturation of fatty acids. In addition, according to Mavrommatis and Tsiplakou [53], MFC has a positive and negative correlation with the contents of C18:0 and C18:2 cis-9, trans-11 in milk, respectively. Therefore, the lower content of C18:0 and the higher proportion of C18:2 cis-9, trans-11 in milk detected in the present study partially explain the reduction in MFC. Likewise, the lower MFY observed could be directly related to the lower MFC because, in dairy cows, there is a positive correlation (r = 0.40) between MFY and MFC [54].

*4.2. Milk Fatty Acid Profile*

According to Hanuš et al. [55], making specific changes in the fatty acid profile of milk can improve the health of a large part of the human population. For example, it has been reported that a lower content of SFAs (C4:0-C20:0) in milk can decrease the risk of CVDs in humans [2]. In the present meta-analysis, MIAs supplementation decreased the content of most SFAs (C4:0, C6:0, C8:0, C10:0, C11:0, C12:0, C15:0, C16, C17:0, C18:0, and C20:0) in milk, which could benefit the health of consumers. In dairy cows or other lactating ruminants, the percentage transfer of C4:0, C6:0, and C8:0 to the blood has not been evaluated. However, in lactating ruminants, dietary supplementation with *Schizochytrium* sp. MIAs decreases between 21 and 76% of the blood levels of C10:0, C11:0, C12:0, C14:0, C15:0, C16:0, C17:0, C18:0, and C20:0 [1,23,53,56]. Therefore, similar effects of *Schizochytrium* sp. MIAs consumption in the present study could reduce the amount of C10:0, C11:0, C12:0, C14:0, C15:0, C16, C17:0, C18:0, and C20:0 reaching the mammary gland and reduce the percentage of these fatty acids in milk.

In dairy cows, vitamin $B_{12}$ deficiency decreases the transformation of propionic acid to succinyl-CoA [47]. According to Pećina and Ivanković [57], this reduction leads to an accumulation of propionic acid and favors the formation of C15:0 and C17:0 FAs. *Schizochytrium* sp. MIAs contain vitamin $B_{12}$, which partially explains the reduction observed in the current study in the contents of FAs C15:0 and C17:0 in milk. On the other hand, dietary supplementation with *Schizochytrium* sp. MIAs decreases the relative abundance of *Butyrivibrio proteoclasticus* bacteria in the ruminal fluid of dairy goats [58]. According to Costa et al. [18], *B. proteoclasticus* has a predominant role in the final step of the ruminal biohydrogenation (RBH) process, which is required for the saturation of 18-carbon unsaturated fatty acids and their conversion into C18:0 [59]. Therefore, a similar effect of consuming *Schizochytrium* sp. MIAs on the ruminal microbiota of dairy cows could explain the reduction observed in the current study in the concentration of C18:0 in milk. Mavrommatis and Tsiplakou [53] indicate that, in the ruminant mammary gland, C18:0 can be used as a substrate by the enzyme stearoyl-CoA desaturase (Δ9-desaturase) to synthesize C18:1 n-9 cis de novo. Therefore, the reduction of C18:1 n-9 cis in milk observed in the present meta-analysis could be related to the lower concentration of C18:0 observed in response to supplementation with *Schizochytrium* sp. MIAs.

In the present study, the use of *Schizochytrium* sp. MIAs as a dietary supplement for dairy cows increased the contents of C18:2 n-6 cis and C18:2 cis-9, trans-11 in milk. Similar results were reported by Orzuna-Orzuna et al. [19] in a meta-analysis of lactating small ruminants supplemented with MIAs. In dairy cows, it has been reported that supplementation with low doses (between 4.5 and 10.6 g/kg DM) of *Schizochytrium* sp. MIAs increases blood levels of vaccenic (C18:1 trans-11) acid to between 17.4 and 45.4% [1,23]. This effect could increase the amount of C18:1 trans-11 reaching the mammary gland and result in increased amounts of C18:2 cis-9, trans-11 in milk because, in the mammary gland, the enzyme $\Delta 9$-desaturase can convert C18:1 trans-11 to C18:2 cis-9, trans-11 [60]. In humans, the consumption of C18:2 n-6 cis decreases serum cholesterol levels, improves glucose metabolism, and is inversely correlated with the incidence of CVDs [61]. Likewise, the consumption of C18:2 cis-9, trans-11 has anti-inflammatory effects and decreases the risk of obesity, diabetes, atherosclerosis, and some types of cancer in humans [62].

The contents of C20:5 n-3, C22:0, and C22:6 n-3 in milk increased in response to the dietary inclusion of *Schizochytrium* sp. MIAs. A previously published meta-analysis [19] also detected a higher ratio of C20:5 n-3, C22:0, and C22:6 n-3 in milk from small ruminants supplemented with MIAs. It has been confirmed that MIAs contain C20:5 n-3 that are partially protected from the effects of the ruminal biohydrogenation (RBH) process [63]. This effect could increase the flow and duodenal absorption of C20:5 n-3 and subsequently increase the incorporation of this fatty acid in milk. Likewise, it has been reported that C22:6 n-3 can be saturated at C22:0 in the rumen [64]. *Schizochytrium* sp. MIAs contain a high concentration (up to 38% of total FAs) of C22:6 n-3 [12,28], and the saturation of a part of this C22:6 n-3 could be related to the higher C22:0 content observed in the current study. In addition, recent studies [1,31] have reported that C22:6 n-3 from *Schizochytrium* sp. MIAs consumed by dairy cows could be transferred to milk with an efficiency of up to 13.2%, which partially explains its increase in milk. Shahidi and Ambigaipalan [8] indicate that the consumption of foods rich in C20:5 n-3 and C22:6 n-3 decreases the risk of CVDs and improves cognitive function in humans. Likewise, a high consumption of C22:0 improves the intestinal absorption of C22:6 n-3 and reduces the incidence of fatty liver [65]. Therefore, including *Schizochytrium* sp. MIAs in diets for dairy cows could help produce milk that benefits consumers' health.

In the present study, the dietary inclusion of *Schizochytrium* sp. MIAs increased the MUFAs, PUFAs, and total $\omega$-3 contents in milk and, at the same time, decreased the total SFAs content and the $\omega$-6/$\omega$-3 ratio. These effects could be directly related to the changes observed in the proportions of individual FAs in the current study. A previous meta-analysis [66] found that MUFA consumption reduces the risk of stroke in humans by 17%. Maki et al. [67] reported that the consumption of PUFAs reduces the incidence of CVD problems and diabetes by 30 and 50%, respectively. Likewise, the consumption of $\omega$-3 reduces mortality in patients with heart problems by up to 20% [7]. In contrast, previous studies [55,68] have indicated that the consumption of foods with a high SFA content and a high $\omega$-6/$\omega$-3 ratio increases the risk of obesity, atherosclerosis, CVDs, and Alzheimer's in humans. Therefore, the changes detected in MUFAs, PUFAs, total $\omega$-3, SFAs, and $\omega$-6/$\omega$-3 ratio in milk suggest that *Schizochytrium* sp. MIAs can be used as a dietary additive to produce healthier cow's milk for humans.

## 5. Conclusions

The results of the present study suggest that the inclusion of *Schizochytrium* sp. microalgae in diets for dairy cows decreases dry matter intake, as well as milk fat content and yield. However, *Schizochytrium* sp. microalgae can be used as a dietary supplement to improve milk yield and fatty acid profile in milk from dairy cows. Furthermore, more studies are needed to evaluate the simultaneous use of *Schizochytrium* sp. microalgae and some other additives that prevent the reduction of milk fat content.

**Author Contributions:** Conceptualization, J.F.O.-O.; methodology, J.F.O.-O.; software, J.F.O.-O., J.E.G.-R. and G.R.-G.; validation, J.E.G.-R., J.R.G.-M., A.L.-B., G.R.-G. and S.J.-C.; formal analysis, J.F.O.-O.; investigation, J.E.G.-R., J.R.G.-M. and G.R.-G.; resources, J.E.G.-R., J.R.G.-M., A.L.-B., G.R.-G. and S.J.-C.; data curation, J.E.G.-R., J.R.G.-M. and S.J.-C.; writing—original draft preparation, J.F.O.-O.;

writing—review and editing, J.E.G.-R., J.R.G.-M., A.L.-B., G.R.-G. and S.J.-C.; visualization, S.J.-C. and A.L.-B.; supervision, A.L.-B.; project administration, A.L.-B.; funding acquisition, A.L.-B. All authors have read and agreed to the published version of the manuscript.

**Funding:** This research received no external funding.

**Institutional Review Board Statement:** Not applicable.

**Data Availability Statement:** The datasets used and analyzed during the current study are available from the corresponding author upon reasonable request.

**Conflicts of Interest:** The authors declare no conflicts of interest.


# References

1. Liu, G.; Yu, X.; Li, S.; Shao, W.; Zhang, N. Effects of Dietary Microalgae (*Schizochytrium* Spp.) Supplement on milk Performance, Blood Parameters, and Milk Fatty Acid Composition in dairy Cows. *Czech J. Anim. Sci.* **2020**, *65*, 162–171. [CrossRef]
2. Pereira, P.C. Milk nutritional composition and its role in human health. *Nutrition* **2014**, *30*, 619–627. [CrossRef] [PubMed]
3. Plata-Pérez, G.; Angeles-Hernandez, J.C.; Morales-Almaráz, E.; Del Razo-Rodríguez, O.E.; López-González, F.; Peláez-Acero, A.; Campos-Montiel, R.G.; Vargas-Bello-Pérez, E.; Vieyra-Alberto, R. Oilseed Supplementation Improves Milk Composition and Fatty Acid Profile of Cow Milk: A Meta-Analysis and Meta-Regression. *Animals* **2022**, *12*, 1642. [CrossRef] [PubMed]
4. Samková, E.; Kalač, P. Rapeseed supplements affect propitiously fatty acid composition of cow milk fat: A meta-analysis. *Livest. Sci.* **2021**, *244*, 104382. [CrossRef]
5. Gallardo, W.B.; Teixeira, I.A.M.A. Associations between Dietary Fatty Acid Profile and Milk Fat Production and Fatty Acid Composition in Dairy Cows: A Meta-Analysis. *Animals* **2023**, *13*, 2063. [CrossRef] [PubMed]
6. Till, B.E.; Huntington, J.A.; Posri, W.; Early, R.; Taylor-Pickard, J.; Sinclair, L.A. Influence of rate of inclusion of microalgae on the sensory characteristics and fatty acid composition of cheese and performance of dairy cows. *J. Dairy Sci.* **2019**, *102*, 10934–10946. [CrossRef] [PubMed]
7. Livingstone, K.M.; Lovegrove, J.A.; Cockcroft, J.R.; Elwood, P.C.; Pickering, J.E.; Givens, D.I. Does dairy food intake predict arterial stiffness and blood pressure in men? Evidence from the caerphilly prospective study. *Hypertension* **2013**, *61*, 42–47. [CrossRef]
8. Shahidi, F.; Ambigaipalan, P. Omega-3 Polyunsaturated Fatty Acids and Their Health Benefits. *Annu. Rev. Food Sci. Technol.* **2018**, *9*, 345–381. [CrossRef]
9. Altomonte, I.; Salari, F.; Licitra, R.; Martini, M. Use of Microalgae in Ruminant Nutrition and Implications on Milk Quality—A Review. *Livest. Sci.* **2018**, *214*, 25–35. [CrossRef]
10. Vanbergue, E.; Peyraud, J.L.; Hurtaud, C. Effects of new n-3 fatty acid sources on milk fatty acid profile and milk fat properties in dairy cows. *J. Dairy Res.* **2018**, *85*, 265–272. [CrossRef]
11. Saadaoui, I.; Rasheed, R.; Aguilar, A. Microalgal-based feed: Promising alternative feedstocks for livestock and poultry production. *J. Anim. Sci. Biotechnol.* **2021**, *12*, 76. [CrossRef] [PubMed]
12. Dineshbabu, G.; Goswami, G.; Kumar, R.; Sinha, A.; Das, D. Microalgae—nutritious, sustainable aqua- and animal feed source. *J. Funct. Foods* **2019**, *62*, 103545. [CrossRef]
13. Lamminen, M.; Halmemies-Beauchet-Filleau, A.; Kokkonen, T.; Simpura, I.; Jaakkola, S.; Vanhatalo, A. Comparison of microalgae and rapeseed meal as supplementary protein in the grass silage based nutrition of dairy cows. *Anim. Feed Sci. Technol.* **2017**, *234*, 295–311. [CrossRef]
14. Da Silva, G.G.; de Jesus, E.F.; Takiya, C.S.; del Valle, T.A.; da Silva, T.H.; Vendramini, T.H.A.; Yu, E.J.; Rennó, F.P. Short communication: Partial replacement of ground corn with algae meal in a dairy cow diet: Milk yield and composition, nutrient digestibility, and metabolic profile. *J. Dairy Sci.* **2016**, *99*, 8880–8884. [CrossRef] [PubMed]
15. Lamminen, M.; Halmemies-Beauchet-Filleau, A.; Kokkonen, T.; Jaakkola, S.; Vanhatalo, A. Different Microalgae Species as a Substitutive Protein Feed for Soya Bean Meal in Grass Silage Based Dairy Cow Diets. *Anim. Feed Sci. Technol.* **2019**, *247*, 112–126. [CrossRef]
16. Lamminen, M.; Halmemies-Beauchet-Filleau, A.; Kokkonen, T.; Vanhatalo, A.; Jaakkola, S. The effect of partial substitution of rapeseed meal and faba beans by Spirulina platensis microalgae on milk production, nitrogen utilization, and amino acid metabolism of lactating dairy cows. *J. Dairy Sci.* **2019**, *102*, 7102–7117. [CrossRef] [PubMed]
17. Valente, L.; Cabrita, A.; Maia, M.; Valente, I.; Engrola, S.; Fonseca, A.; Freire, J. *Microalgae as Feed Ingredients for Livestock Production and Aquaculture. Microalgae—Cultivation, Recovery of Compounds and Applications*; Academic Press: Cambridge, MA, USA, 2021; pp. 239–312.
18. Costa, D.F.A.; Castro-Montoya, J.M.; Harper, K.; Trevaskis, L.; Jackson, E.L.; Quigley, S. Algae as Feedstuff for Ruminants: A Focus on Single-Cell Species, Opportunistic Use of Algal By-Products and On-Site Production. *Microorganisms* **2022**, *10*, 2313. [CrossRef] [PubMed]
19. Orzuna-Orzuna, J.F.; Chay-Canul, A.J.; Lara-Bueno, A. Performance, milk fatty acid profile and oxidative status of lactating small ruminants supplemented with microalgae: A meta-analysis. *Small Rumin. Res.* **2023**, *226*, 107031. [CrossRef]

20. Glover, E.K.; Budge, S.; Rose, M.; Rupasinghe, H.P.V.; MacLaren, L.; Green-Johnson, J.; Fredeen, A.H. Effect of Feeding Fresh Forage and Marine Algae on the Fatty Acid Composition and Oxidation of Milk and Butter. *J. Dairy Sci.* **2012**, *95*, 2797–2809. [CrossRef] [PubMed]

21. Marques, J.A.; Del Valle, T.A.; Ghizzi, L.G.; Zilio, E.M.C.; Gheller, L.S.; Nunes, A.T.; Silva, T.B.P.; Dias, M.S.d.S.; Grigoletto, N.T.S.; Koontz, A.F.; et al. Increasing dietary levels of docosahexaenoic acid-rich microalgae: Ruminal fermentation, animal performance, and milk fatty acid profile of mid-lactating dairy cows. *J. Dairy Sci.* **2019**, *102*, 5054–5065. [CrossRef]

22. Moran, C.A.; Morlacchini, M.; Keegan, J.D.; Warren, H.; Fusconi, G. Dietary supplementation of dairy cows with a docosahexaenoic acid-rich thraustochytrid, *Aurantiochytrium limacinum*: Effects on milk quality, fatty acid composition and cheese making properties. *J. Anim. Feed Sci.* **2019**, *28*, 3–14. [CrossRef]

23. Till, B.E.; Huntington, J.A.; Kliem, K.E.; Taylor-Pickard, J.; Sinclair, L.A. Long term dietary supplementation with microalgae increases plasma docosahexaenoic acid in milk and plasma but does not affect plasma 13,14-dihydro-15-keto PGF2α concentration in dairy cows. *J. Dairy Res.* **2020**, *87*, 14–22. [CrossRef] [PubMed]

24. Vahmani, P.; Fredeen, A.H.; Glover, K.E. Effect of supplementation with fish oil or microalgae on fatty acid composition of milk from cows managed in confinement or pasture systems. *J. Dairy Sci.* **2013**, *96*, 6660–6670. [CrossRef] [PubMed]

25. Littell, J.H.; Corcoran, J.; Pillai, V. *Systematic Reviews and Meta-Analysis*, 1st ed.; Oxford University Press: Oxford, UK, 2008; pp. 111–132.

26. Nishikawa-Pacher, A. Research questions with PICO: A universal mnemonic. *Publications* **2022**, *10*, 21. [CrossRef]

27. Moher, D.; Liberati, A.; Tetzlaff, J.; Altman, D.G.; Group, P. Preferred reporting items for systematic reviews and meta-analyses: The PRISMA statement. *PLoS Med.* **2009**, *6*, e1000097. [CrossRef] [PubMed]

28. Chi, G.; Xu, Y.; Cao, X.; Li, Z.; Cao, M.; Chisti, Y.; He, H. Production of polyunsaturated fatty acids by *Schizochytrium* (*Aurantiochytrium*) spp. *Biotechnol. Adv.* **2022**, *55*, 107897. [CrossRef]

29. Dorantes-Iturbide, G.; Orzuna-Orzuna, J.F.; Lara-Bueno, A.; Mendoza-Martínez, G.D.; Miranda-Romero, L.A.; Lee-Rangel, H.A. Essential oils as a dietary additive for small ruminants: A meta-analysis on performance, rumen parameters, serum metabolites, and product quality. *Vet. Sci.* **2022**, *9*, 475. [CrossRef]

30. Moate, P.J.; Williams, S.R.O.; Hannah, M.C.; Eckard, R.J.; Auldist, M.J.; Ribaux, B.E.; Jacobs, J.L.; Wales, W.J. Effects of feeding algal meal high in docosahexaenoic acid on feed intake, milk production, and methane emissions in dairy cows. *J. Dairy Sci.* **2013**, *96*, 3177–3188. [CrossRef]

31. Moran, C.A.; Morlacchini, M.; Keegan, J.D.; Fusconi, G. The Effect of Dietary Supplementation with *Aurantiochytrium limacinum* on Lactating Dairy Cows in Terms of Animal Health, Productivity and Milk Composition. *J. Anim. Physiol. Anim. Nutr.* **2018**, *102*, 576–590. [CrossRef]

32. Vanbergue, E.; Hurtaud, C.; Peyraud, J.L.; Beuvier, E.; Duboz, G.; Buchin, S. Effects of n-3 fatty acid sources on butter and hard cooked cheese; technological properties and sensory quality. *Int. Dairy J.* **2018**, *82*, 35–44. [CrossRef]

33. Schwarzer, G.; Carpenter, J.R.; Rücker, G. *Meta-Analysis with R*; Springer: Berlin/Heidelberg, Germany, 2015; Volume 4784, ISBN 978-3-319-21415-3.

34. Viechtbauer, W. Conducting Meta-Analyses in R with the metafor Package. *J. Stat. Softw.* **2010**, *36*, 1–48. [CrossRef]

35. Takeshima, N.; Sozu, T.; Tajika, A.; Ogawa, Y.; Hayasaka, Y.; Furukawa, T.A. Which Is more generalizable, powerful and interpretable in meta-analyses, mean difference or standardized mean difference? *BMC Med. Res. Methodol.* **2014**, *14*, 30. [CrossRef]

36. Der-Simonian, R.; Laird, N. Meta-analysis in clinical trials. *Control Clin. Trials* **1986**, *7*, 177–188. [CrossRef] [PubMed]

37. Borenstein, M.; Hedges, L.V.; Higgins, J.P.T.; Rothstein, H.R. *Introduction to Meta-Analysis*, 1st ed.; John Wiley and Sons, Ltd.: Chichester, UK, 2009; p. 413.

38. Borenstein, M.; Higgins, J.P.T.; Hedges, L.V.; Rothstein, H.R. Basics of Meta-Analysis: I2 Is Not an Absolute Measure of Heterogeneity. *Res. Synth. Methods* **2017**, *8*, 5–18. [CrossRef] [PubMed]

39. Egger, M.; Smith, G.D.; Schneider, M.; Minder, C. Bias in meta-analysis detected by a simple, graphical test. *Br. Med. J.* **1997**, *315*, 629–634. [CrossRef]

40. Begg, C.B.; Mazumdar, M. Operating characteristics of a rank correlation test for publication bias. *Biometrics* **1994**, *50*, 1088–1101. [CrossRef]

41. Orzuna-Orzuna, J.F.; Hernández-García, P.A.; Chay-Canul, A.J.; Galván, C.D.; Ortiz, P.B.R. Microalgae as a dietary additive for lambs: A meta-analysis on growth performance, meat quality, and meat fatty acid profile. *Small Rumin. Res.* **2023**, *226*, 107072. [CrossRef]

42. Mavrommatis, A.; Skliros, D.; Flemetakis, E.; Tsiplakou, E. Changes in the Rumen Bacteriome Structure and Enzymatic Activities of Goats in Response to Dietary Supplementation with *Schizochytrium* spp. *Microorganisms* **2021**, *9*, 1528. [CrossRef]

43. Mavrommatis, A.; Sotirakoglou, K.; Skliros, D.; Flemetakis, E.; Tsiplakou, E. Dose and time response of dietary supplementation with *Schizochytrium* sp. on the abundances of several microorganisms in the rumen liquid of dairy goats. *Livest. Sci.* **2021**, *247*, 104489. [CrossRef]

44. Relling, A.; Reynolds, C. Feeding rumen-inert fats differing in their degree of saturation decreases intake and increases plasma concentrations of gut peptides in lactating dairy cows. *J. Dairy Sci.* **2007**, *90*, 1506–1515. [CrossRef]

45. Zanton, G.I.; Bowman, G.R.; Vázquez-Añón, M.; Rode, L.M. Meta-analysis of lactation performance in dairy cows receiving supplemental dietary methionine sources or postruminal infusion of methionine. *J. Dairy Sci.* **2014**, *97*, 7085–7101. [CrossRef] [PubMed]
46. Moghimi-Kandelousi, M.; Alamouti, A.A.; Imani, M.; Zebeli, Q. A meta-analysis and meta-regression of the effects of vitamin E supplementation on serum enrichment, udder health, milk yield, and reproductive performance of transition cows. *J. Dairy Sci.* **2020**, *103*, 6157–6166. [CrossRef] [PubMed]
47. Duplessis, M.; Lapierre, H.; Sauerwein, H.; Girard, C.L. Combined biotin, folic acid, and vitamin B12 supplementation given during the transition period to dairy cows: Part I. Effects on lactation performance, energy and protein metabolism, and hormones. *J. Dairy Sci.* **2022**, *105*, 7079–7096. [CrossRef] [PubMed]
48. Wang, L.; Li, Z.; Lei, X.; Yao, J. Effect of folic acid supplementation on lactation performance of Holstein dairy cows: A meta-analysis. *Anim. Feed Sci. Technol.* **2023**, *296*, 115551. [CrossRef]
49. De Oca-Flores, M.E.; Espinoza-Ortega, A.; Arriaga-Jordán, C.M. Technological and physicochemical properties of milk and physicochemical aspects of traditional Oaxaca cheese. *Rev. Mex. Cienc. Pecu.* **2019**, *10*, 367–378. [CrossRef]
50. Yang, Q.; Lu, T.; Yan, J.; Li, J.; Zhou, H.; Pan, X.; Lu, Y.; He, N.; Ling, X. Regulation of polyunsaturated fatty acids synthesis by enhancing carotenoid-mediated endogenous antioxidant capacity in *Schizochytrium* sp. *Algal Res.* **2021**, *55*, 102238. [CrossRef]
51. Hussein, M.; Harvatine, K.H.; Weerasinghe, W.M.P.B.; Sinclair, L.A.; Bauman, D.E. Conjugated linoleic acid-induced milk fat depression in lactating ewes is accompanied by reduced expression of mammary genes involved in lipid synthesis. *J. Dairy Sci.* **2013**, *96*, 3825–3834. [CrossRef] [PubMed]
52. Vahmani, P.; Glover, K.E.; Fredeen, A.H. Effects of pasture versus confinement and marine oil supplementation on the expression of genes involved in lipid metabolism in mammary, liver, and adipose tissues of lactating dairy cows. *J. Dairy Sci.* **2014**, *97*, 4174–4183. [CrossRef] [PubMed]
53. Mavrommatis, A.; Tsiplakou, E. The impact of the dietary supplementation level with *Schizochytrium* sp. on milk chemical composition and fatty acid profile, of both blood plasma and milk of goats. *Small Rumin. Res.* **2020**, *193*, 106252. [CrossRef]
54. Wongpom, B.; Koonawootrittriron, S.; Elzo, M.A.; Suwanasopee, T. Milk yield, fat yield and fat percentage associations in a Thai multibreed dairy population. *Agric. Nat. Res.* **2017**, *51*, 218–222. [CrossRef]
55. Hanuš, O.; Samková, E.; Křížová, L.; Hasoňová, L.; Kala, R. Role of Fatty Acids in Milk Fat and the Influence of Selected Factors on Their Variability—A Review. *Molecules* **2018**, *23*, 1636. [CrossRef] [PubMed]
56. Christodoulou, C.; Kotsampasi, B.; Dotas, V.; Simoni, M.; Righi, F.; Tsiplakou, E. The effect of Spirulina supplementation in ewes' oxidative status and milk quality. *Anim. Feed Sci. Technol.* **2023**, *295*, 115544. [CrossRef]
57. Pećina, M.; Ivanković, A. Candidate genes and fatty acids in beef meat, a review. *Ital. J. Anim. Sci.* **2021**, *20*, 1716–1729. [CrossRef]
58. Mavrommatis, A.; Skliros, D.; Sotirakoglou, K.; Flemetakis, E.; Tsiplakou, E. The effect of forage-to-concentrate ratio on *Schizochytrium* spp.-supplemented goats: Modifying rumen microbiota. *Animals* **2021**, *11*, 2746. [CrossRef] [PubMed]
59. Dewanckele, L.; Vlaeminck, B.; Hernandez-Sanabria, E.; Ruiz-González, A.; Debruyne, S.; Jeyanathan, J.; Fievez, V. Rumen Biohydrogenation and Microbial Community Changes Upon Early Life Supplementation of 22:6n-3 Enriched Microalgae to Goats. *Front. Microbiol.* **2018**, *9*, 573. [CrossRef] [PubMed]
60. Póti, P.; Pajor, F.; Bodnár, Á.; Penksza, K.; Köles, P. Effect of micro-alga supplementation on goat and cow milk fatty acid composition. *Chil. J. Agric. Res.* **2015**, *75*, 259–263. [CrossRef]
61. Marangoni, F.; Agostoni, C.; Borghi, C.; Catapano, A.L.; Cena, H.; Ghiselli, A.; Poli, A. Dietary linoleic acid and human health: Focus on cardiovascular and cardiometabolic effects. *Atherosclerosis* **2020**, *292*, 90–98. [CrossRef] [PubMed]
62. Yang, B.; Chen, H.; Stanton, C.; Ross, R.P.; Zhang, H.; Chen, Y.Q.; Chen, W. Review of the roles of conjugated linoleic acid in health and disease. *J. Funct. Foods* **2015**, *15*, 314–325. [CrossRef]
63. Vítor, A.C.; Francisco, A.E.; Silva, J.; Pinho, M.; Huws, S.A.; Santos-Silva, J.; Alves, S.P. Freeze-dried *Nannochloropsis oceanica* biomass protects eicosapentaenoic acid (EPA) from metabolization in the rumen of lambs. *Sci. Rep.* **2021**, *11*, 21878. [CrossRef]
64. Aldai, N.; Delmonte, P.; Alves, S.P.; Bessa, R.J.; Kramer, J.K. Evidence for the initial steps of DHA biohydrogenation by mixed ruminal microorganisms from sheep involves formation of conjugated fatty acids. *J. Agric. Food Chem.* **2018**, *66*, 842–855. [CrossRef]
65. Moreira, D.K.T.; Santos, P.S.; Gambero, A.; Macedo, G.A. Evaluation of structured lipids with behenic acid in the prevention of obesity. *Food Res. Int.* **2017**, *95*, 52–58. [CrossRef] [PubMed]
66. Schwingshackl, L.; Hoffmann, G. Monounsaturated fatty acids, olive oil and health status: A systematic review and meta-analysis of cohort studies. *Lipids Health Dis.* **2014**, *13*, 154. [CrossRef] [PubMed]
67. Maki, K.C.; Eren, F.; Cassens, M.E.; Dicklin, M.R.; Davidson, M.H. $\omega$-6 polyunsaturated fatty acids and cardiometabolic health: Current evidence, controversies, and research gaps. *Adv. Nutr.* **2018**, *9*, 688–700. [CrossRef]
68. Patterson, E.; Wall, R.; Fitzgerald, G.F.; Ross, R.P.; Stanton, C. Health implications of high dietary omega-6 polyunsaturated fatty acids. *J. Nutr. Metab.* **2012**, *2012*, 539426. [CrossRef] [PubMed]

# Effect of Different Levels of Calcium and Addition of Magnesium in the Diet on Garden Snails' (*Cornu aspersum*) Condition, Production, and Nutritional Parameters

Anna Rygało-Galewska [1,*], Klara Zglińska [1], Mateusz Roguski [1], Kamil Roman [2], Wiktor Bendowski [1], Damian Bień [3] and Tomasz Niemiec [1]

[1] Division of Animal Nutrition, Institute of Animal Sciences, Warsaw University of Life Sciences, 166 Nowoursynowska St., 02-787 Warsaw, Poland; klara_zglinska@sggw.edu.pl (K.Z.); mateusz_roguski@sggw.edu.pl (M.R.); wiktor_bendowski@sggw.edu.pl (W.B.); tomasz_niemiec@sggw.edu.pl (T.N.)

[2] Institute of Wood Sciences and Furniture, Warsaw University of Life Sciences, 166 Nowoursynowska St., 02-787 Warsaw, Poland; kamil_roman@sggw.edu.pl

[3] Department of Animal Breeding, Institute of Animal Sciences, Warsaw University of Life Sciences, 166 Nowoursynowska St., 02-787 Warsaw, Poland; damian_bien@sggw.edu.pl

* Correspondence: anna_rygalo-galewska@sggw.edu.pl

**Abstract:** Edible snails are an attractive protein source due to their high growth rate, cost-efficiency, and nutritional value. Calcium is crucial for snail growth, reproduction, and shell formation, while magnesium plays a role in enzyme function and muscle tone. This study aimed to optimise calcium and magnesium levels in *Cornu aspersum* diets to optimise the production and technological characteristics of the derived animal products. Snails were fed specific diets in controlled conditions with varying calcium and magnesium levels (44.3, 66.1, 88.7, 103.5 Ca g/kg feed and 3.3, 5.6, 7.2 Mg g/kg feed) for four months. Their growth, shell characteristics, and meat composition were evaluated. As calcium in the feed increased, carcass and shell weights were higher. Also, the crushing force of the shells was higher with increasing amount of calcium in the feed. In the group with 10.35% calcium and 0.72% magnesium, snail growth significantly slowed down after three months, with lower mortality. It is suggested that a shortened fattening cycle by 3–4 weeks compared to the magnesium-free diet is possible. However, based on meat, shell, mortality, and feed intake analysis, a 0.56% magnesium concentration in the feed seems to give better results, as magnesium content at 0.72% might be toxic to snails. Further investigation is to confirm the possibility of neutralising the negative effects of magnesium in the diet through increasing calcium and phosphorus intake.

**Keywords:** edible snails; *Cornu aspersum*; snail nutrition; magnesium addition; calcium addition

**Citation:** Rygało-Galewska, A.; Zglińska, K.; Roguski, M.; Roman, K.; Bendowski, W.; Bień, D.; Niemiec, T. Effect of Different Levels of Calcium and Addition of Magnesium in the Diet on Garden Snails' (*Cornu aspersum*) Condition, Production, and Nutritional Parameters. *Agriculture* **2023**, *13*, 2055. https://doi.org/10.3390/agriculture13112055

Academic Editors: Petru Alexandru Vlaicu, Arabela Elena Untea and Mihaela Saracila

Received: 28 September 2023
Revised: 23 October 2023
Accepted: 25 October 2023
Published: 26 October 2023

## 1. Introduction

New animal protein sources are being investigated to meet the growing human population's demands. Edible snails are an interesting source because they are easy to maintain, have a fast fattening cycle, are cost-effective, and are relatively environmentally friendly in production [1–3]. Snails are also valued for their low-calorie meat and caviar. In many European coastal countries (Italy, Spain, France, Greece), land snails are an integral part of the national cuisine, and the demand for their production is constantly high [4,5]. Snail meat is a good source of protein (59.53–67.42% of dry matter—DM), exogenous amino acids and unsaturated fatty acids. However, snail meat is relatively low in fat (4.24–7.30% DM). It is also a good source of vitamins and minerals [6].

One of the biggest problems in edible snail farming is achieving adequate mechanical strength in the shells. More shell brittleness hampers pre-processing—preventing mechanical cleaning and sorting of shells as well as long-distance transport, causing financial losses, as shells that are damaged, even if they only result in minor losses, are unsuitable for further use. Shell quality also affects the value of the snails for export, as a garden snail (*Cornu aspersum*) is traditionally served in the shell [7].

The shell makes up about a third of a snail's body weight. The function of the shell is to protect the snail's internal organs, prevent water loss, provide shelter from the cold, and protect from predators and microorganisms [8,9]. The outer part of the shell—the periostracum—can be divided into three layers: the outer conchiolin layer, made up of organic matter, which gives the shell its colour; the crystalline middle layer, made up mainly of calcium carbonate (about 95–99% of the total shell composition); and the inner layer, known as the pearl layer [10,11]. The shell ends with an opening called a peristomium. In adult specimens, the edge of the opening forms a characteristic upturned lip [12]. The quality of the shell is closely linked to the formation of its internal structure. This process involves the formation of an organic matrix of amino acids and glycoproteins in which calcium carbonate crystallises in the form of aragonite. The level of bioavailable calcium (Ca) forms in the feed influences the correct crystallisation and formation of aragonite layers under farm conditions. The mature shell exhibits species-specific microstructural variations in the arrangement of crystals and organic elements. The degree of mineral saturation of successive layers is highest in the inner layers and decreases outwards through the layers [9,13].

The shell also stores Ca, which the snails use to produce eggshells during the laying period. Excessive drainage of minerals from the shell leads to animal death or cannibalism (feeding on the shells of other living or dead individuals) [14].

Ca influences the adequate mineralisation of the skeleton in vertebrate organisms. It also affects muscle contraction, enzyme activation, cell differentiation, immune response, programmed cell death, and neuronal activity [15]. It is an essential element for the proper growth and development of animals [16,17].

In snails, Ca is crucial for shell formation [18–20] and during egg-laying by adults [7,21,22]. Ca deficiency can retard snail growth and render their shells more susceptible to damage. Conversely, excess Ca can reduce their body weight and cause shell thinning, increasing their vulnerability [7,23]. The Ca content in their diet influences the snails' nutritional preferences [24], and its availability can be constrained by shell growth [25,26]. Snails acquire Ca ions through dietary consumption, water filtration, or passive diffusion across their bodies. Once absorbed, these ions may be temporarily stored in the mantle's connective tissues or the calcifying mantle epithelium [27]. As previously noted, soluble Ca and bicarbonate ions circulate in the hemolymph, and Ca can be stored or transported as amorphous granules [28]. These granules are a Ca reserve readily available for release, especially during urgent situations like shell repair [29]. In minor quantities, Ca is deposited in the digestive gland, connective tissue cells, foot tissue, and surrounding major blood vessels [30,31].

Both Ca deficiency and its excess in the diet can lead to a decrease in snail body weight [7,32]. Adequate Ca intake enhances the reproductive outcomes of these animals [26,33,34], and a subsequent rise in their survival rate has been observed [7,35,36]. Depending on the snail species and the Ca source, 6.0 to 20.9% dry matter in the feed is advised [8,37,38].

Magnesium (Mg) is distinguished by its pronounced chemical activity [39,40], which influences various physiological functions, such as improving energy metabolism, protein synthesis, and replication prior to cell division [41] as well as cell proliferation and wound healing [42–44]. Magnesium supplementation has also been shown to have a beneficial effect on reproduction in animals [45,46]. Over 90% of intracellular Mg is associated with ribosomes, lysosomes, phospholipids, nucleotides, and mitochondria functions and structures. It aids in the structural stabilisation of proteins, nucleic acids, and membrane surface bonds [47,48]. Mg's attributes related to membrane stabilisation have implications for generating reactive oxygen species and lipid peroxidation and releasing and modulating neurotransmitters [49]. This mineral orchestrates the function of approximately 300 enzymes pivotal in the metabolic transformation of carbohydrates, fats, proteins, nucleic acids, and redox reactions. It facilitates the synthesis of high-energy compounds, namely ATP and ADP, within cells and is involved in phosphorylation reactions, forming complexes with ATP ($Mg^{2+}$ ATP) [43,50]. A myriad of enzymes, directly or indirectly, are influenced by fluctuations in Mg concentrations, encompassing those engaged in cell cycle regulation [51] and glucose utilisation [52].

Furthermore, Mg partially influences smooth muscle tone by counteracting Ca in potassium and Ca channels [53–55]. The metabolism of Mg is intricately intertwined with that of Ca and phosphorus [56–58]. The transport mechanisms of Mg are interconnected with other elements, predominantly via shared proteins that oversee their exchange and translocation across cellular membranes. These elements are Na [59–61], Cu, Fe, Mn [62–64], and K [53,65,66]. Particularly strong associations have been found between Mg and Na, K, and Ca [67–69]. Antagonistic interactions with Ca and zinc have also been demonstrated [70,71]. The interplay between Mg and phosphate absorption appears to be due to Mg–Ca interactions, mainly driven by renal phosphorus conservation [72].

Despite the absence of mucosal injury, Mg deficiency causes subclinical inflammation in the small intestine, leading to functional changes in local and distant organs and increasing susceptibility to oxidative stress [73]. Mg deficiency primarily results in elevated cellular $Ca^{2+}$, which acts as a signal for priming cells to initiate an inflammatory response. The primary effects of Mg deficiency result from changes in extracellular Mg concentrations due to direct membrane effects and Mg–Ca interactions [74]. Mg deficiency also disrupts bone metabolism, impairing bone growth [75] and leading to bone loss [76].

Mg is absorbed primarily in the small intestine at an efficiency of approximately 60%, mostly via passive transport. In this site, potassium, Ca, and ammonia are its antagonists [77]. Mg is transported to various tissues, and its uptake occurs in response to physiological demand [78]. Analogous to Ca, the primary site of Mg retention is the kidneys [79]. In dietary Mg insufficiency, diminished plasma Mg concentrations can precipitate a swift release from the bone surface [80]. Consequently, it has been established that deficiencies in Mg can culminate in osteoporosis [81].

In snails, Mg constitutes a vital component of their shells [18,19,82]. Considering the literature addressing the impact of dietary Mg on fattening metrics or the quality of snail shells [83], this research endeavours to elucidate these associations.

Determining the optimal amount of Ca and Mg in snail feed can increase animal survival and improve reproductive performance. Farmers may also benefit from better feed utilisation by snails, increased mechanical shell strength, and accelerated animal maturation, reducing fattening time. These benefits may translate into direct economic benefits, allowing savings in the fattening period. Snails' price depends on their maturity level, based on the appearance of their shell lip. Also, an increased final body weight of snails and animal survival rate result in higher earnings for the farmer. Snail meat's enhanced quality and dietary benefits, including mineral enrichment, may be crucial to the end consumer.

The objective of this study was to determine the optimal Ca concentration and assess the influence of different levels of Mg addition in the diet for *Cornu aspersum*. That was done to optimise the production and technological characteristics of the derived animal products.

## 2. Materials and Methods

### 2.1. Animals and Experimental Design

In Experiments I and II, *Cornu aspersum* snails were housed at the Warsaw University of Life Sciences animal facility (52.15890, 21.04554). The animal facility was maintained under strict, controlled conditions: ambient temperature was consistently held at $22\,°C \pm 1\,°C$, with a relative humidity of $60\% \pm 10\%$. A 12:12 h light cycle was implemented. The room was provided with a mechanical-gravity ventilation system. The snails were housed in 10 L ventilated plastic containers with sterilised soil at the base to regulate humidity (refer to Figure 1). The mobile racks on which the containers were placed were repositioned twice a week. Additionally, a weekly rotation was carried out on the containers located on individual shelves.

A complete feed mixture, formulated for experiments at the Division of Animal Nutrition (detailed composition provided in Table 1), was administered ad libitum. This feed was presented in a flour form, with corn meal as its primary component. The composition of this compound feed was determined based on prior research and the standards set by the French INRA [84]. Per these standards, the protein content should range between 16–18% (and reach up to 24% [25]), while fat and fibre contents should not exceed 4% of the dry matter. Mayaki et al. [85] suggest that the optimal Ca content in snail feed should be 8%. However, other research recommends values ranging from 6% [37] to 17.5% [86] to achieve optimal production outcomes. Ca carbonate was selected as the Ca source primarily because of its prevalent use in snail production as fodder chalk and its distinction as the purest and most efficient form of Ca supplement in feed [7,87,88]. The premix included only vitamins and amino acids, devoid of any minerals.

**Figure 1.** (**a**) Experimental container; (**b**) Setting of containers in the animal house; (**c**) Adult snails foraging on experimental feed.

Residual feed post-consumption was gathered to determine the FCR (Feed Conversion Ratio)—a metric representing the quantity of feed ingested in kg (DM) by the snails for every kg of body weight gain.

In both experiments, snails were procured commercially as hatchlings and introduced into the experimental containers when they were 2–3 days old. Upon their allocation to the containers, which was done randomly, both their body weight and shell width were ascertained using a scale and a caliper (AOS Absolute Digimatic Caliper, Mitutoyo, Japan). This instrument was consistently employed for all subsequent measurements. At the commencement of the experiment, each experimental group consisted of 120 snails, with

40 snails allocated to each experimental container within the group. The initial density was determined in alignment with the guidelines established for the commercial rearing of snails [12,89,90].

**Table 1.** Ingredients and determined chemical composition of snail diets in Experiments I and II.

| Ingredients (%) | Experiment I | | | | | Experiment II | | |
|---|---|---|---|---|---|---|---|---|
| | C I | C II | C III | C IV | CON | M I | M II | M III |
| Corn meal | 53.0 | 50.0 | 45.0 | 43.3 | 43.8 | 43.5 | 43.3 | 43.1 |
| Post-extraction soybean meal 460 | 18.0 | 19.0 | 21.0 | 21.0 | 21.0 | 21.0 | 21.0 | 21.0 |
| Wheat bran | 10.0 | 8.0 | 6.0 | 5.0 | 5.0 | 5.0 | 5.0 | 5.0 |
| Fodder yeast | 4.0 | 4.0 | 4.0 | 4.0 | 4.0 | 4.0 | 4.0 | 4.0 |
| Rapeseed oil | 2.0 | 2.0 | 2.5 | 2.5 | 2.0 | 2.0 | 2.0 | 2.0 |
| Monocalcium phosphate | 2.0 | 2.0 | 2.0 | 2.0 | 2.0 | 2.0 | 2.0 | 2.0 |
| NaCl | 0.2 | 0.2 | 0.2 | 0.2 | 0.2 | 0.2 | 0.2 | 0.2 |
| Vitamin premix | 2.0 | 2.0 | 2.0 | 2.0 | 2.0 | 2.0 | 2.0 | 2.0 |
| Ca carbonate | 8.9 | 13.2 | 17.3 | 20.0 | 20.0 | 20.0 | 20.0 | 20.0 |
| Mg oxide (g) | - | - | - | - | - | 0.8 | 3.1 | 4.7 |
| Composition (% DM) | | | | | | | | |
| Crude protein | $18.22 \pm 0.53$ | $18.86 \pm 0.76$ | $18.72 \pm 0.44$ | $18.30 \pm 0.68$ | $18.40 \pm 0.58$ | $18.74 \pm 0.63$ | $19.03 \pm 0.71$ | $18.82 \pm 0.42$ |
| Ether extracts | $2.93 \pm 0.09$ | $3.11 \pm 0.12$ | $3.32 \pm 0.11$ | $3.31 \pm 0.07$ | $3.01 \pm 0.10$ | $3.33 \pm 0.13$ | $3.15 \pm 0.10$ | $3.12 \pm 0.08$ |
| Crude fibre | $3.62 \pm 0.09$ | $4.05 \pm 0.09$ | $3.47 \pm 0.09$ | $3.30 \pm 0.09$ | $3.33 \pm 0.09$ | $3.62 \pm 0.09$ | $3.59 \pm 0.09$ | $3.81 \pm 0.09$ |
| Crude ash | $21.15 \pm 0.62$ | $23.48 \pm 0.73$ | $25.48 \pm 0.93$ | $27.30 \pm 1.03$ | $27.24 \pm 0.79$ | $26.30 \pm 0.81$ | $27.00 \pm 0.53$ | $27.60 \pm 0.86$ |
| Nitrogen-free extract | $54.08 \pm 2.09$ | $50.50 \pm 1.83$ | $49.01 \pm 1.12$ | $47.78 \pm 1.06$ | $48.02 \pm 1.66$ | $48.01 \pm 0.67$ | $47.23 \pm 1.34$ | $46.65 \pm 1.12$ |
| Ca | $4.43 \pm 0.13$ | $6.61 \pm 0.21$ | $8.87 \pm 0.32$ | $10.35 \pm 0.39$ | $10.35 \pm 0.22$ | $10.35 \pm 0.31$ | $10.35 \pm 0.40$ | $10.35 \pm 0.25$ |
| Mg | $0.26 \pm 0.01$ | $0.25 \pm 0.01$ | $0.26 \pm 0.01$ | $0.24 \pm 0.01$ | $0.25 \pm 0.01$ | $0.33 \pm 0.01$ | $0.56 \pm 0.02$ | $0.72 \pm 0.02$ |
| Calculated BE (MJ) | $13.21 \pm 0.52$ | $12.79 \pm 0.44$ | $12.59 \pm 0.49$ | $12.32 \pm 0.37$ | $12.26 \pm 0.42$ | $12.43 \pm 0.50$ | $12.28 \pm 0.38$ | $12.14 \pm 0.45$ |

Experiment I was designed to explore the influence of varying Ca concentrations on snail production metrics, shell morphometry, and the nutritional attributes of the mollusc meat (refer to Figure 2). Each month, fifty individuals were randomly selected from each experimental group and their weight and shell width were measured, following the same procedure used at the start of the experiment. Mortality rates for animals within the group were determined, and feed consumption was calculated. Lastly, at the end of the experiment, fifty randomly selected individuals from each group were euthanised through freezing and utilised for further laboratory analyses pertaining to the quality of their carcasses and shells.

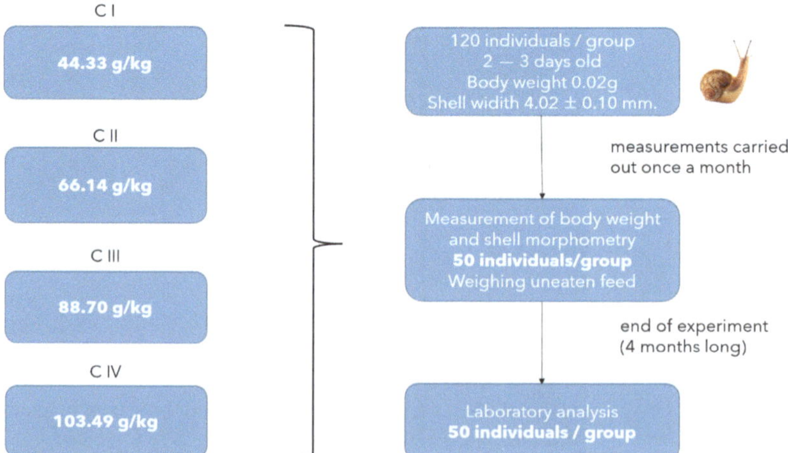


**Figure 2.** Scheme illustrating the design of Experiment I: categorisation of the groups based on Ca content, details of the snails' initial body parameters, the methodology for measuring the snails, and the end of the research.

Given the extensive range of recommended dietary Ca levels cited in the literature, the decision was made to evaluate four distinct levels in this study. The C I group contained Ca levels of 44.22 g/kg feed, the C II group, 66.14 g/kg feed; the C III group, 88.70 g/kg feed; and the C IV group, 103.49 g/kg feed.

Experiment II, as depicted in Figure 3, aimed to assess the impact of Mg concentrations on snail production results, shell morphometry, and the nutritional attributes of the mollusc meat. The procedure in Experiment II mirrored that of Experiment I, except that forty-six individuals, chosen at random, underwent laboratory analysis instead of fifty. The diet from Experiment I, which yielded the most favourable results regarding snail fattening, was adopted as the benchmark for the control diet in this study. Mg oxide was chosen as the Mg supplement for the diet, given its status as the most prevalent mineral source with the highest available Mg concentration for feed [77]. Furthermore, Mg oxide typically ensures efficient absorption of Mg ions.

The control group (CG) had a Mg content of 2.45 g/kg feed (with no additional Mg). This level was determined based on prior research on poultry [91]. The M I group had a Mg content of 2.36 g/kg feed, the M II group was at 5.56 g/kg feed, and the M III group was set at 7.15 g/kg feed. The Mg concentrations in these rations were chosen based on studies in poultry, assuming that the nutritional requirements of laying hens may be similar to those of snails [92,93].

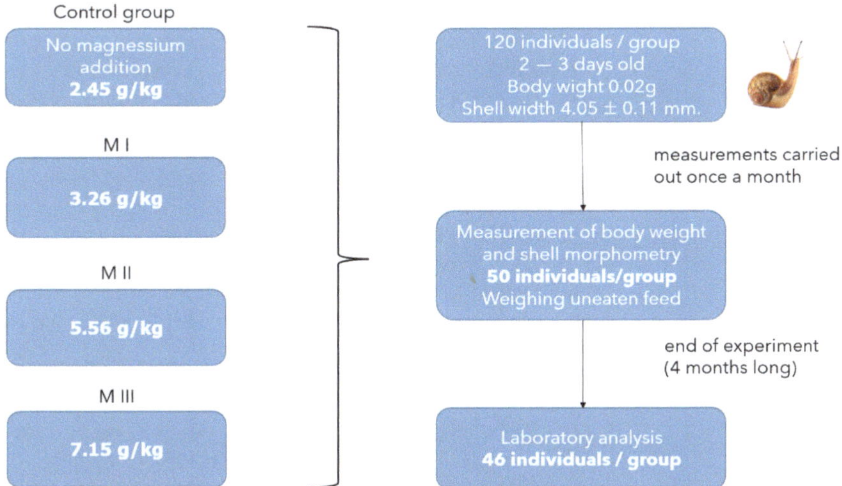


**Figure 3.** Scheme illustrating the design of Experiment II: categorisation of the groups based on Mg content, details of the snails' initial body parameters, the methodology for measuring the snails, and the end of the research.

## 2.2. Experimental and Analytical Procedures

In the laboratory, carcasses were removed from shells. Carcass weight and the proportion of carcass to total body weight were calculated using data from fifty randomly chosen snails from Experiment I and forty-six from Experiment II. Then, further analyses of the chemical composition of snail meat, TBARS (thiobarbituric acid reactive substances) and mineral composition of the snail meat and shells were conducted in 6 replicates.

The chemical composition of the snail carcasses was determined according to AOAC (Association of Official Analytical Chemists) [94]: moisture and dry matter by drying at 105 °C to constant weight, crude ash by incineration at 550 °C for six h, crude protein ($N \times 6.25$) using the micro-Kjeldahl technique (Kjeltec System 1026 Distilling Unit, Foss Tecator, Sweden), and crude fat after extraction with petroleum ether by the Soxhlet method.

Thiobarbituric acid reactive substances (TBARS) were expressed as equivalents of malondialdehyde (MDA), and MDA precursor—1.2.3.3-tetraethoxypropane (TEP) was used as a standard to plot a standard curve. The methodology presented by Uchiyam and Mihar [95] was used. The test tissues of snail carcasses were homogenised in 1% potassium chloride solution and then centrifuged at 2000 rpm for 15 min at 4 °C. Then, 1% phosphoric acid, 1% potassium chloride, 2% butyl hydroxyanisole (BHA), and 0.4% TBA were added to the resulting filtrate. After adding the reagents and mixing, the sample was capped and placed in a water bath at 100 °C for 60 min. After cooling, 4 mL of butanol was added to the sample and shaken for 2 min. The absorbance of compounds dissolved in the butanol phase was read at 532 nm. The results were read in nmol/mL from a standard curve against TEP.

Inductively coupled plasma optical emission spectrometry (ICP-OES) determined the mineral components of carcasses and snail shells. First, test samples were ground and mineralised in a chamber furnace at 450 °C for 12 h. Then, the acid digestion (hydrochloric acid 37%, AnalaR NORMAPUR®, VWR Chemicals, Radnor, PA, USA) occurred on a hotplate. The material was then filtered through blotting paper and diluted to 50 mL with distilled water. The total contents of Ca, Cu (copper), Fe (iron), Mg, Mn (manganese), P (phosphorus), K (potassium), Zn (zinc), Si (silicone), and Na (sodium) were determined with ICP-OEC equipment (Perkin Elmer Avio 200; Waltham, MA, USA).

During shell analysis, the share of matured individuals was determined by inspecting the presence of solid shell lip, indicating that the individual has reached maturity and ended the growth cycle [9]. Also, shell measurements were taken to establish the shell shape index—shell width-to-height ratio [96] and solidity index. This method is based on one used on poultry eggs [97], previously used in snail shell estimation by Ligaszewski et al. [98]. The formula follows [shell weight $\times$ (height width)$^{-1}$] $\times$ 100.

An Instron 3382 testing device (Norwood, MA, USA) was employed to evaluate the crushing force exerted on the shells. This assessment aimed to discern the potential influence of varying Mg levels in the feed on shell robustness. The testing setup encompassed several components: a device designated for compression tests, plates marked for compression testing, and a computer system for data acquisition.

The device was calibrated to administer loads ranging from 0 to 100 kN (kilonewton). Static tests were conducted at a consistent speed of 5 mm per minute. The compressive strength was ascertained based on the orientation of the applied materials and the resistance they generated.

*2.3. Statistical Analysis*

The results were analysed using one-way analysis of variance (ANOVA), using Statistica 13.3 (TIBCO Software Inc., Palo Alto, CA, USA). A difference of $p < 0.05$ between means was significant. Further, the post-hoc LSD and repeated measures ANOVA tests were performed.

The used mathematical model followed Yij = $\mu$ + $\alpha$i + eij, where Yij is the observation value, $\mu$ is the overall mean, $\alpha$i is the effect of the experimental diet (calcium or magnesium level), and eij is random error.

Statistical analysis for the parameters of body weight, shell width, and mortality was performed with the PS IMAGO PRO 9.0 software, using ANOVA tests for repeated measures. The pairwise comparison method with Bonferroni correction was used for significant interaction effects. All study variables met the sphericity condition. A $p < 0.05$ was taken as statistically significant.

## 3. Results

The development of snails is characterised by three stages that can be distinguished: slow growth of the body and rapid development of the internal organs in the first month, rapid increase in body weight in the second and third months, and stabilisation of body weight and a slowdown in growth after 13 weeks of age [12]. The results of the measurements carried out in each month of Experiment I and Experiment II are presented in Figures 4–9.

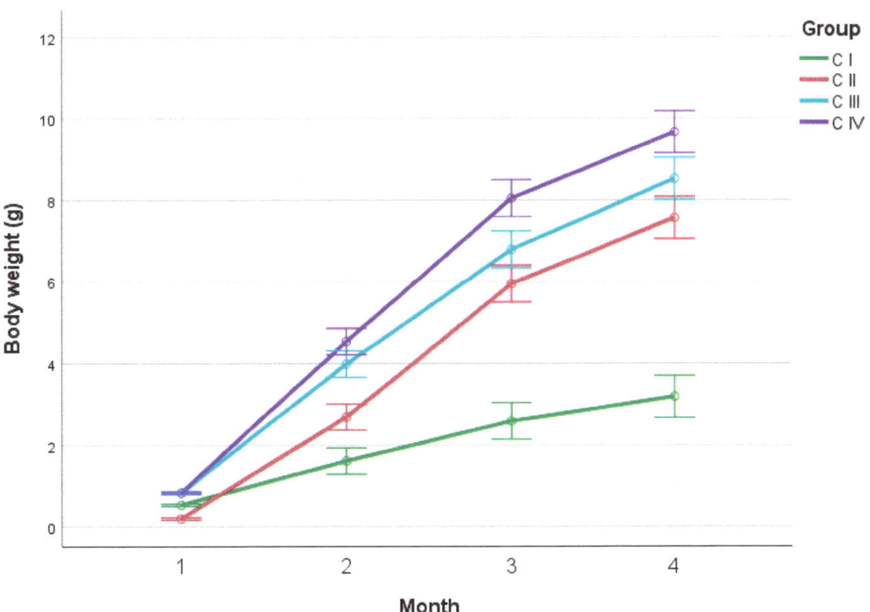

**Figure 4.** Results of monthly body weight (g) measurements of snails in Experiment I. Data are given as mean ± SD (n = 50). C I, C II, C III, C IV—diet with Ca dose respectively: 44.3, 66.1, 88.7, 103.5 g/kg feed.

There was no significant difference ($p > 0.05$) in the body weights of the snails between C III (0.85 g) and C IV (0.68 g) during the first month. Snails in group C II had the lowest weight gain, and those in group C III had the highest (0.18 g vs. 0.83 g). Throughout the second and third months, snails in group C I showed the lowest mean weight gains (1.08 g and 1.65 g), while animals in group C IV showed the greatest (3.87 g and 3.51 g). In the second month, the weights of snails in groups C III (3.99 g) and C IV (4.54 g) did not differ significantly ($p > 0.05$). Group C I showed the lowest body weight (1.61 g). During the third month, no significant differences were found ($p > 0.05$), only between C II (5.95 g) and C III (6.79 g). Group C I showed the lowest body weight (3.26 g), and group C IV showed the highest (8.05 g). In the final measurement, there was no significant difference ($p > 0.05$) in the body weights of snails between groups C II (7.56 g) and C III (8.52 g). At the end of the experiment, snails in the C I group had the lowest mean body weight (3.19 g) and those in C IV, the highest (9.67 g).

In the first month of the experiment, the snails belonging to group C III exhibited the largest shell width (13.58 mm). In comparison, those from C II had the smallest (8.94 mm). Significant shell width differences ($p \leq 0.01$) were observed in all tested groups. However, during the second measurement, groups C III (22.37 mm) and C IV (23.47 mm) did not show significant differences in shell width ($p > 0.05$). Group C I showed the lowest shell width at 14.79 mm. At month 3, there were no significant differences ($p > 0.05$) in is-total between groups C II (28.17 mm) and C III (27.56 mm). Group C I had the lowest shell width throughout the Experiment I duration. In group C II, snails showed the highest shell width increases among all experimental groups in months 2–4 (11.95 mm, 7.28 mm and 2.37 mm, respectively). Finally, animals in group C I showed the smallest average shell width at 17.51 mm in the final month of the experiment, while snails in group C IV showed the largest average shell width at 30.63 mm. By the end of the experiment, there were no significant differences ($p > 0.05$) in shell width between snails in groups CII (30.54 mm) and CIII (29.77 mm), CII and CIV, and CIII and CIV. The most significant overall increase in shell width ($p \leq 0.01$) over time was observed in CII (+21.60 mm), whereas the smallest was in CI (+5.92 mm).

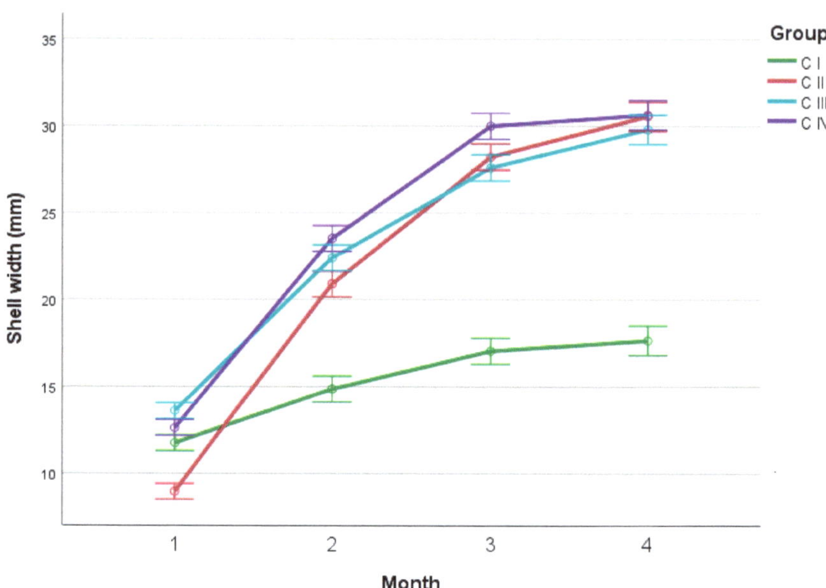

**Figure 5.** Results of monthly shell width (mm) measurements of snails in Experiment I. Data are given as mean ± SD (n = 50). C I, C II, C III, C IV—diet with Ca dose respectively: 44.3, 66.1, 88.7, 103.5 g/kg feed.

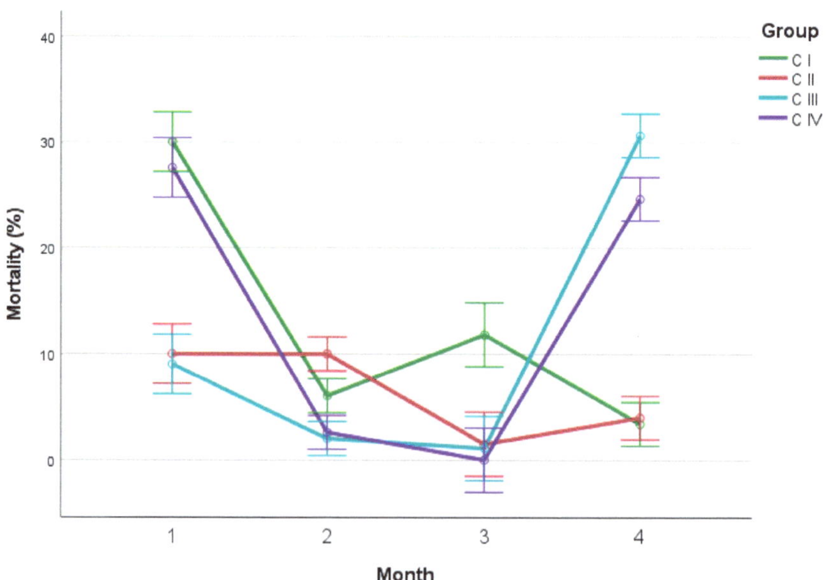

**Figure 6.** Results of monthly mortality (%) records of snails in Experiment I. Data are given as mean ± SD (n = 3). C I, C II, C III, C IV—diet with Ca dose respectively: 44.3, 66.1, 88.7, 103.5 g/kg feed.

During the first month, the mortality rate was the lowest ($p \leq 0.01$) for group C III at 9.03%, while group C I registered the highest rate at 30%. For the first measurement, the number of deaths did not differ significantly ($p > 0.05$) between groups C I (30%) and C IV (27.17%), and between groups C II (10%) and C III (9.83%). For the second measurement, the number of deaths did not differ between groups C III (2.10%) and C IV (2.60%). For the third measurement, it did not differ between groups C II (1%) and C III (1.50%) and between groups C II and C IV (0%). By the fourth month of the experiment, the mortality rates for groups C I and C II were relatively low at 3.40% and 4.00%, respectively. However, group C III recorded a significantly higher mortality rate of 30.90%, while group C IV had a mortality rate of 25.00%. The number of deaths did not differ significantly ($p > 0.05$) between groups CI and CII.

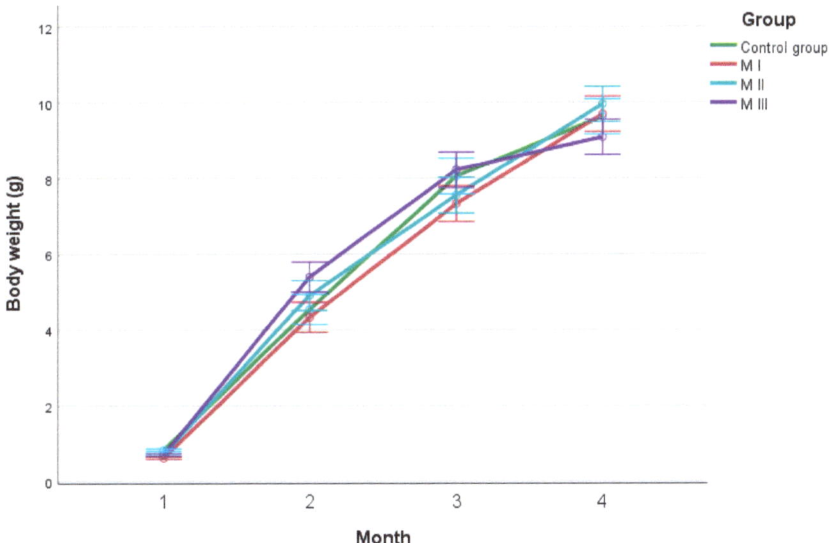


**Figure 7.** Results of monthly body weight measurements (g) of snails in Experiment I. Data are given as mean $\pm$ SD (n = 50). No-Mg control diet; M I, M II, M III—diet with Mg dose respectively: 3.3 g/kg, 5.6, 7.2 g/kg feed.

Magnesium supplementation in the diet of snails in Experiment II had a significant effect ($p \leq 0.01$) on their mean body weight. During the first measurement, there was no significant difference ($p > 0.05$) in mean body weight between the M II and M III groups. The control group had the highest body weight (0.83 g) whilst M I had the lowest (0.62 g). In the second month of measurements, only the CG (4.40 g) and M III (5.39 g), and M I (4.32 g) and M III groups had significantly different ($p \leq 0.05$) body weights. Differences ($p \leq 0.05$) in snail body weight were observed only between the M I (7.32 g) and M III (8.21 g) groups during the third measurement. No significant differences ($p > 0.05$) in the snails' body weights were evident during the last month of measurements. At the end of the experiment, the snails had gained body weights of CG = 9.62 g; M I = 9.68 g; M II = 9.95 g; M III = 9.08 g. Animals in the M II group consistently showed the most significant differences in weight gain compared to the control group, specifically, in the first month (0.70 g vs. 0.40 g), in the third month (4.70 g vs. 3.71 g) and the fourth month (2.41 g vs. 1.49 g). Conversely, snails in the M III group showed the lowest weight gains in the mentioned months: 0.69 g, 2.82 g, and 0.87 g, respectively.

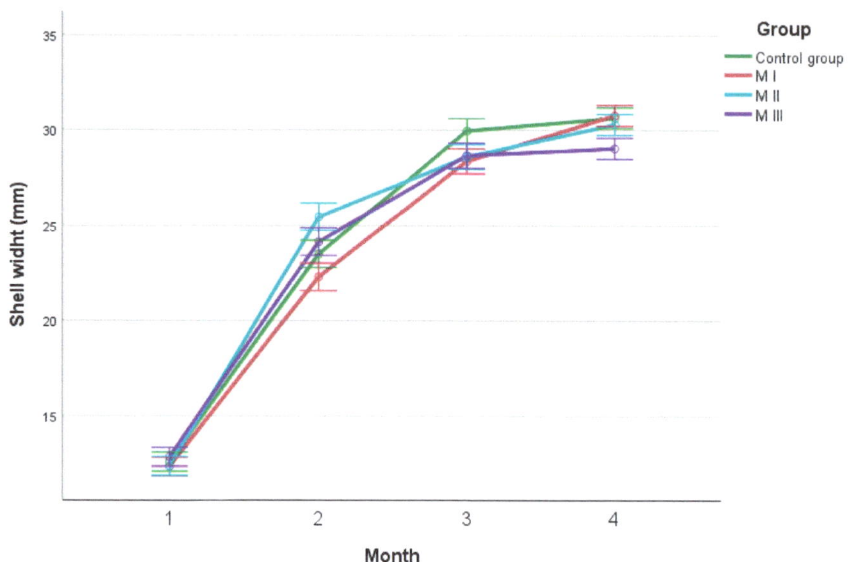


**Figure 8.** Results of monthly shell width (mm) measurements of snails in Experiment I. Data are given as mean ± SD (n = 50). No-Mg control diet; M I, M II, M III—diet with Mg dose respectively: 3.3 g/kg, 5.6, 7.2 g/kg feed.

No significant difference was observed in shell width during the first month of the experiment ($p > 0.05$). Among all groups, M I had the smallest shell width (12.34 mm), while MIII had the largest (12.84 mm). However, shell width varied significantly ($p \leq 0.05$) between CG and M II (25.41 mm), M I (22.30 mm) and M II, as well as between M I and M III (24.14 mm) during the second month of measurements. Similarly, during the third measurement round, significant differences were found ($p \leq 0.05$) between the CG (29.98 mm) and M I (28.39 mm), CG and M II (28.72 mm), and CG and M III (28.68 mm) groups. Significant differences in shell width ($p \leq 0.01$) were observed between groups CG (30.66 mm) and M III (29.03 mm), M I (30.79 mm) and MIII, and between M II (30.30 mm) and MIII by the end of the measurements. The largest shell width of 30.79 mm was observed in M I, while the smallest width of 29.06 mm was observed in M III. Regarding shell width gains, the M II group had significantly more substantial gains: 18.25% greater than the control group in the first month (13.04 mm vs. 11.03 mm). In contrast, in the third month, the M II group had significantly smaller gains: 47.77% less than the control group (3.31 mm vs. 6.36 mm). In the last month, the M I group had the highest increase in shell width—2.40 mm. Furthermore, the largest shell width change over time was found in MI (+18.45 mm) and the smallest in MIII (+16.21 mm).

During the initial measurement, mortality exhibited a substantial difference ($p \leq 0.01$) exclusively between the CG (27.33%) and MII (13.50%) groups. In the second measurement, no significant differences ($p > 0.05$) were found while comparing CG (2.6%) and M II (0%), and CG and M III (2.9%). No significant differences in mortality were observed ($p > 0.05$) between the M I (7.6%) and M II (5.9%) groups after three months of measurements. The mortality rate of group M III was the highest at 13.6%. At the end of the study, mortality rates were noted as follows: CG = 25.00%; MI = 30.50%; MII = 35.28%; and MIII = 28.00%. Only the CG and MI groups showed significant differences ($p \leq 0.01$).

During the examination of carcass and shell traits in Table 2, animals in group C IV in Experiment I had the highest mean carcass weight (8.27 g). The lowest body weight was noted in group C I at 2.89 g. The trend in shell mass was similar, with group C IV displaying the heaviest shells at 1.40 g and group C I at 0.29 g.

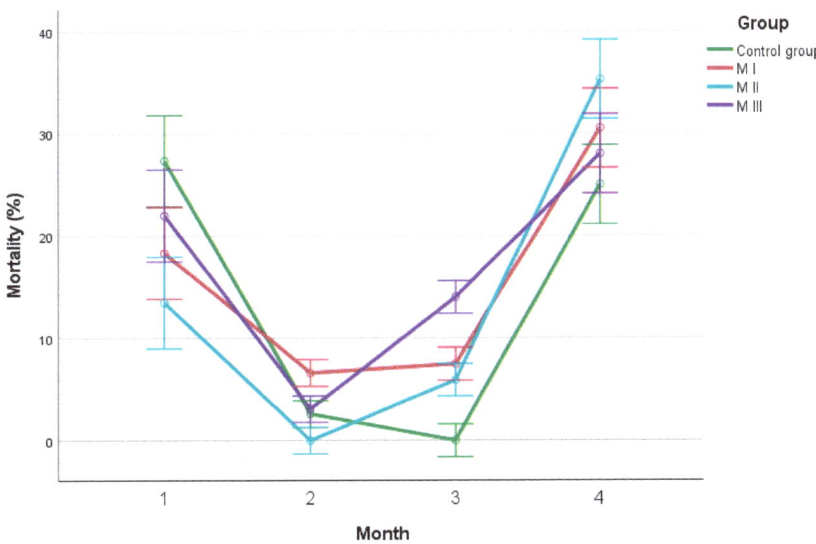



**Figure 9.** Results of monthly mortality (%) records of snails in Experiment I. Data are given as mean $\pm$ SD (n = 3). No-Mg control diet; M I, M II, M III—diet with Mg dose respectively: 3.3 g/kg, 5.6, 7.2 g/kg feed.

**Table 2.** Mean ($\pm$SD) results of carcass and shell characteristics and feed intake at the end of Experiment I and Experiment II.

| Indices | n | Experiment I Experimental Groups | | | | SEM | *p*-Value |
|---|---|---|---|---|---|---|---|
| | | C I | C II | C III | C IV | | |
| Carcass weight (g) | 50 | 2.89 $\pm$ 1.95 [a] | 6.87 $\pm$ 1.40 [b] | 7.41 $\pm$ 1.39 [b] | 8.27 $\pm$ 1.68 [c] | 0.23 | <0.0001 |
| Shell weight (g) | 50 | 0.29 $\pm$ 0.22 [a] | 0.70 $\pm$ 0.34 [b] | 1.12 $\pm$ 0.21 [c] | 1.40 $\pm$ 0.24 [d] | 0.03 | <0.0001 |
| Share of the carcass in total body weight (%) | 50 | 91.24 $\pm$ 0.93 [c] | 90.78 $\pm$ 1.12 [c] | 86.87 $\pm$ 2.32 [b] | 85.46 $\pm$ 2.20 [a] | 0.24 | <0.0001 |
| Shell shape index | 50 | 1.44 $\pm$ 0.36 | 1.39 $\pm$ 0.08 | 1.44 $\pm$ 0.07 | 1.43 $\pm$ 0.12 | 0.03 | 0.5572 |
| Solidity index (g/cm$^2$) $\times$ 100 | 50 | 11.19 $\pm$ 2.43 [a] | 10.25 $\pm$ 2.10 [a] | 18.08 $\pm$ 2.85 [b] | 21.15 $\pm$ 3.55 [c] | 0.39 | <0.0001 |
| Crushing force of the shells (N) | 15 | 12.44 $\pm$ 6.68 [a] | 17.47 $\pm$ 8.01 [a] | 32.85 $\pm$ 13.46 [b] | 54.83 $\pm$ 17.39 [c] | 3.38 | <0.0001 |
| TBARS (nmol/g lyophilisate) | 6 | 56.94 $\pm$ 0.65 [ab] | 58.87 $\pm$ 1.47 [c] | 58.02 $\pm$ 1.84 [bc] | 55.59 $\pm$ 0.38 [a] | 0.45 | <0.0001 |
| Mature individuals (%) | 50 | 4 $\pm$ 0.02 [a] | 20 $\pm$ 0.04 [b] | 20 $\pm$ 0.04 [b] | 46 $\pm$ 0.05 [c] | 0.06 | <0.0001 |
| Total FCR (kg DM/kg) | 3 | 0.86 $\pm$ 0.32 [a] | 1.10 $\pm$ 0.14 [ab] | 0.94 $\pm$ 0.10 [a] | 1.28 $\pm$ 0.05 [b] | 0.09 | 0.0009 |
| Total feed intake (g/individual) | 3 | 11.09 $\pm$ 0.31 | 8.34 $\pm$ 0.03 | 8.75 $\pm$ 0.51 | 10.64 $\pm$ 0.25 | 0.92 | 0.1647 |

| Indices | n | Control Group | Experiment II Experimental Groups | | | SEM | *p*-Value |
|---|---|---|---|---|---|---|---|
| | | | M I | M II | M III | | |
| Carcass weight (g) | 46 | 8.23 $\pm$ 1.30 [b] | 8.18 $\pm$ 1.51 [b] | 8.45 $\pm$ 1.53 [b] | 7.47 $\pm$ 1.65 [a] | 0.22 | 0.0134 |
| Shell weight (g) | 46 | 1.39 $\pm$ 0.23 [a] | 1.50 $\pm$ 0.35 [ab] | 1.49 $\pm$ 0.28 [a] | 1.61 $\pm$ 0.27 [b] | 0.04 | 0.0047 |
| Share of the carcass in total body weight (%) | 46 | 85.45 $\pm$ 2.12 [c] | 84.52 $\pm$ 2.07 [b] | 84.92 $\pm$ 1.87 [bc] | 81.98 $\pm$ 3.04 [a] | 0.30 | <0.0001 |
| Shell shape index | 46 | 1.43 $\pm$ 0.10 [c] | 1.29 $\pm$ 0.08 [b] | 1.27 $\pm$ 0.08 [ab] | 1.24 $\pm$ 0.11 [a] | 0.02 | <0.0001 |
| Solidity index (g/cm$^2$) $\times$ 100 | 46 | 21.17 $\pm$ 3.13 [a] | 20.26 $\pm$ 3.52 [a] | 20.51 $\pm$ 3.03 [a] | 23.58 $\pm$ 3.84 [b] | 0.51 | <0.0001 |
| Crushing force of the shells (N) | 15 | 57.29 $\pm$ 17.31 | 59.38 $\pm$ 23.48 | 61.41 $\pm$ 22.61 | 49.73 $\pm$ 10.73 | 5.77 | 0.5123 |
| TBARS (nmol/g lyophilisate) | 6 | 55.59 $\pm$ 0.35 [a] | 55.33 $\pm$ 0.32 [a] | 55.90 $\pm$ 1.36 [a] | 57.65 $\pm$ 1.96 [b] | 0.39 | 0.0246 |
| Mature individuals (%) | 46 | 46.00 $\pm$ 0.41 | 63.00 $\pm$ 0.49 | 65.00 $\pm$ 0.48 | 59.00 $\pm$ 0.52 | 0.07 | 0.2287 |
| Total FCR (kg DM/kg) | 3 | 1.27 $\pm$ 0.06 [b] | 1.10 $\pm$ 0.25 [ab] | 0.94 $\pm$ 0.17 [a] | 1.61 $\pm$ 0.12 [c] | 0.09 | 0.0049 |
| Total feed intake (g/individual) | 3 | 11.09 $\pm$ 0.29 | 8.34 $\pm$ 0.23 | 8.75 $\pm$ 0.88 | 10.64 $\pm$ 3.04 | 0.91 | 0.1643 |

C I, C II, C III, C IV—diet with Ca dose respectively: 44.3, 66.1, 88.7, 103.5 g/kg feed; No-Mg control diet; M I, M II, M III—diet with Mg dose respectively: 3.3 g/kg, 5.6, 7.2 g/kg feed. The means indicated with different superscripts (a, b, c, d) are significantly different ($p < 0.05$).

The share of carcass in total body weight was highest among snails in group C I at 91.24. Animals in group C IV showed the lowest share at 85.46%.

The shell shape index demonstrated minimal variation, with all groups exhibiting values around 1.44, signifying a similarity in shell shape among the experimental groups. The solidity index shows that animals in group C IV had the highest value of 21.15, while in group C I, the lowest score was recorded at 11.19 g/cm$^2$.

Snails in group C IV displayed the highest crushing force at 54.83 N, and group C I had the lowest, at 12.44 N. The shells of the C I group showed unique characteristics, such as softness, flexibility and translucency. They exhibited a delayed fracture pattern characterised by stretching and planar flattening.

Thiobarbituric acid reactive substance (TBARS) results provide essential information in studies related to lipid oxidation processes in biological samples and indicate changes in health status and metabolism, or the effects of external factors on the studied organisms. The results showed that snails in group C III had the highest levels at 58.87 nmol/g lyophilizate in carcasses and snails in group C IV had the lowest, at 55.59 nmol/g.

Furthermore, group C IV had the highest percentage of mature individuals at 46%. In contrast, in groups C I, C II, and C III, the results were 10 times (C I) and 1.3 times (CII and C III) lesser, correspondingly.

The feed conversion ratio (FCR) in group C IV showed the highest value at 1.28 kg of feed (DM) per kg of snail body weight. Results of groups C I and C III showed comparable FCR values.

Regarding feed intake, individuals in group C II consumed the least amount of feed in the first month, with an average of 4.92 g per individual. Contrarily, group C IV had the highest feed intake, averaging 11.10 g per individual. The trend remained constant throughout the experiment, whereby Group C IV consistently displayed the greatest feed intake, and the largest discrepancy was noted in the final month. On the other hand, Group C I constantly demonstrated the lowest feed intake, with the minimum value of 2.06 g per specimen documented in the first month.

When carcass and shell characteristics were examined in Experiment II, the snails in group M II exhibited the highest mean carcass weight at 8.45 g compared with other groups. The lowest weight was recorded in group M II at 7.47 g.

Animals in group M III exhibited the heaviest shells, weighing 1.61 g, whereas the control group showed shells weighing 1.39 g. The mean proportion of the carcass of snails in total body weight was the highest in the control group, at 85.45%, while group M III had the lowest proportion, at 81.98%.

The control group displayed a shell shape index of snails of 1.43, the highest among the groups, while group M III recorded the lowest at 1.24. Group M I had the lowest solidity index of 20.26, whereas group M III had the highest solidity index of 23.58 g/cm$^2$. The crushing force of the shells showed differences, but these were not statistically significant.

The TBARS levels in snail carcass tissues from group M III were the highest at 57.65 nmol/g lyophilisate, whereas those in snails from group M I were the lowest at 55.33 nmol/g.

The proportion of mature snails did not exhibit a distinct trend of maximum and minimum values (46–65%) between the groups but demonstrated fluctuation.

The FCR records revealed that group M III displayed the most efficient feed conversion, with the lowest FCR recorded at 0.94, while group M I reflected the highest FCR at 1.27. Also, group M III presented the highest feed intake at 11.09 g/individual, whereas group M I showed the lowest intake (8.34 g/individual).

The investigation into the percentage differences in proximate composition among the experimental groups in Experiment I and Experiment II demonstrates significant variations in the nutritional profiles of snail carcasses (Table 3).

**Table 3.** Mean ($\pm$SD; n = 6) proximate composition of snail carcasses (% of DM) in Experiment I and Experiment II.

| Item | Experiment I Experimental Groups | | | | SEM | *p*-Value |
|---|---|---|---|---|---|---|
| | C I | C II | C III | C IV | | |
| Crude protein | $60.26 \pm 1.33$ [c] | $58.06 \pm 1.56$ [b] | $53.38 \pm 1.79$ [a] | $59.97 \pm 1.96$ [bc] | 0.69 | <0.0001 |
| Ether extracts | $4.66 \pm 0.62$ [c] | $3.14 \pm 2.41$ [b] | $1.49 \pm 0.26$ [a] | $1.48 \pm 0.24$ [a] | 0.51 | 0.0008 |
| Crude ash | $14.07 \pm 0.72$ [c] | $14.30 \pm 1.38$ [c] | $11.09 \pm 1.26$ [b] | $9.20 \pm 0.62$ [a] | 0.43 | <0.0001 |

| Item | Control Group | Experiment II Experimental Groups | | | SEM | *p*-Value |
|---|---|---|---|---|---|---|
| | | M I | M II | M III | | |
| Crude protein | $59.97 \pm 1.92$ [b] | $52.16 \pm 1.72$ [a] | $52.48 \pm 1.00$ [a] | $52.14 \pm 0.76$ [a] | 0.56 | <0.0001 |
| Ether extracts | $1.56 \pm 0.28$ [a] | $1.68 \pm 0.72$ [a] | $2.29 \pm 1.51$ [ab] | $2.86 \pm 0.25$ [b] | 0.33 | 0.0442 |
| Crude ash | $9.20 \pm 0.58$ | $9.22 \pm 0.24$ | $9.01 \pm 0.39$ | $8.79 \pm 0.97$ | 0.25 | 0.6056 |

C I, C II, C III, C IV—diet with Ca dose respectively: 44.3, 66.1, 88.7, 103.5 g/kg feed; No-Mg control diet; M I, M II, M III—diet with Mg dose respectively: 3.3, 5.6, 7.2 g/kg feed. The means indicated with different superscripts (a, b, c) are significantly different ($p < 0.05$).

The study found that snails in group C I exhibited the highest crude protein content in meat, at 60.26%. Group C III showed the lowest crude protein content of 53.38%.

The lipid content exhibited comparable changes across the experimental groups. Animals in group C I had the highest ether extract content, measuring 4.66%. Both groups C III and C IV exhibited the lowest ether extract content, measuring 1.49%.

Groups C I and C II exhibited the highest crude ash content values, registering at 14.07%. Snails in group C IV showed the lowest crude ash content among all experimental groups, measuring 9.20%.

The proximate composition analysis of the snail carcasses conducted in Experiment II illustrates substantial disparities in crude protein and ether extract content. Animals in the control group displayed the highest crude protein content at 59.97% in their carcasses. In contrast, the M III group showed the lowest content, a 13.16% reduction compared to the control group.

In contrast, the control group displayed the lowest ether extract content of 1.56%. In contrast, the M III group exhibited the highest ether extract content, with a significant increase of 83.33% compared to the control group. These results emphasise the substantial variations in these two components between the study groups.

Crude ash content was similar across all groups, ranging from 8.79% (group M III) to 9.22% (group M II).

Additional analyses were conducted concerning the mineral composition (in DM) of the snails' meat and shells (in FM). The results are presented in Table 4.

In Experiment I, animals in group C III had the highest Ca content in meat (1.45%), with the lowest being in group C II (1.31%). In the shells, the highest Ca content was found in group C IV (38.96%), with the lowest being in group C II (36.16%). Snails in group C II recorded the highest concentration of Cu in meat (56.99 mg/kg), while group C IV presented the lowest (36.70 mg/kg). In the shells, group C I had the highest Cu content (6.12 mg/kg), and group C IV showed the lowest (1.88 mg/kg).

The Fe content in snail meat was highest in group C I (357.52 mg/kg) and lowest in group C IV (127.82 mg/kg). Regarding the shells, group C III had the highest Fe content (202.39 mg/kg), and group C IV had the lowest (135.82 mg/kg). While analysing potassium content, animals in group C II had the highest K content in meat (7718.49 mg/kg), while group C IV had the lowest (6220.12 mg/kg). There were no significant differences in K content in the shells among the groups.

**Table 4.** The mean ($\pm$SD; n = 6) mineral content of snail carcasses (in DM) and shells (in FM) in Experiment I and Experiment II.

| Item | | Experiment I Experimental Groups | | | | SEM | *p*-Value |
| | | C I | C II | C III | C IV | | |
|---|---|---|---|---|---|---|---|
| Ca (%) | Meat | 1.36 ± 0.23 | 1.31 ± 0.09 | 1.45 ± 0.05 | 1.40 ± 0.12 | 0.07 | 0.5847 |
| | Shell | 37.99 ± 0.96 [bc] | 36.16 ± 0.97 [a] | 37.05 ± 1.15 [ab] | 38.96 ± 0.22 [c] | 0.45 | 0.0048 |
| Cu (mg/kg) | Meat | 36.70 ± 1.19 [a] | 56.99 ±4.04 [c] | 45.33 ±1.58 [b] | 60.33 ± 3.90 [c] | 1.49 | <0.0001 |
| | Shell | 6.12 ± 2.60 [b] | 3.47 ± 0.47 [a] | 3.51 ± 0.72 [a] | 1.88 ± 0.02 [a] | 0.68 | 0.0072 |
| Fe (mg/kg) | Meat | 357.52 ± 31.75 [d] | 312.01 ± 28.83 [c] | 218.19 ± 30.51 [b] | 127.82 ± 9.70 [a] | 13.38 | <0.0001 |
| | Shell | 196.76 ± 35.77 [bc] | 129.26 ± 50.56 [a] | 202.39 ±23.17 [c] | 135.82 ± 26.76 [ab] | 17.41 | 0.0181 |
| K (mg/kg) | Meat | 7097.36 ± 430.28 [b] | 7718.49 ± 275.80 [c] | 6960.40 ± 529.03 [b] | 6220.12 ± 214.21 [a] | 191.53 | 0.0012 |
| | Shell | 806.19 ± 184.37 | 791.41 ±289.89 | 726.69 ± 126.45 | 1096.67 ± 48.90 | 92.06 | 0.0624 |
| Mg (mg/kg) | Meat | 2948.14 ± 425.59 [a] | 3532.52 ± 122.51 [b] | 2670.22 ± 308.84 [a] | 2684.13 ± 186.15 [a] | 142.78 | 0.0026 |
| | Shell | 508.48 ± 32.46 [b] | 322.13 ± 9.54 [a] | 356.97 ± 26.00 [a] | 320.12 ±5.89 [a] | 10.74 | <0.0001 |
| Mn (mg/kg) | Meat | 25.98 ± 2.90 | 23.31 ± 2.65 | 24.57 ± 2.08 | 21.82 ± 0.94 | 1.14 | 0.1148 |
| | Shell | 14.58 ± 0.68 [c] | 10.98 ± 0.78 [b] | 9.99 ± 0.56 [a] | 9.76 ± 0.20 [a] | 0.30 | <0.0001 |
| Na (mg/kg) | Meat | 5915.44 ± 886.60 [a] | 7934.27 ± 175.55 [b] | 5467.21 ± 783.89 [a] | 5299.06 ± 298.35 [a] | 308.26 | <0.0001 |
| | Shell | 1198.09 ± 169.30 [c] | 806.95 ± 242.83 [ab] | 752.42 ± 117.97 [a] | 1050.68 ± 27.87 [bc] | 79.87 | 0.0057 |
| P (mg/kg) | Meat | 11,109.90 ± 1289.60 [b] | 11,616.93 ± 265.68 [b] | 10,415.35 ± 880.25 [ab] | 9271.07 ± 415.49 [a] | 409.35 | 0.0089 |
| | Shell | 896.51 ± 55.47 [c] | 558.19 ± 14.21 [b] | 532.45 ± 47.52 [b] | 428.37 ± 16.78 [a] | 18.92 | <0.0001 |
| Si (mg/kg) | Meat | 57.49 ± 15.46 [bc] | 36.40 ± 9.59 [a] | 45.59 ± 14.02 [ab] | 66.00 ± 10.35 [c] | 6.30 | 0.0293 |
| | Shell | 182.30 ± 45.08 [c] | 93.11 ± 27.66 [b] | 190.42 ± 36.81 [c] | 24.98 ± 3.69 [a] | 16.13 | <0.0001 |
| Zn (mg/kg) | Meat | 105.27 ± 11.43 [b] | 104.58 ± 8.85 [b] | 95.25 ± 7.30 [b] | 57.26 ± 3.06 [a] | 4.12 | <0.0001 |
| | Shell | 6.40 ± 1.04 [a] | 6.28 ± 2.25 [a] | 5.26 ± 1.16 [a] | 17.91 ± 1.75 [b] | 0.77 | <0.0001 |

| Item | | Experiment II Experimental Groups | | | | SEM | *p*-Value |
| | | Control Group | M I | M II | M III | | |
|---|---|---|---|---|---|---|---|
| Ca (%) | Meat | 1.38 ± 0.11 [b] | 1.16 ± 0.05 [a] | 1.29 ± 0.09 [ab] | 1.26 ± 0.23 [ab] | 0.05 | 0.0274 |
| | Shell | 38.96 ± 0.21 | 33.78 ± 6.48 | 38.70 ± 2.39 | 37.64 ± 1.08 | 1.48 | 0.0977 |
| Cu (mg/kg) | Meat | 60.33 ± 3.88 [b] | 48.69 ± 4.43 [a] | 57.20 ± 2.94 [b] | 48.86 ± 2.72 [a] | 1.78 | 0.0013 |
| | Shell | 1.90 ± 0.02 | 3.06 ± 1.53 | 3.84 ±1.45 | 4.98 ± 2.66 | 0.82 | 0.1048 |
| Fe (mg/kg) | Meat | 127.82 ± 9.72 [a] | 228.16 ± 8.39 [c] | 223.68 ± 18.68 [c] | 174.40 ± 7.05 [b] | 5.93 | <0.0001 |
| | Shell | 135.80 ± 26.76 [a] | 184.22 ± 2.28 [b] | 200.70 ± 25.37 [b] | 131.80 ± 10.40 [a] | 8.78 | <0.0001 |
| K (mg/kg) | Meat | 6220.12 ± 214.23 | 6102.19 ± 233.64 | 5829.50 ± 443.64 | 5759.94 ± 228.28 | 147.78 | 0.1424 |
| | Shell | 1096.70 ± 27.92 [b] | 750.43 ± 90.35 [a] | 882.32 ± 138.52 [a] | 827.05 ± 140.74 [a] | 53.64 | 0.0043 |
| Mg (mg/kg) | Meat | 2684.13 ± 186.13 [c] | 2279.38 ± 44.22 [a] | 2500.01 ± 154.35 [bc] | 2381.87 ± 92.34 [ab] | 65.65 | 0.0055 |
| | Shell | 320.10 ± 5.78 [b] | 271.84 ± 2.85 [a] | 367.77 ± 42.97 [c] | 408.85 ± 16.86 [d] | 11.62 | <0.0001 |
| Mn (mg/kg) | Meat | 21.82 ± 0,92 [a] | 22.77 ± 1,10 [a] | 25.34 ± 1.80 [b] | 21.17 ± 1.30 [a] | 0.66 | 0.0043 |
| | Shell | 9.80 ± 0.22 [b] | 9.18 ± 0.47 [ab] | 9.25 ± 0.58 [b] | 8.42 ± 0.46 [a] | 0.21 | 0.0058 |
| Na (mg/kg) | Meat | 5299.06 ± 298,35 [c] | 4899.43 ± 142,00 [b] | 4644.27 ± 262,32 [ab] | 4497.35 ± 112,38 [a] | 109.15 | 0.0013 |
| | Shell | 1050.70 ± 26.98 [c] | 738.78 ± 32.79 [a] | 849.92 ± 101.69 [ab] | 941.13 ± 180.83 [bc] | 52.60 | 0.0082 |
| P (mg/kg) | Meat | 9271.07 ± 415.42 [b] | 7905.68 ± 332.83 [a] | 8164.64 ± 330.69 [a] | 7958.51 ± 327.05 [a] | 176.73 | <0.0001 |
| | Shell | 428.40 ± 16.79 [b] | 244.33 ± 24.08 [a] | 259.35 ± 39.55 [a] | 296.59 ± 38.89 [a] | 15.10 | <0.0001 |
| Si (mg/kg) | Meat | 66.00 ± 10.32 [a] | 102.26 ± 19.74 [b] | 119.27 ± 10.67 [b] | 104.25 ± 3.47 [b] | 6.24 | <0.0001 |
| | Shell | 25.00 ± 3.66 [a] | 179.36 ± 24.08 [b] | 178.05 ± 39.55 [b] | 149.24 ± 38.89 [b] | 29.65 | 0.0094 |
| Zn (mg/kg) | Meat | 57.26 ± 3.09 [c] | 49.63 ± 1.24 [a] | 53.86 ± 1.85 [b] | 53.35 ± 1.71 [b] | 1.04 | 0.0022 |
| | Shell | 17.90 ± 1.72 [c] | 4.52 ± 0.96 [ab] | 4.42 ± 0.62 [a] | 6.63 ± 1.74 [b] | 0.62 | <0.0001 |

C I, C II, C III, C IV—diet with Ca dose respectively: 44.3, 66.1, 88.7, 103.5 g/kg feed; No-Mg control diet; M I, M II, M III—diet with Mg dose respectively: 3.3, 5.6, 7.2 g/kg feed. The means indicated with different superscripts (a, b, c, d) are significantly different ($p < 0.05$).

Snails in group C II had the highest Mg content in meat at 3532.52 mg/kg, while the lowest was detected in group C IV at 2670.22 mg/kg. In the shells, group C I exhibited the highest Mg content at 508.48 mg/kg, and group C IV had the lowest at 320.12 mg/kg. Animals in group C I displayed the highest manganese concentration in meat (25.98 mg/kg) and shells (14.58 mg/kg). Group C IV had the lowest manganese content in both meat and shells, with measurements of 21.82 mg/kg and 9.76 mg/kg, respectively.

Animals in group C II showed the highest Na content in meat at 7934.27 mg/kg, whereas group C IV showed the lowest at 5299.06 mg/kg. In the shells, inversely, group C IV had the highest Na content at 1050.68 mg/kg, and group C II had the lowest at 806.95 mg/kg. The highest phosphorus content in snail meat was observed in group C II

(11,616.93 mg/kg), whereas the lowest was in group C IV (9271.07 mg/kg). In the shells, group C I had the highest phosphorus content (896.51 mg/kg), whereas group C IV had the lowest (428.37 mg/kg).

The meat from group C I exhibited the highest silicon content (66.00 mg/kg), while the lowest was in group C II (36.40 mg/kg). In the shells, group C I demonstrated the greatest Si concentration (182.30 mg/kg), whereas group C IV presented the lowest (24.98 mg/kg). Group C I exhibited the highest zinc content in meat (105.27 mg/kg); conversely, group C IV had the lowest (57.26 mg/kg). In the shells, the Zn content was highest in group C IV (17.91 mg/kg) and lowest in group C I (6.40 mg/kg).

In Experiment II, animals in the control group had the highest Ca content in the meat at 1.40% and in shells (38.96%), while M I had the lowest content in meat at 1.16% and in the shell at 33.78%. Copper content displayed differences across the groups, with M I having the lowest at 48.69 mg/kg and the control group having the highest at 60.33 mg/kg in snail meat. Cu content increased across the groups in the shell, with the control group having the lowest (1.90 mg/kg) and M III having the highest (4.98 mg/kg).

The iron content in the snail meat exhibited a significant increase in M I (228.16 mg/kg) compared with the control group (127.82 mg/kg). Regarding the shell, the Fe content was the highest in group M II (200.70 mg/kg) and the lowest in group M III (131.80 mg/kg). There were slight, not statistically significant, differences in the potassium content of the snail meat. However, in the shell, the control group exhibited a high concentration of 1096.70 mg/kg, while M I showed the lowest concentration at 750.43 mg/kg.

Differences across the groups in the Mg content in the snail meat were noted, with M I recording the minimum value (2279.38 mg/kg) and the control group showing the maximum value (2684.13 mg/kg). However, the Mg content in the shell varied significantly, with the highest value recorded in M III (408.85 mg/kg) and the lowest in M I (271.84 mg/kg). The Mn content in the meat in group M III was the lowest (21.17 mg/kg); in group M II, the Mn content recorded was the highest (25.34 mg/kg). The M III group shells exhibited the lowest Mn content (8.42 mg/kg), in contrast to M II, which contained the highest Mn content (9.25 mg/kg).

The sodium levels in the meat were most elevated in the control group (5299.06 mg/kg) and least in M III (4497.35 mg/kg). Na levels varied in the shells, with M I exhibiting the lowest (738.78 mg/kg) and the control group the highest (1050.70 mg/kg). Phosphorus content in the meat was the lowest in group M I (7905.68 mg/kg) and the highest in the control group (9271.07 mg/kg). In shell analysis, the highest P content was found in the control group (428.40 mg/kg) and the lowest in the M I group (244.33 mg/kg).

The silicon content in the snails' meat was highest in the M II group, measuring 119.27 mg/kg, while the lowest value was recorded in the control group at 66.00 mg/kg. Regarding the shell, Si content in the control group was the lowest (25.00 mg/kg), and M I had the highest (179.36 mg/kg) values. The lowest zinc content in the meat was detected in the M I group (49.63 mg/kg), with the highest value found in M II (53.86 mg/kg). Zn content in the shell varied among the groups, with the highest content in the control group (17.90 mg/kg) and the lowest in group C II (4.42 mg/kg).

## 4. Discussion

### 4.1. Growth Rates during Fattening

#### 4.1.1. Experiment I

As anticipated, the body weights of the snails increased concerning the Ca concentration in the mixtures. As an indicator of shell development, the shell width parameter showed similar results across all groups from the second month until the fattening period's end, except for the group receiving the lowest Ca supplement (C I). Based on the findings, it can be inferred that the C I group (44.3 g Ca/kg) lacked sufficient calcium for the adequate growth and development of snails. Individuals in this group displayed indications of arrested development at approximately 1 month of age compared to the other experimental groups. From day 30 onwards, the low Ca content was deficient and impeded proper shell development. This finding corroborates previous research linking low Ca levels in their diet to the development of dwarfism in these animals [99–101]. However, it is worth mentioning that in the first fattening stage, a level of 4.4% satisfies the animals' Ca requirements. Although the shell widths in groups C II, C III, and C IV during the second and third months differ considerably, they eventually equalise by the end of the fattening period. That implies that shell growth is sustained regardless of the Ca level in the mixtures.

Previous studies on *Archachatina marginata* demonstrated that increasing the proportion of Ca in the feed by up to 20% (above the standard level in snail nutrition) increased body weight and shell width [32]. Also, Oluokun et al. [25] conducted studies that indicated body weight and shell width of snails were positively influenced by Ca levels, showing that animals fed with an 8% added Ca mixture feed noted an increase in shell width of approx. 30%, and an approx. 25% increase in body weight compared with animals fed with a feed composition containing 4% Ca.

The experimental results show that the C IV group had the highest total FCR results compared to other experimental groups, yet it was not significantly different. Also, group C IV showed minimal growth during the final month of the fattening period, which was attributed to the onset of reproduction. As a result, this group may have had an overestimated FCR compared to other groups, as the snails' growth ceased despite their continuous feed consumption. The study results show that total feed intake was rising along with Ca levels in the feed, suggesting that feed containing higher doses of Ca is more palatable or encourages animals to eat in larger quantities, supported by previous research [24]. It was previously shown that increased Ca supply likely affects digestive processes and the absorption of other nutrients, which may affect metabolic efficiency and feed energy utilisation [7]. However, current results contrast with those of Oluokun et al. [25], who demonstrated that FCR improved with higher percentages of Ca in animal feed. Experimental groups, except C IV, exhibited FCRs comparable to those documented in the literature for *Cornu aspersum* [102].

The results show an association between Ca in the diet at 4.43% and 8.87% and higher mortality rates in the early fattening stages. A different trend is observed in the 6.61% Ca group, which can reduce mortality throughout the entire fattening period. It is assumed that low calcium levels in snails' feed increase mortality in the first month of rearing due to deficiencies. In contrast, high levels contribute to high mortality in the final phase of fattening due to the exhaustion of animals during laying, which commences earlier due to the accelerated maturation process. Maintaining sufficient calcium levels in the snails' feed is recommended to reduce mortality rates.

Studies by Tchakounte et al. [36] on *Archachatina marginata* showed reduced snail mortality with increasing Ca concentration (14–18%) in the feed, valid for animals in the middle fattening period in the current study. Notably, snails experience their highest mortality during the first month of development and when entering the breeding and egg-laying phase [21,36]. Notably, none of the experimental groups achieved a mortality rate of 40% during the first month of fattening, which has been reported in the literature as the standard [12]. The results were closer to the 30% obtained by Desbuquois (1997) [103]. Studies on *Helix Aspersa Maxima* reported mortality rates of 8–10% [102], and comparable results, below 12% mortality, were observed in all experimental groups in the current study in the second and third month of fattening.

### 4.1.2. Experiment II

It was shown that Mg at 0.72% promoted increased animal weight in the second and third months of rearing. Therefore, it appears justifiable to incorporate magnesium supplementation during the intermediate stage of snail feeding (second and third month). Adding Mg to the diet did not significantly affect the final body weight of the animals, yet it was observed that it accelerated the snails' growth in time. Previous studies on farm animals and shrimps confirm the beneficial effect of Mg supplementation on animal weight gain [104–106]. Compared to the Berillis et al. [8] study (20% Ca in feed mixture), snails' body weights were higher, and the widths of the shells were comparable to the current research.

In addition to compound feed, results of Experiment II showed that total FCR and feed consumption were lower in the groups with Mg at 3.3 g and 5.6 g/kg feed (approx. 1 kg DM/kg and approx. 8.5 g/individual), which indicates better feed utilization by animals and the possibility of savings for the farmer due to lower feed expenses. The high feed conversion rate observed in both the control and M III groups, coupled with the weight gain and shell width measurements in the last two months of fattening and the number of mature individuals in each group, suggests that the snails' growth was nearly complete. The snails began to reproduce after the development and fattening phase ended. Consequently, the fattening period of the snails could be completed 2–3 weeks earlier, thereby positively impacting the economic factors of snail fattening. It was shown that higher Mg content had a beneficial effect on feed conversion in farm animals and shrimps, which is, to some extent, consistent with the study's findings [104,105]. All experimental groups exhibited FCRs better than those documented in the literature for *Cornu aspersum*, with values up to 1.85 [9,12,107].

Groups with additional Mg in their diet exhibited a comparatively lower mortality rate in the initial month of fattening than CG. Mg at 0.56% in the diet increased mortality in the last month. Results below 14% of mortality were observed in all experimental groups in the current study in the second and third month of fattening. This study's findings suggest that adding magnesium to the diet at levels of 3.3 g/kg and 5.6 g/kg in the first month of fattening and 5.6 g/kg in the second month resulted in favourable survival of the snails. However, during the following two months, the addition of magnesium to the diet showed no better results than its absence in the control group. All groups receiving Mg supplementation exhibited more significant mortality in the final month of fattening, potentially due to the snails laying eggs and maturing faster than the control group. This faster maturation is believed to be the primary reason for the higher mortality rate observed [108]. A study on shrimps indicated higher animal survival rates with increasing Mg supplementation. However, this association was not observed in the present snail study [105,106]. The mortality results were close to 20–30% obtained in previous studies on snails [109,110] in the first and fourth month of fattening. Earlier studies on *Helix Aspersa Maxima* reported mortality rates of up to 10% [111].

*4.2. Carcass Characteristics*

4.2.1. Experiment I

Increasing dietary Ca levels were associated with increased carcass weight from 2.89 g to 8.27 g. The increased calcium content in the diet is thought to have a positive effect on the development of the snails and their growth, not only in terms of total body weight, but also in terms of carcass weight. Calcium is speculated to enhance the rapidity and effectiveness of fattening these animals. At the same time, it was noted that under the influence of increased dietary Ca intake, the percentage of carcass weight in the total body weight of the snail was reduced from 91.24% to 85.46%. Previous studies corroborate the present findings [31]. It can be assumed that this effect is due to the maturation and saturation of the shell with minerals and increasing its weight [9,25,32]. Therefore, calcium supplementation has a positive impact not only on the weight of the snail's body but also on the weight of its shell and overall weight.

The proximate composition analyses showed that introducing 4.43% Ca in the diet contributes to increased snail meat protein. In contrast, the presence of 8.87% Ca in the diet appears to have a negative effect on the protein content of the meat. As the amount of Ca in the feed increased, a decrease in the fat and ash content of the snail meat was observed (down to 1.48% and 9.20%, respectively). The findings indicate that an elevated calcium content in the diet correlates with a heightened proportion of carbohydrates in snail carcasses. Such carbohydrates mostly exist in the form of glycogen and galactose, which enable the animals to prepare for hibernation, among other functions [112,113]. Current findings show comparable crude protein and fat levels and higher crude ash content than previous snail studies [102,114]. The current research shows higher amounts of crude ash in snail meat and lower levels of crude protein than previous studies on *Helix Aspersa Maxima* [111,115].

Mineral analyses of the carcasses revealed that Ca at 4.43% in the diet influences higher iron content and that Ca at 6.61% contributes to the meat's potassium, Mg, sodium, and phosphorus content while reducing the amount of silicon. In addition, a higher amount of Ca in feed was linked to higher silicon content and lower potassium, Mg, sodium, and phosphorus content. Ca interacts with these elements, frequently dislodging them from their binding and transportation positions on the exterior of cell membranes. Based on the study results, it can be assumed that Ca at a level above 8.87% starts to have negative influence on these elements' absorption. Also, Ca at 10.35% increased snail carcasses' copper. It is worth noting that copper is a component of haemocyanin, which is responsible for transporting oxygen in the snail's haemolymph throughout the body [12,116]. This relationship may be attributed to the presence of larger and heavier carcasses, which require a greater supply of nutrients to a larger body surface area. Consequently, an increased amount of hemolymph is necessary for proper body function. An association was observed among vertebrates (rats, dogs, chickens) between increased Ca supply and Mg deficits [117–119]. In the current study, no such relationship was shown.

As a result of the experiments related to the mineral analysis of snail meat, significant differences were observed in the carcasses' Ca, iron, phosphorus, Mg, sodium, potassium, manganese, and zinc contents. The values of these minerals were significantly different compared with the results of previous studies [120,121], where higher Ca, iron, and phosphorus contents were found in the carcasses studied. In contrast, a lower Ca content was recorded compared with the results of Çağıltay (2011) [115]. Moreover, copper and Mg contents were lower compared with the study of Gomot (1998) [120]. Compared with *A. achatina*, it was proven that the *Cornu aspersum* snails' meat contained more Ca, iron, copper, potassium, and phosphorus and less zinc [122]. It was established that *Cornu aspersum* snails can accumulate copper in their foot tissues [123]. The snails in the current experiment exhibited decreased Ca, Mg, and phosphorus levels and elevated iron and zinc levels in their carcasses, except in the C IV group, compared with the Niemiec et al. [124] study. On the other hand, the potassium and manganese levels were higher than in previous studies [115,125]. Fagbuaro et al. [126] showed lower iron, sodium and potassium levels than in the current study. Thus, the available literature confirms variations in the mineral composition of meat.

The increased TBARS concentration in the C II and C III groups, compared with the C I Ca group, could indicate a possible increase in lipid oxidation processes or biological tissue reactivity within these groups. A decrease in TBARS values in the C IV group might signify a particular reduction in these oxidation processes. The results indicate that the effect of calcium in the diet at a level of 10.35% is beneficial from the point of view of lipid peroxidation in snail tissues.

### 4.2.2. Experiment II

Post-mortem analyses showed that Mg at 0.72% decreased the proportion of carcass weight in the body weight of the animals along with carcass weight, favouring an increase in shell weight. It can be assumed that Mg in the feed above this level begins to have a negative effect on the weight gain of snails, especially carcass weight. Compared with the Ligaszewski et al. [127] study, lower carcass weights were demonstrated, along with a higher share of the carcass in the total body weight of animals.

It was proven that Mg supplementation increases protein utilisation in foods due to increased protein absorption and synthesis [128]. Stronger protein digestion may be related to increased sodium–potassium triphosphatase activity in enterocytes and changes in membrane transport systems [129]. That could explain the weight gain observed during the second and third month of fattening of the snails. At 2–3 months, snails achieve complete development of their internal organs, marking the beginning of their fattening period. However, this did not result in an increase in final protein quantity in snail carcasses, as analyses of the proximate composition showed that adding Mg to the diet contributed to a linear decrease in protein and ash while increasing carcass fat. It is possible that increasing magnesium levels in the diet may result in increased deposition of elements in the shell (thus reducing the ash content of the meat) or reduce the digestibility and absorption of nutrients such as protein and minerals. This may be related to the possibility of diarrhea when taking high doses of magnesium in the diet [40]. A study on shrimps confirms Mg supplementation's positive impact on animal carcasses' fat content [105]. It was also shown that *Cornu aspersum* carcasses contained five times less crude fat compared with the research conducted by Niemiec et al. [124] on reared snails fed a diet without added Mg. The crude protein levels were comparable across the groups.

Experiment II meat mineral assays showed that Mg at 0.72% reduced the meat's manganese content. Mg at 0.33% increased the iron and decreased the carcass's Ca, copper, Mg, potassium and zinc content. Mg at 0.56% increased manganese and silicone in the meat. Adding Mg to the diet also resulted in a linear decrease in the sodium and potassium content of the snail meat. It is worth pointing out that sodium and potassium are the main positive ions responsible for maintaining animals' physiological pH levels. Objectively, the ions' presence affects muscle contractility and tension. Proper sodium concentrations in the body also aid the absorption of monosaccharides and amino acids [16]. Reducing the percentage of these constituents in carcasses may suggest the potential for several disorders in the organism. The current findings challenge preceding research demonstrating an inverse relationship between Ca concentrations and Mg and copper concentrations [31]. Reductions in the content of elements like K, Mg, Mn, Na, Fe, and P in the meat of snails fed with feed mix containing 0.72% Mg can indicate the antagonistic effects between these elements.

These results partially align with the study's findings on Roman snails, which showed an association between Mg, Cu, and Zn in soft tissues [130]. The study by Beeby and Richmond (2015) displayed that Mg levels did not impact Mg deposition in *Cantareus Aspersus* snails' bodies, contradicting the present study's results [131]. However, the element concentrations used in the cited research were notably low (720–2263 μg/g). In contrast to studies on poultry [118,132], the present study demonstrates that Mg addition affects the mineral composition of meat by increasing iron levels and decreasing Ca and Mg levels. Similarly, studies on shrimps that received an elevated Mg dose (254.36 μg/g) in the feed yielded conflicting outcomes to the current findings. The heightened Mg dosage raised the copper, zinc, Ca, sodium, and potassium concentration in the organisms' carcasses. The sole shared result was the rise in iron levels [105]. The study's findings corroborate previously reported links between elevated Mg intake and reduced absorption of manganese [133,134], zinc [135,136], and Ca [137,138] into body tissues. Both Mg and zinc were proven to minimise oxidative stress and free radicals [139,140]. Like Mg, zinc is a component or an activator of nearly 300 enzymes and plays various biological roles in the body [127,128], like regulating hormonal functions [141–143] and the immune response [144,145]. Higher iron content in Mg groups can indicate increasing immunity because immune responses are, among others, aided by enzymes containing iron, which promote the improvement of non-specific immunity. [146,147]. Mg supplementation also enhances protein utilisation in food by increasing protein absorption and synthesis [128]. It was shown that excessive Mg results in increased phosphate resorption [148], which promotes the formation of Ca salts that may favour Fe retention in the body [149].

In Experiment II, it was observed that the Mg content of snail meat was higher compared with the results of Çağıltay (2011) [115] and Fagbuaro et al. [126] but lower than in the study by Özogul et al. [125]. In contrast, potassium and manganese contents were higher than in the last-mentioned study [125]. Snail meat's zinc content was also lower than previous studies' results [120,125,126]. Furthermore, it was noted that the Mg content was lower than in the results of Özogul et al. [125]. The levels of potassium and manganese in Experiment II were similar to the results of Çağıltay (2011) [115].

The current study showed that Mg content at 0.72% increased TBARS levels compared with other experimental groups, indicating an increased incidence of oxidative stress.

Significant reductions in the content of a number of elements in the meat of snails fed with feed mix containing 0.72% Mg can confirm antagonistic effects between these elements, as shown in previous studies [59,63,66]. Alongside a statistically significant increase in TBARS, these findings suggest that a high Mg level may be toxic to snails. Generally, the lack of this element is associated with oxidative stress, inflammation, cytokine production, and increased free radicals [150–152]. It was proven that the negative impact of Mg on the body can be neutralised by increasing the availability of Ca and phosphorus in the diet [153].

### 4.3. Characteristics of Shells

#### 4.3.1. Experiment I

The study showed that increasing Ca levels positively affected shell weight, as it also positively affects shell width, and shells may have a higher mineral saturation. It was proven that increasing Ca content in feed has a positive effect on snails' development and growth. Also, the results of Ireland and Marigomez (1992) confirm the beneficial impact of an increased Ca proportion in the feed on the weight of shells [31]. Yet, the results were lower than those of the Berillis et al. [8] study.

Mineral analyses of the shells showed that Ca at 4.43% increased the Mg and sodium in shells. For Ca at 6.61%, a reduction in the Ca and iron content of the shells was observed. Ca at 8.87% contributed to an increase in iron and silicon while decreasing sodium and zinc. In contrast, Ca at 10.35% increased the Ca content and reduced the amount of silicon in the shells. As the proportion of Ca in the diet increased, a decrease in copper, Mg, manganese, and phosphorus and an increase in zinc was observed in the snail shells. The mechanisms responsible for the deposition of individual minerals in snail shells are still poorly understood and require further research.

A solidity index increase in response to higher dietary Ca levels was observed. This change may have implications for shell structural resistance. As such, the study showed that snail shell crushing force also increased (from 12.44 N to 54.83 N). In addition, an association was observed between shell crushing force, solidity index, and shell weight (from 0.29 g to 1.40 g). The current shells' crushing force was lower than in previous studies [84,98]. Also, in the present study, if differences in the shell shape index were statistically significant, they had little impact on crush resistance, and besides the Ca-deficient group, a higher Ca content enhanced the shell's crush resistance, unlike the results of the previous research on *Cornu aspersum* [84,98]. Also, less Ca was present in the shells: 37–38%, compared to 41.5%.

These relationships may indicate significant interactions between dietary Ca and physical characteristics and shell strength and the potential influence of individual maturity on its strength [26,154,155]. Studies on marine gastropods have demonstrated that the low Ca content of the water causes a substantial decrease in shell hardness [156]. Previous studies noted that shells of snails fed with insufficient Ca exhibited discolouration, thin walls, and reduced resistance to crushing. That was particularly evident in the group with a Ca level of 4.43% [34].

In the current study, there is evidence of earlier maturation of animals under the influence of increased dietary Ca, like higher body weight, a higher percentage of matured individuals, and smaller weight and shell size gains in the last month of the experiment.

#### 4.3.2. Experiment II

The study showed that the addition of Mg increased snail shell weight from 1.39 g to 1.61 g. Considering the absence of considerable variations in shell dimensions, it is likely that mineral saturation within the shell structure may cause an increase in their mass. The results are comparable to those of the Berillis et al. [8] experiment.

It was also observed that Mg at 0.72% promoted a higher solidity index of shells (23.58 g/cm$^2$) and higher shell weight than other experimental groups. In addition, a reduction in the shell shape index was observed, meaning that the shells became flatter. Shell crush strength increased up to a level of 0.56% of the dietary Mg content (61.41 N) and then decreased significantly (to 49.73 N), which is even lower than that of the control group (57.29 N). The experimental findings have established that the augmentation of Mg to the extent of 0.56% has a beneficial effect on the strength of the shell, as well as on its hardness and density (strength index). The shell crushing force can also be linked to the Fe amount in shell composition. Interestingly, a study on commercially raised *Cornu aspersum* in outdoor plots demonstrated more than twice the shell-crushing force compared to the best-observed result in the current experiment [157]. On the other hand, in a study by Berillis et al. 2013 [8], the shell-crushing force was up to five times lower than in the current study. It is plausible that dissimilar measurement techniques account for the disparity.

Mineral analyses of the shells showed that Mg at 0.33% in the diet reduced the Mg, potassium, Na, and P while increasing the Si content of the shell. Mg at 0.56% increased iron in the shell while decreasing Zn. In contrast, Mg at 0.72% resulted in a decrease in iron in the shell. Adding Mg to the diet also contributes to increased Mg in the shell while reducing the levels of manganese. The antagonistic interaction between magnesium and other elements is also apparent in the elemental composition of shells, although it is not as evident as it is in the composition of meat.

Assuming a similarity between hen eggshells and snail shells, the relationship between increased Mg in the feed and increased Mg in the shell composition is confirmed. However, this is not the case for Ca—studies on poultry showed that a higher proportion of Mg in the feed reduced the amount of Ca in the eggshell. The Ca level remained statistically unchanged in snail shells [158,159]. A study on eggshells found that adding Mg to feed increased zinc and reduced manganese levels [160]. This study affirms the association between Mg addition and decreased manganese levels in the shell. Still, the zinc content was lower in the Mg-added group than in the control group. However, there was an observable trend of increasing zinc concentration with increasing Mg levels within the experimental groups. Substantial decreases in the elemental levels, including Fe, K, and Mn, highlight the antagonistic interactions between these minerals and Mg. It is worth mentioning that the deposition of Mg and copper within the chicken eggshells increases along with their respective concentrations in the feed [53,61,62].

It was noted that Mg addition helped accelerate the snails' maturation process (from 46% to 65% mature individuals), potentially allowing for a shorter fattening period.

*4.4. Impact of Ca and Mg*

It is crucial to consider that snails excrete some nutrients absorbed from their feed, which ultimately affects the soil. According to Gomot et al. [83], consuming Ca and Mg-rich soil, which is a natural behaviour of snails, positively impacts their growth and enhances the resilience and hardness of their shells. Soils richer in Ca and Mg were also shown to reduce oxidative stress physiology parameters in snails *Pila globose* [161]. In addition, a previous study showed that these kinds of soils are preferred by a wide range of gastropod snail species [162].

When considering the effect of soil Ca on plants, it should be noted that Ca deficiency results in reduced plant growth, impairs root system growth, and causes leaf discolouration. Mg plays a role in the structure of chlorophyll, and thus in photosynthesis, and is also required for many enzymatic reactions in plant organisms [163]. It was proven that it is Mg deficiency that is related to the occurrence of plant diseases (17) rather than its excess (6 linked diseases), which may be linked to the impact of these ion elements on the pH of the soil and plant immune mechanisms [164]. However, the intended Mg level application is unlikely to impact soil pH or the availability of Ca to plants (Mg-induced Ca deficiency), as it is comparably low in the feed compared with Ca levels [165,166]. Infiltration of Ca and Mg into groundwater is possible. Nevertheless, the augmented presence of these minerals in drinking water has been shown to have beneficial effects on human health [167,168].

Typically, to mitigate low pH, toxic ions, and other ionic imbalances and their harmful effects on plant roots, soil solution concentrations of Ca ranging from 1 to 5 mM are necessary. Ca and Mg requirements for plants vary with plant species, from 100 to 2000 µg/g [169]. A study on maize showed that ratios of 2:1 to 3:1 of Ca:Mg gave optimum yield of plants without symptoms of Mg deficiency. Ratios of up to 6:1 also gave satisfactory results [170]. It can be assumed that additional amounts of Ca in soil can be beneficial for plants as long as the Ca:Mg ratio is kept at the right level.

### 5. Conclusions

In the group featuring 10.35% Ca and 0.72% Mg, snail growth was observed to slow down after three months, accompanied by a decrease in mortality. However, based on the analysis of total animal mortality and feed intake, it seems that a feed containing 0.56% magnesium would be the safer option, as high Mg (0.72%) could potentially exhibit slight toxicity towards snails. Shortening the fattening period by 3–4 weeks could increase profitability for snail farmers.

**Author Contributions:** Conceptualisation, A.R.-G. and T.N.; methodology, T.N.; investigation, A.R.-G., K.Z., M.R., W.B. and K.R.; formal analysis: A.R.-G.; data curation, A.R.-G. and D.B.; resources, K.Z., K.R., M.R. and W.B.; writing—original draft preparation A.R.-G.; writing—review T.N.; visualisation, A.R.-G. and D.B.; supervision, T.N. All authors have read and agreed to the published version of the manuscript.

**Funding:** This research received no external funding.

**Institutional Review Board Statement:** Not applicable.

**Data Availability Statement:** Not applicable.

**Acknowledgments:** This manuscript is part of a PhD thesis.

**Conflicts of Interest:** The authors declare no conflict of interest.

## References

1. Zucaro, A.; Forte, A.; De Vico, G.; Fierro, A. Environmental Loading of Italian Semi-Intensive Snail Farming System Evaluated by Means of Life Cycle Assessment. *J. Clean. Prod.* **2016**, *125*, 56–67. [CrossRef]
2. Hatziioannou, M.; Issari, A.; Neofitou, C.; Aifadi, S.; Matsiori, S. Economic Analysis and Production Techniques of Snail Farms in Southern Greece. *World J. Agric. Res.* **2014**, *2*, 276–279. [CrossRef]
3. Ogunniyi, L.T. Economic Analysis of Snail Production IN Ibadan, Oyo Stat. *Int. J. Agric. Econ. Rural. Dev.* **2009**, *2*, 26–34.
4. Zagata, L.; Sutherland, L.A. Deconstructing the 'Young Farmer Problem in Europe': Towards a Research Agenda. *J. Rural Stud.* **2015**, *38*, 39–51. [CrossRef]
5. Elmslie, L. Snail Collection and Small-Scale Production in Africa and Europe. In *Ecological Implications of Minilivestock: Potential of Insects, Rodents, Frogs, and Snails*; Paoletti, M.G., Ed.; Science Publishers: New Hampshire, UK, 2005; pp. 93–121. ISBN 9781482294439.
6. Rygało-Galewska, A.; Zglińska, K.; Niemiec, T. Edible Snail Production in Europe. *Animals* **2022**, *12*, 2732. [CrossRef]
7. Ireland, M.P. The Effect of Dietary Calcium on Growth, Shell Thickness and Tissue Calcium Distribution in the Snail *Achatina fulica*. *Comp. Biochem. Physiol. A Physiol.* **1991**, *98*, 111–116. [CrossRef]
8. Berillis, P.; Hatziioannou, M.; Panagiotopoulos, N.; Neofitou, C. Similar Shell Features between Rear and Wild *Cornu aspersum* Snails. *World J. Agric. Res.* **2013**, *1*, 1–4.
9. Murphy, B. *Breeding and Growing Snails Commercially in Australia*; Rural Industries Research Development Corporation: Wagga Wagga, Australia, 2001; ISBN 0-642-58219-X.
10. Lehmann, U.; Hillmer, G. *Bezkręgowce Kopalne*; Wydawnictwo Geologiczne: Warszawa, Poland, 1991.
11. Zhang, C.; Zhang, R. Matrix Proteins in the Outer Shells of Molluscs. *Mar. Biotechnol.* **2006**, *8*, 572–586. [CrossRef]
12. Sowiński, G.; Wąsowski, R. *Chów Ślimaków. Pielęgnacja, Żywienie, Zarys Chorób z Profilaktyką Oraz Kulinaria*; Wydawnictwo Uniwersytetu Warmińsko-Mazurskiego: Olsztyn, Poland, 2000; ISBN 83-88343-40-8.
13. Ligaszewski, M. *Opracowanie Metody Oceny Jakości Produkcji Ślimaków Jadalnych z Gatunku Helix Aspersa z Różnych Warunków Hodowlanych i Środowiskowych na Podstawie Analizy cech Jakościowych I Ilościowych ich Muszli*; Instytut Zootechniki Pib: Balice, Poland, 2008.
14. Fournié, J.; Chétail, M. Calcium Dynamics in Land Gastropods. *Am. Zool.* **1984**, *24*, 857–870. [CrossRef]

15. Pu, F.; Chen, N.; Xue, S. Calcium Intake, Calcium Homeostasis and Health. *Food Sci. Hum. Wellness* **2016**, *5*, 8–16. [CrossRef]
16. Radwińska, J.; Zarczyńska, K. Effects of Mineral Deficiency on the Health of Young Ruminants. *J. Elem.* **2014**, *19*, 915–928.
17. Jaiswal, J.K. Calcium—How and Why? *J. Biosci.* **2001**, *26*, 357–363. [CrossRef] [PubMed]
18. Jatto, O.E.; Asia, I.O.; Medjor, W.E. Proximate and Mineral Composition of Different Species of Snail Shell. *Pac. J. Sci. Technol.* **2010**, *11*, 416–419.
19. Beeby, A.; Richmond, L. Magnesium and the Deposition of Lead in the Shell of Three Populations of the Garden Snail *Cantareus aspersus*. *Environ. Pollut.* **2011**, *159*, 1667–1672. [CrossRef] [PubMed]
20. Hotopp, K.P. Land Snails and Soil Calcium in Central Appalachian Mountain Forest. *Southeast. Nat.* **2002**, *1*, 27–44. [CrossRef]
21. Tompa, A.S.; Wilbur, K.M. Calcium Mobilisation during Reproduction in Snail *Helix aspersa*. *Nature* **1977**, *270*, 53–54. [CrossRef]
22. Emelue, G.U.; Omonzogbe, E.A. Growth Performance of African Giant Land Snails (*Archachatina marginata*) Fed with Feed Formulated with Different Calcium Sources. *Malays. J. Sustain. Agric.* **2018**, *2*, 1–4. [CrossRef]
23. Voelker, J. Der Chemische Einfluss von Kalzimkarbonat auf Wachstum, Entwicklung und Gehausebau von *Achatina fulica* Bowdich (Pulmonata). *Mitteilungen Hambg. Zool. Mus. Inst. Univ. Hambg.* **1959**, *57*, 37–78.
24. Chevalier, L.; Desboquois, C.; Papineau, J.; Charrier, M. Influence of Inorganic Compounds on Food Selection by the Brown Garden Snail *Cornu aspersum* (Müller) (Gastropoda: Pulmonata). *J. Moll. Stud.* **2000**, *66*, 61–68. [CrossRef]
25. Oluokun, J.A.; Omole, A.J.; Fapounda, O. Effect of Increasing the Level of Calcium Supplementation in the Diets of Growing Snail on Performance Characteristics. *Res. J. Agric. Biol. Sci.* **2005**, *1*, 76–79.
26. Gouveia, A.R.; Pearce-Kelly, P.; Quicke, D.L.J.; Leather, S.R. Effects of Different Calcium Concentrations Supplemented on the Diet of *Partula gibba* on Their Morphometric Growth Parameters, Weight and Reproduction Success. *Malacologia* **2011**, *54*, 139–146. [CrossRef]
27. Marin, F.; Roy, N.L.; Marie, B. The Formation and Mineralization of Mollusk Shell. *Front. Biosci.* **2012**, *4*, 1099–1125. [CrossRef] [PubMed]
28. Roinel, N.; Morel, F.; Istin, M. Etude des Granules Calcifiés du Manteau des Lamellibranches à L'aide de la Microsonde Électronique. *Calcif. Tissue Res.* **1973**, *11*, 163–170. [CrossRef] [PubMed]
29. Fleury, C.; Marin, F.; Marie, B.; Luquet, G.; Thomas, J.; Josse, C.; Serpentini, A.; Lebel, J.M. Shell Repair Process in the Green Ormer *Haliotis tuberculata*: A Histological and Microstructural Study. *Tissue Cell* **2008**, *40*, 207–218. [CrossRef] [PubMed]
30. Simkiss, K.; Watkins, B. The Influence of Gut Microorganisms on Zinc Uptake in *Helix aspersa*. *Environ. Pollut.* **1990**, *66*, 263–271. [CrossRef]
31. Ireland, M.P.; Marigomez, I. The Influence of Dietary Calcium on the Tissue Distribution of Cu, Zn, Mg & P and Histological Changes in the Digestive Gland Cells of the Snail *Achatina fulica* Bowdichm. *J. Molluscan Stud.* **1992**, *58*, 157–168. [CrossRef]
32. Kouassi Kouadio, D.; Jean-Baptiste, A.; Mamadou, K. Growth Performance of *Archachatina marginata* Bred on the Substrate Amended with Industrial Calcium: Mikhart. *Int. J. Sci. Res.* **2016**, *5*, 582–586. [CrossRef]
33. Aman, J.B.; Adou, C.F.D.; Karamoko, M.; Otchoumou, A. Effect of Source and Amendment Rate of Rearing Substrate on the Growth and Yield of *Archachatina marginata*. *J. Res. Ecol.* **2019**, *7*, 2546–2554.
34. Crowell, H.H. Laboratory Study of Calcium Requirements of the Brown Garden Snail, *Helix aspersa* Müller. *J. Molluscan Stud.* **1973**, *40*, 491–503. [CrossRef]
35. Wacker, A.; Baur, B. Effects of Protein and Calcium Concentrations of Artificial Diets on the Growth and Survival of the Land Snail *Arianta arbustorum*. *Invertebr. Reprod. Dev.* **2004**, *46*, 47–53. [CrossRef]
36. Tchakounte, F.M.; Kana, J.R.; Azine, P.C.; Meffowoet, C.P.; Djuidje, V.P. Growth and Reproductive Performances of African Giant Snail (*Archachatina marginata*) as Affected by Dietary Calcium Levels. *Sci. J. Vet. Adv.* **2019**, *8*, 263–271.
37. Otchoumou, A.; Dupont-Nivet, M.; Dosso, H. Effects of Diet Quality and Dietary Calcium on Reproductive Performance in *Archachatina ventricosa* (Gould 1850), Achatinidae, under Indoor Rearing Conditions. *Invertebr. Reprod. Dev.* **2012**, *56*, 14–20. [CrossRef]
38. Lobo, H.L.; Santos, A.G. Níveis de Cálcio em Rações para Escargot (*Cornu aspersum*). Bachelor's Thesis, Federal University of São João del-Rei, São João del-Rei, Brazil, 2017; p. 32.
39. Drapała, T. *Chemia Ogólna Nieorganiczna*; Pwn: Warszawa, Poland, 1986.
40. Karmanska, A.; Stanczak, A.; Karwowski, B. Magnez Aktualny Stan Wiedzy. *Bromatol. Chem. Toksykol.* **2015**, *48*, 677–689.
41. Rubin, H. Central Role for Magnesium in Coordinate Control of Metabolism and Growth in Animal Cells. *Proc. Natl. Acad. Sci. USA* **1975**, *72*, 3551–3555. [CrossRef] [PubMed]
42. Banai, S.; Haggroth, L.; Epstein, S.E.; Casscells, W. Influence of Extracellular Magnesium on Capillary Endothelial Cell Proliferation and Migration. *Circ. Res.* **1990**, *67*, 645–650. [CrossRef]
43. Wolf, F.I.; Cittadini, A. Magnesium in Cell Proliferation and Differentiation. *Front. Biosci.* **1999**, *4*, 607–617. [CrossRef]
44. Maier, J.A.M.; Bernardini, D.; Rayssiguier, Y.; Mazur, A. High Concentrations of Magnesium Modulate Vascular Endothelial Cell Behaviour In Vitro. *Biochim. Biophys. Acta-Mol. Basis Dis.* **2004**, *1689*, 6–12. [CrossRef]
45. Chandra, A.K.; Sengupta, P.; Goswami, H.; Sarkar, M. Effects of Dietary Magnesium on Testicular Histology, Steroidogenesis, Spermatogenesis and Oxidative Stress Markers in Adult Rats. *Indian J. Exp. Biol.* **2013**, *51*, 37–47.
46. Gaál, K.K.; Sáfár, O.; Gulyás, L.; Stadler, P. Magnesium in Animal Nutrition. *J. Am. Coll. Nutr.* **2004**, *23*, 754s–757s. [CrossRef]
47. Cowan, J.A. Introduction to the Biological Chemistry of Magnesium Ion. In *The Biological Chemistry of Magnesium*; Cowan, J.A., Ed.; Vch Publishers: New York, NY, USA, 1995; pp. 1–23, ISBN 1560816279.

48. Rubin, H. The Logic of the Membrane, Magnesium, Mitosis (MMM) Model for the Regulation of Animal Cell Proliferation. *Arch. Biochem. Biophys.* **2007**, *458*, 16–23. [CrossRef]

49. Vink, R.; Cernak, I. Regulation of Intracellular Free Magnesium in Central Nervous System Injury. *Front. Biosci.* **2000**, *5*, A541. [CrossRef] [PubMed]

50. Pasternak, K.; Kocot, J.; Horecka, A. The Biochemistry of Magnesium. *J. Elementol.* **2010**, *15*, 601–616. [CrossRef]

51. Cefaratti, C.; Romani, A.; Scarpa, A. Differential Localization and Operation of Distinct $Mg^{2+}$ Transporters in Apical and Basolateral Sides of Rat Liver Plasma Membrane. *J. Biol. Chem.* **2000**, *275*, 3772–3780. [CrossRef]

52. Garfinkel, D.; Garfinkel, L. Magnesium and Regulation of Carbohydrate Metabolism at the Molecular Level. *Magnesium* **1988**, *7*, 249–261.

53. Tammaro, P.; Smith, A.L.; Crowley, B.L.; Smirnov, S.V. Modulation of the Voltage-Dependent $K^+$ Current By Intracellular $Mg^{2+}$ in Rat Aortic Smooth Muscle Cells. *Cardiovasc. Res.* **2005**, *65*, 387–396. [CrossRef] [PubMed]

54. Shi, J.; Krishnamoorthy, G.; Yang, Y.; Hu, L.; Chaturvedi, N.; Harilal, D.; Qin, J.; Cui, J. Mechanism of Magnesium Activation of Calcium-Activated Potassium Channels. *Nature* **2002**, *418*, 876–880. [CrossRef] [PubMed]

55. Agus, Z.S.; Morad, M. Modulation of Cardiac Ion Channels by Magnesium. *Annu. Rev. Physiol.* **1991**, *53*, 299–307. [CrossRef]

56. Mcdonald, P.; Edwards, R.A.; Greenhalgh, J.F.D.; Morgan, C.A.; Sinclair, L.A.; Wilkinson, R.G. *Animal Nutrition*, 7th ed.; Longman Scientific and Technical: Madison, WI, USA, 2011; ISBN 9781408204238.

57. Horky, P.; Sochor, J.; Skladanka, J.; Klusonova, I.; Nevrkla, P. Effect of Selenium, Vitamins E and C on Antioxidant Potential and Quality of Boar Ejaculate. *J. Anim. Feed Sci.* **2016**, *25*, 29–36. [CrossRef]

58. Freitag, J.J.; Martin, K.J.; Conrades, M.B.; Bellorin-Font, E.; Teitelbaum, S.; Klahr, S.; Slatopolsky, E. Evidence for Skeletal Resistance to Parathyroid Hormone in Magnesium Deficiency: Studies in Isolated Perfused Bone. *J. Clin. Investig.* **1979**, *64*, 1238–1244. [CrossRef]

59. Schlingmann, K.P.; Gudermann, T. A Critical Role of TRPM Channel-Kinase for Human Magnesium Transport. *J. Physiol.* **2005**, *566*, 301–308. [CrossRef]

60. Monteilh-Zoller, M.K.; Hermosura, M.C.; Nadler, M.J.S.; Scharenberg, A.M.; Penner, R.; Fleig, A. TRPM7 Provides an Ion Channel Mechanism for Cellular Entry of Trace Metal Ions. *J. Gen. Physiol.* **2003**, *121*, 49–60. [CrossRef] [PubMed]

61. Okorodudu, A.O.; Yang, H.; Elghetany, M.T. Ionized Magnesium in the Homeostasis of Cells: Intracellular Threshold for Mg in Human Platelets. *Clin. Chim. Acta* **2001**, *303*, 147–154. [CrossRef] [PubMed]

62. Féray, J.C.; Garay, R. A One-to-One $Mg^{2+}$:$Mn^{2+}$ Exchange in Rat Erythrocytes. *J. Biol. Chem.* **1987**, *262*, 5763–5768. [CrossRef] [PubMed]

63. Goytain, A.; Quamme, G.A. Functional Characterization of the Human Solute Carrier, SLC41A2. *Biochem. Biophys. Res. Commun.* **2005**, *330*, 701–705. [CrossRef]

64. Schmitz, C.; Perraud, A.-L.; Johnson, C.O.; Inabe, K.; Smith, M.K.; Penner, R.; Kurosaki, T.; Fleig, A.; Scharenberg, A.M. Regulation of Vertebrate Cellular $Mg^{2+}$ Homeostasis by TRPM7. *Cell* **2003**, *114*, 191–200. [CrossRef]

65. Matsuda, H. Magnesium Gating of the Inwardly Rectifying $K^+$ Channel. *Annu. Rev. Physiol.* **1991**, *53*, 289–298. [CrossRef]

66. Gröber, U.; Schmidt, J.; Kisters, K. Magnesium in Prevention and Therapy. *Nutrients* **2015**, *7*, 8199–8226. [CrossRef]

67. Shils, M.E. Experimental Human Magnesium Depletion. *Medicine* **1969**, *48*, 61–85. [CrossRef]

68. Whang, R.; Flink, E.B.; Dyckner, T.; Wester, P.O.; Aikawa, J.K.; Ryan, M.P. Magnesium Depletion as a Cause of Refractory Potassium Repletion. *Arch. Intern. Med.* **1985**, *145*, 1686–1689. [CrossRef]

69. Chang, C.; Bloom, S. Interrelationship of Dietary Mg Intake and Electrolyte Homeostasis in Hamsters: I. Severe Mg Deficiency, Electrolyte Homeostasis, and Myocardial Necrosis. *J. Am. Coll. Nutr.* **1985**, *4*, 173–185. [CrossRef]

70. Goytain, A.; Quamme, G.A. Functional Characterization of ACDP2 (Ancient Conserved Domain Protein), a Divalent Metal Transporter. *Physiol. Genom.* **2005**, *22*, 382–389. [CrossRef] [PubMed]

71. Toba, Y.; Kajita, Y.; Masuyama, R.; Takada, Y.; Suzuki, K.; Aoe, S. Nutrient Metabolism Dietary Magnesium Supplementation Affects Bone Metabolism and Dynamic Strength of Bone in Ovariectomized Rats. *J. Nutr.* **2000**, *130*, 216–220. [CrossRef] [PubMed]

72. Clark, I. Effects of Magnesium Ions on Calcium and Phosphorus Metabolism. *Am. J. Physiol.* **1968**, *214*, 348–356. [CrossRef] [PubMed]

73. Scanlan, B.J.; Tuft, B.; Elfrey, J.E.; Smith, A.; Zhao, A.; Morimoto, M.; Chmielinska, J.J.; Tejero-Taldo, M.I.; Mak, I.T.; Weglicki, W.B.; et al. Intestinal Inflammation Caused by Magnesium Deficiency Alters Basal and Oxidative Stress-Induced Intestinal Function. *Mol. Cell. Biochem.* **2007**, *306*, 59–69. [CrossRef] [PubMed]

74. Gunther, T.; Hollriegl, V.; Vormann, J.; Bubeck, J.; Classen, H.G. Increased Lipid Peroxidation in Rat Tissues by Magnesium Deficiency and Vitamin E Depletion. *Magnes. Bull.* **1994**, *16*, 38–43.

75. Rude, R.K.; Gruber, H.E.; Wei, L.Y.; Frausto, A.; Mills, B.G. Magnesium Deficiency: Effect on Bone and Mineral Metabolism In the Mouse. *Calcif. Tissue Int.* **2003**, *72*, 32–41. [CrossRef]

76. Rude, R.K.; Kirchen, M.E.; Gruber, H.E.; Stasky, A.A.; Meyer, M.H. Magnesium Deficiency Induces Bone Loss in the Rat. *Miner. Electrolyte Metab.* **1998**, *24*, 314–320. [CrossRef]

77. Lipiński, K.; Stasiewicz, M.; Purwin, C.; Zuk-Gołaszewska, K. Effects of Magnesium on Pork Quality. *J. Elem.* **2011**, *16*, 325–337. [CrossRef]

78. Vormann, J.; Günther, T.; Höllriegl, V.; Schumann, K. Pathobiochemical Effects of Graded Magnesium Deficiency in Rats. *Z. Ernahrungswissenschaft* **1998**, *37* (Suppl. S1), 92–97.

79. Quamme, G.A.; De Rouffignac, C. Epithelial Magnesium Transport and Regulation by the Kidney. *Front. Biosci.* **2000**, *5*, D694–D711. [CrossRef]
80. Elin, R.J. Magnesium: The Fifth but Forgotten Electrolyte. *Am. J. Clin. Pathol.* **1994**, *102*, 616–622. [CrossRef] [PubMed]
81. Weaver, V.M.; Welsh, J. 1,25-Dihydroxycholecalciferol Supplementation Prevents Hypocalcemia in Magnesium-Deficient Chicks. *J. Nutr.* **1993**, *123*, 764–771. [CrossRef]
82. Yoshimura, T.; Tamenori, Y.; Kawahata, H.; Suzuki, A. Fluctuations of Sulfate, S-Bearing Amino Acids and Magnesium in a Giant Clam Shell. *Biogeosciences* **2014**, *11*, 3881–3886. [CrossRef]
83. Gomot, A.; Gomot, L.; Boukraa, S.; Bruckert, S. Influence of Soil on the Growth of the Land Snail *Helix aspersa*. An Experimental Study of the Absorption Route for the Stimulating Factors. *J. Molluscan Stud.* **1989**, *55*, 1–7. [CrossRef]
84. Ligaszewski, M.; Pol, P. *Wybrane Zagadnienia z Dziedziny Helikultury*; Zespół Wydawnictw i Poligrafii iz Pib: Kraków, Poland, 2019; ISBN 9788376073927.
85. Mayaki, O.M.; Ozumba, A.U.; Aderele, A.A.; Daramola, A.O. Effect of Sources of Fibre on Performance of Growing Snail. *Niger. Food J.* **2013**, *31*, 28–32. [CrossRef]
86. Sampelayo, M.R.S.; Fonolla, J.; Extremera, F.G. Factors Affecting the Food Intake, Growth and Protein Utilization in the Helix Aspersa Snail. Protein Content of the Diet and Animal Age. *Lab. Anim.* **1991**, *25*, 291–298. [CrossRef]
87. Bonnet, J.-C.; Vrillon, J.-L.; Aupinel, P. *L'escargot Helix aspersa*; Inra: Paris, France, 1992; ISBN 9782759201976.
88. García, A.; Perea, J.; Martín, R.; Acero, R.; Mayoral, A.; Peña, F.; Luque, M. Effect of Two Diets on the Growth of the *Helix aspersa* Müller During the Juvenile Stage. In Proceedings of the 56th Annual Meeting EAAP, Uppsala, Sweden, 5–8 June 2005; pp. 1–9.
89. Cameron, R.A.D.; Carter, M.A. Intra- and Interspecific Effects of Population Density on Growth and Activity in Some Helicid Land Snails (Gastropoda: Pulmonata). *J. Anim. Ecol.* **1979**, *48*, 237. [CrossRef]
90. Staikou, A.; Dimitriadou, L. Effect of Crowding on Growth and Mortality in the Edible Snail *Helix lucorum* (Gastropoda: Pulmonata) in Greece. *Isr. J. Zool.* **1989**, *36*, 1–9.
91. Guo, Y.; Zhang, G.; Yuan, J.; Nie, W. Effects of Source and Level of Magnesium and Vitamin E on Prevention of Hepatic Peroxidation and Oxidative Deterioration of Broiler Meat. *Anim. Feed Sci. Technol.* **2003**, *107*, 143–150. [CrossRef]
92. Lee, S.; Britton, W.M. Magnesium Toxicity: Effect on Phosporus Utilization by Broiler Chicks. *Poult. Sci.* **1980**, *59*, 1989–1994. [CrossRef]
93. Hess, J.; Britton, W. Effects of Dietary Magnesium Excess in White Leghorn Hens. *Poult Sci* **1997**, *76*, 703–710. [CrossRef] [PubMed]
94. AOAC. *Official Methods of Analysis of the Association of Official Analytical Chemists*; AOAC Intl.: Gaithersburg, MD, USA, 2005.
95. Uchiyama, M.; Mihara, M. Determination of Malonaldehyde Precursor in Tissues by Thiobarbituric Acid Test. *Anal. Biochem.* **1978**, *86*, 271–278. [CrossRef]
96. Chevallier, H. La Variabilité de L'escargot Petit-Gris *Hélix aspersa* Muller. *Bull. Muséum Natl. D'histoibe Nat.* **1977**, *3*, 425–442.
97. Cooke, A.S. Shell Thinning in Avian Eggs by Environmental Pollutants. *Environ. Pollut.* **1973**, *4*, 85–152. [CrossRef]
98. Ligaszewski, M.; Surówka, K.; Stekla, J. The Shell Features of *Cornu aspersum* (Synonym *Helix aspersa*) And *Helix pomatia*: Characteristics and Comparison. *Am. Malacol. Bull.* **2009**, *27*, 173–181. [CrossRef]
99. Fournie, J.; Chetail, M. Evidence for a Mobilization of Calcium Reserves for Reproduction Requirements in *Deroceras reticulatum* (Syn: *Agriolimax reticulatus*) (Gastropoda: Pulmonata). *Malacologia* **1980**, *22*, 285–291.
100. Cowie, R.H.; Cain, A.J. Laboratory Maintenance and Breeding of Land Snails, with an Example from *Helix aspersa*. *J. Molluscan Stud.* **1983**, *49*, 176–177. [CrossRef]
101. Dan, N.A. Studies on the Growth and Ecology of *Helix asposa* Muller. Ph.D. Thesis, University of Manchester, Manchester, UK, 1978.
102. Milinsk, M.C.; Padre, R.D.G.; Hayashi, C.; De Oliveira, C.C.; Visentainer, J.V.; De Souza, N.E.; Matsushita, M. Effects of Feed Protein and Lipid Contents on Fatty Acid Profile of Snail (*Helix aspersa maxima*) Meat. *J. Food Compos. Anal.* **2006**, *19*, 212–216. [CrossRef]
103. Desbuquois, C. Influence of Egg Cannibalism on Growth, Survival and Feeding in Hatchlings of the Land Snail *Helix aspersa* Müller (Gastropoda, Pulmonata, Stylommatophora). *Reprod. Nutr. Dev.* **1997**, *37*, 191–202. [CrossRef]
104. Suttle, N. *Mineral Nutrition of Livestock*; Suttle, N., Ed.; Cabi: Cambridge, UK, 2010; ISBN 9781845934736.
105. Srinivasan, V.; Bhavan, P.S.; Rajkumar, G.; Satgurunathan, T.; Muralisankar, T. Dietary Supplementation of Magnesium Oxide (MgO) Nanoparticles for Better Survival and Growth of the Freshwater Prawn *Macrobrachium rosenbergii* Post-Larvae. *Biol. Trace Elem. Res.* **2017**, *177*, 196–208. [CrossRef]
106. Jahan, I.; Reddy, A.K.; Sudhagar, S.A.; Harikrishna, V.; Singh, S.; Varghese, T.; Srivastava, P.P. The Effect of Fortification of Potassium and Magnesium in the Diet and Culture Water on Growth, Survival and Osmoregulation of Pacific White Shrimp, *Litopenaeus vannamei* Reared in Inland Ground Saline Water. *Turk. J. Fish Aquat. Sci.* **2018**, *18*, 1235–1243. [CrossRef] [PubMed]
107. Wacker, A. Lipids in the Food of a Terrestrial Snail. *Invertebr. Reprod. Dev.* **2005**, *47*, 205–212. [CrossRef]
108. Egonmwan, R.I. Effects of Dietary Calcium on Growth and Oviposition of the African Land Snail *Limicolaria flammea* (Pulmonata: Achatinidae). *Rev. Biol. Trop.* **2008**, *56*, 333–343. [CrossRef] [PubMed]
109. Jess, S.; Marks, R.J. Population Density Effects on Growth in Culture of the Edible Snail *Helix aspersa* Var. *Maxima*. *J. Molluscan Stud.* **1995**, *61*, 313–323. [CrossRef]
110. Daguzan, J.; Bonnet, J.C.; Perrin, Y.; Perrin, E.; Rouet, H. Contribution à L'élevage de L'escargot Petit-Gris: *Helix aspersa* Müller (Mollusque Gastéropode Pulmoné Stylommatophore). I.—Reproduction et Éclosion des Jeunes, en Bâtiment et en Conditions Thermohygrométriques Contrôlées. *Ann. Zootech.* **1981**, *30*, 249–272. [CrossRef]

111. Milinsk, M. Influence of Diets Enriched with Different Vegetable Oils on the Fatty Acid Profiles of Snail *Helix aspersa maxima*. *Food Chem.* **2003**, *82*, 553–558. [CrossRef]

112. Nicolai, A.; Filser, J.; Lenz, R.; Bertrand, C.; Charrier, M. Adjustment of Metabolite Composition in the Haemolymph to Seasonal Variations in the Land Snail *Helix pomatia*. *J. Comp. Physiol. B* **2011**, *181*, 457–466. [CrossRef]

113. Dimitriadis, V.K.; Domouhtsidou, G.P. Carbohydrate Cytochemistry of the Intestine and Salivary Glands of the Snail *Helix lucorum*: Effect of Starvation and Hibernation. *J. Molluscan Stud.* **1995**, *61*, 215–224. [CrossRef]

114. Baby, R.L.; Hasan, I.; Kabir, K.A.; Naser, M.N. Nutrient Analysis of Some Commercially Important Molluscs of Bangladesh. *J. Sci. Res.* **2010**, *2*, 390–396. [CrossRef]

115. Çağıltay, F. Amino Acid, Fatty Acid, Vitamin and Mineral Contents of the Edible Garden Snail (*Helix aspersa*). *J. Fishscicom.* **2011**, *5*, 354–363. [CrossRef]

116. Mikkelsen, F.F.; Weber, R.E. Oxygen Transport and Hemocyanin Function in the Pulmonate Land Snail, *Helix pomatia*: Physiological and Molecular Implications of Polyphasic Oxygen-Binding Curves. *Physiol. Zool.* **1992**, *65*, 1057–1073. [CrossRef]

117. Tufts, E.V.; Greenberg, D.M. The Biochemistry of Magnesium Deficiency. *J. Biol. Chem.* **1938**, *122*, 693–714. [CrossRef]

118. Bunce, G.E.; Chiemchaisri, Y.; Phillips, P.H. The Mineral Requirements of the Dog IV. Effect of Certain Dietary and Physiologic Factors upon the Magnesium Deficiency Syndrome. *J. Nutr.* **1962**, *76*, 23–29. [CrossRef]

119. Nugara, D.; Edwards, H.M. Influence of Dietary Ca and P Levels on the Mg Requirement of the Chick. *J. Nutr.* **1963**, *80*, 181–184. [CrossRef]

120. Gomot, A. Biochemical Composition of Helix Snails: Influence of Genetic and Physiological Factors. *J. Molluscan Stud.* **1998**, *64*, 173–181. [CrossRef]

121. Çelik, M.Y.; Duman, M.B.; Sariipek, M.; Uzun Gören, G.; Kaya Öztürk, D.; Kocatepe, D.; Karayücel, S. Comparison of Fatty Acids and Some Mineral Matter Profiles of Wild and Farmed Snails, *Cornu aspersum* Müller, 1774. *Molluscan Res.* **2019**, *39*, 234–240. [CrossRef]

122. Engmann, F.; Afoakwah, N.A.; Darko, P.O.; Sefah, W. Proximate and Mineral Composition of Snail (*Achatina achatina*) Meat; any Nutritional Justification for Acclaimed Health Benefits? *J. Basic Appl. Sci. Res.* **2013**, *3*, 8–15.

123. Gomot, A.; Pihan, F. Comparison of the Bioaccumulation Capacities of Copper and Zinc in Two Snail Subspecies (*Helix*). *Ecotoxicol. Environ. Saf.* **1997**, *38*, 85–94. [CrossRef]

124. Niemiec, T.; Łozicki, A.; Pietrasik, R.; Pawęta, S.; Rygało-Galewska, A.; Matusiewicz, M.; Zglińska, K. Impact of Ag Nanoparticles (AgNPs) and Multimicrobial Preparation (EM) on the Carcass, Mineral and Fatty Acid Composition of *Cornu aspersum aspersum* Snails. *Animals* **2021**, *11*, 1926. [CrossRef]

125. Özogul, Y.; Özogul, F.; Olgunoglu, A.I. Fatty Acid Profile and Mineral Content of the Wild Snail (*Helix pomatia*) from the Region of the South of the Turkey. *Eur. Food Res. Technol.* **2005**, *221*, 547–549. [CrossRef]

126. Fagbuaro, O.; Oso, J.A.; Edward, J.B.; Ogunleye, R.F. Nutritional Status of Four Species of Giant Land Snails in Nigeria. *J. Zhejiang Univ. Sci. B* **2006**, *7*, 686–689. [CrossRef] [PubMed]

127. Ligaszewski, M.; Pol, P. Ocena Wpływu Różnych Systemów Chowu Ślimaka Szarego (*Helix aspersa*) na Wartość Odżywczą i Wydajność Jego Mięsa. *Wiadomości Zootech.* **2016**, *54*, 18–34.

128. Rico, M.C.; Lerma, A.; Planells, E.; Aranda, P.; Gonzalez, J.L. Changes in the Nutritive Utilization of Protein Induced by Mg Deficiency in Rats. *Int. J. Vitam. Nutr. Res.* **1995**, *65*, 122–126.

129. Bara, M.; Guiet-Bara, A.; Durlach, J. Regulation of Sodium and Potassium Pathways by Magnesium in Cell Membranes. *Magnes. Res.* **1993**, *6*, 167–177. [PubMed]

130. Nica, D.V.; Bordean, D.M.; Hărmănescu, M.; Bura, M.; Gergen, I. Interactions among Heavy Metals (Cu, Cd, Zn, Pb) and Metallic Macroelements (K, Ca, Na, Mg) in Roman Snail (*Helix pomatia*) Soft Tissues. *Acta Met.–MEEMB* **2014**, *11*, 65.

131. Beeby, A.; Richmond, L. Magnesium and the Regulation of Lead in Three Populations of the Garden Snail *Cantareus aspersus*. *Environ. Pollut.* **2010**, *158*, 2288–2293. [CrossRef]

132. Liu, Y.X.; Guo, Y.M.; Wang, Z. Effect of Magnesium on Reactive Oxygen Species Production in the Thigh Muscles of Broiler Chickens. *Br. Poult. Sci.* **2007**, *48*, 84–89. [CrossRef]

133. Woerpel, H.R.; Balloun, S.L. Effect of Iron and Magnesium on Manganese Metabolism. *Poult. Sci.* **1964**, *43*, 1134–1142. [CrossRef]

134. Sanchez-Morito, N.; Planells, E.; Aranda, P.; Llopis, J. Magnesium-Manganese Interactions Caused by Magnesium Deficiency in Rats. *J. Am. Coll. Nutr.* **1999**, *18*, 475–480. [CrossRef]

135. Lease, J.G.; Williams, W.P. The Effect of Added Magnesium on the Availability of Zinc with Some High-Protein Feedstuffs. *Poult. Sci.* **1967**, *46*, 242–248. [CrossRef]

136. Planells, E.; Aranda, P.; Lerma, A.; Llopis, J. Changes in Bioavailability and Tissue Distribution of Zinc Caused by Magnesium Deficiency in Rats. *Br. J. Nutr.* **1994**, *72*, 315–323. [CrossRef]

137. Planells, E.; Llopis, J.; Peran, F.; Aranda, P. Changes in Tissue Calcium and Phosphorus Content and Plasma Concentrations of Parathyroid Hormone and Calcitonin after Long-Term Magnesium Deficiency in Rats. *J. Am. Coll. Nutr.* **1995**, *14*, 292–298. [CrossRef] [PubMed]

138. O'dell, B.L. Magnesium Requirement and Its Relation to other Dietary Constitunents. *Fed. Proc.* **1960**, *19*, 648–654. [PubMed]

139. Afanas'ev, I.B.; Suslova, T.B.; Cheremisina, Z.P.; Abramova, N.E.; Korkina, L.G. Study of Antioxidant Properties of Metal Aspartates. *Analyst* **1995**, *120*, 859–862. [CrossRef] [PubMed]

140. Sahin, K.; Onderci, M.; Sahin, N.; Gulcu, F.; Yildiz, N.; Avci, M.; Kucuk, O. Responses of Quail to Dietary Vitamin E and Zinc Picolinate at Different Environmental Temperatures. *Anim. Feed Sci. Technol.* **2006**, *129*, 39–48. [CrossRef]
141. Al-Saad, K.M.; Al-Sadi, H.I.; Abdul-Majeed, M.O. Clinical, Hematological, Biochemical and Pathological Studies on Zinc Deficiency (Hypozincemia) in Sheep. *Vet. Res.* **2010**, *3*, 14–20.
142. Miao, X.; Sun, W.; Fu, Y.; Miao, L.; Cai, L. Zinc Homeostasis in the Metabolic Syndrome and Diabetes. *Front. Med.* **2013**, *7*, 31–52. [CrossRef]
143. Alves, C.X.; Vale, S.H.L.; Dantas, M.M.G.; Maia, A.A.; Franca, M.C.; Marchini, J.S.; Leite, L.D.; Brandao-Neto, J. Positive Effects of Zinc Supplementation on Growth, Gh, Igf1, and Igfbp3 in Eutrophic Children. *J. Pediatr. Endocrinol. Metab.* **2012**, *25*, 881–887. [CrossRef]
144. Kincaid, R.L. Assessment of Trace Mineral Status of Ruminants: A Review. *J. Anim. Sci.* **2000**, *77*, 1. [CrossRef]
145. El-Far, A.H. Biochemical Alterations in Zinc Deficient Sheep Associated by Hyperlactatemia. *Am. J. Anim. Vet. Sci.* **2013**, *8*, 112–116. [CrossRef]
146. Walter, T.; Olivares, M.; Pizarro, F.; Muñoz, C. Iron, Anemia, and Infection. *Nutr. Rev.* **1997**, *55*, 111–124. [CrossRef]
147. Benito, P.; Miller, D. Iron Absorption and Bioavailability: An Updated Review. *Nutr. Res.* **1998**, *18*, 581–603. [CrossRef]
148. Planells, E.; Aranda, P.; Peran, F.; Llopis, J. Changes in Calcium and Phosphorus Absorption and Retention during Long-Term Magnesium Deficiency in Rats. *Nutr. Res.* **1993**, *13*, 691–699. [CrossRef]
149. Sanchez-Morito, N.; Planells, E.; Aranda, P.; Llopis, J. Influence of Magnesium Deficiency on the Bioavailability and Tissue Distribution of Iron in the Rat. *J. Nutr. Biochem.* **2000**, *11*, 103–108. [CrossRef] [PubMed]
150. Yang, Y.; Wu, Z.; Chen, Y.; Qiao, J.; Gao, M.; Yuan, J.; Nie, W.; Guo, Y. Magnesium Deficiency Enhances Hydrogen Peroxide Production and Oxidative Damage in Chick Embryo Hepatocyte In Vitro. *Biometals* **2006**, *19*, 71–81.
151. Bussière, F.I.; Zimowska, W.; Gueux, E.; Rayssiguier, Y.; Mazur, A. Stress Protein Expression cDNA Array Study Supports Activation of Neutrophils During Acute Magnesium Deficiency in Rats. *Magnes. Res.* **2002**, *15*, 37–42. [PubMed]
152. Weglicki, W.B.; Dickens, B.F.; Wagner, T.L.; Chmielinska, J.J.; Phillips, T.M. Immunoregulation by Neuropeptides in Magnesium Deficiency: Ex Vivo Effect of Enhanced Substance P Production on Circulating T Lymphocytes from Magnesium-Deficient Mice. *Magnes. Res.* **1996**, *9*, 3–11.
153. Haag, J.R.; Palmer, L.S. The Effect of Variations in the Proportions of Calcium, Magnesium, and Phosphorus Contained in the Diet. *J. Biol. Chem.* **1928**, *76*, 367–389. [CrossRef]
154. Ligaszewski, M. Kształtowanie się Wskaźników Wartości Użytkowej Muszli Ślimaków Jadalnych z Rodzaju Helix w Zróżnicowanych Warunkach Hodowlanych i Środowiskowych. *Rocz. Nauk. Zootech.* **2005**, *19*, 1–104.
155. Ligaszewski, M.; Pol, P. Edible Snails Breeding in Poland. *Zagadnienia Doradz. Rol.* **2021**, *1*, 67–85.
156. Glass, N.H.; Darby, P.C. The Effect of Calcium and Ph on Florida Apple Snail, *Pomacea paludosa* (Gastropoda: Ampullariidae), Shell Growth and Crush Weight. *Aquat. Ecol.* **2009**, *43*, 1085–1093. [CrossRef]
157. Łozicki, A.; Niemiec, T.; Pietrasik, R.; Pawęta, S.; Rygało-Galewska, A.; Zglińska, K. The Effect of Ag Nanoparticles and Multimicrobial Preparation as Factors Stabilizing the Microbiological Homeostasis of Feed Tables for *Cornu aspersum* (Müller) Snails on Snail Growth and Quality Parameters of Carcasses and Shells. *Animals* **2020**, *10*, 2260. [CrossRef] [PubMed]
158. Atteh, J.O.; Leeson, S. Influence of Increasing Dietary Calcium and Magnesium Levels on Performance, Mineral Metabolism, and Egg Mineral Content of Laying Hens. *Poult. Sci.* **1983**, *62*, 1261–1268. [CrossRef] [PubMed]
159. Shastak, Y.; Rodehutscord, M. A Review of the Role of Magnesium in Poultry Nutrition. *World Poult. Sci. J.* **2019**, *71*, 125–137. [CrossRef]
160. Holder, D.P.; Huntley, D.M. Influence of Added Manganese, Magnesium, Zinc, and Calcium Level on Egg Shell Quality. *Poult. Sci.* **1978**, *57*, 1629–1634. [CrossRef]
161. Pati, S.G.; Panda, F.; Samanta, L.; Paital, B. Spatio-Temporal Changes in Oxidative Stress Physiology Parameters in Apple Snail *Pila globosa* as a Function of Soil Mg, Ca, Organic Carbon and Aquatic Physico-Chemical Factors. *Environ. Geochem. Health* **2023**, *45*, 2591–2610. [CrossRef]
162. Ondina, P.; Mato, S.; Hermida, J.; Outeiro, A. Importance of Soil Exchangeable Cations and Aluminium Content on Land Snail Distribution. *Appl. Soil Ecol.* **1998**, *9*, 229–232. [CrossRef]
163. Schwartzkopf, C. Potassium, Calcium, Magnesium—How They Relate to Plant Growth. *Usga Green Sect. Rec.* **1972**, *10*, 1–2.
164. Huber, D.M.; Jones, J.B. The Role of Magnesium in Plant Disease. *Plant Soil* **2013**, *368*, 73–85.
165. Kobayashi, H.; Masaoka, Y.; Sato, S. Effects of Excess Magnesium on the Growth and Mineral Content of Rice and *Echinochloa*. *Plant Prod. Sci.* **2005**, *8*, 38–43. [CrossRef]
166. Franklin, W.T.; Olsen, J.S.; Soltanpour, P.N. Effects of Excessive Magnesium in Irrigation Waters on Wheat and Corn Growth. *Commun. Soil Sci. Plant Anal.* **1991**, *22*, 49–61. [CrossRef]
167. Razowska-Jaworek, L. *Calcium and Magnesium in Groundwater Occurrence and Significance for Human Health*, 1st ed.; Razowska-Jaworek, L., Ed.; CRC Press Taylor & Francis Group: Boca Raton, FL, USA, 2019.
168. Rapant, S.; Cvečková, V.; Fajčíková, K.; Sedláková, D.; Stehlíková, B. Impact of Calcium and Magnesium in Groundwater and Drinking Water on the Health of Inhabitants of the Slovak Republic. *Int. J. Environ. Res. Public Health* **2017**, *14*, 278. [CrossRef]

169. Clark, R.B. Physiological Aspects of Calcium, Magnesium, and Molybdenum Deficiencies in Plants. *Soil Acidity Liming* **2015**, *12*, 99–170. [CrossRef]
170. Osemwota, I.O.; Omueti, J.A.I.; Ogboghodo, A.I. Effect of Calcium/Magnesium Ratio in Soil on Magnesium Availability, Yield, and Yield Components of Maize. *Commun. Soil Sci. Plant Anal.* **2007**, *38*, 2849–2860. [CrossRef]

*Article*

# Performance, Carcass Composition, and Meat Quality during Frozen Storage in Male Layer-Type Chickens

Teodora Popova [1,*], Evgeni Petkov [1], Krasimir Dimov [2], Desislava Vlahova-Vangelova [3], Nikolay Kolev [3], Desislav Balev [3], Stefan Dragoev [3] and Maya Ignatova [1]

1   Agricultural Academy, Institute of Animal Science-Kostinbrod, Pochivka St., 2232 Kostinbrod, Bulgaria; e_petkov@ias.bg (E.P.); m_ignatova@ias.bg (M.I.)
2   Agricultural Academy, Institute of Cryobiology and Food Technologies, 53 Cherni Vrah Blvd., 1407 Sofia, Bulgaria; krasimir.dimov@ikht.bg
3   Department of Meat and Fish Technology, University of Food Technologies, 26 Maritsa Blvd., 4002 Plovdiv, Bulgaria; d_vangelova@uft-plovdiv.bg (D.V.-V.); n_kolev@uft-plovdiv.bg (N.K.); d_balev@uft-plovdiv.bg (D.B.); s_dragoev@uft-plovdiv.bg (S.D.)
*   Correspondence: t_popova@ias.bg

**Abstract:** An experiment was carried out in the Institute of Animal Science—Kostinbrod, Bulgaria, to investigate the growth performance of male layer-type chickens (Lohmann Brown Classic), raised to 6 and 9 weeks of age, to evaluate the economic aspects of this rearing, as well as to present changes in the quality characteristics of the meat during frozen storage. The chickens were reared in a controlled microclimate with an initial stocking density of 9 birds/m$^2$. After 6 weeks of age, fragmentation of the stocking density was applied, and then it diminished to 3 birds/m$^2$. The chickens were slaughtered at 6 and 9 weeks of age. Ten 9-week-old chickens were subjected to carcass analysis. Meat quality parameters (pH, color), degree of proteolysis (free amino groups), and lipid oxidation (content of peroxides and TBARS) were assessed in fresh breast and thigh meat (0 d) and in samples stored for 60 and 120 days at −18 °C in chickens slaughtered at 6 and 9 weeks old. The mean live weight of the male layer-type chickens at 6 weeks was 608.81 g, while the 9-week-old chickens reached 1115.93 g. The feed conversion ratio (FCR) for the whole period of rearing was 2.75. There were no considerable deviations in the meat traits, indicating quality deterioration over the course of the frozen storage. There was a significant increase in the pH of the breast and thighs, reaching maximum values for 60 days of storage in the 6-week-old chicks, while in the 9-week-old birds, pH peaked in the samples stored for 120 days. The changes in the dynamics of pH corresponded to those of proteolysis. There was an increase in lightness (L*), allowing for higher values in the samples stored for 60 days to be reached regardless of the type of meat and age of the chickens. The content of the peroxides increased considerably for 60 days of frozen storage and decreased afterwards. During storage, there was a constant increase in the secondary products of lipid oxidation. Our results indicated that the application of practices such as the fragmentation of stocking density and finding the suitable age for slaughter have significant importance for the profitability of producing meat product from male layer-type chickens. We found that rearing this type of bird until 9 weeks of age resulted in lower costs and higher economic efficiency.

**Keywords:** male layer-type chickens; growth performance; economics; meat; frozen storage

**Citation:** Popova, T.; Petkov, E.; Dimov, K.; Vlahova-Vangelova, D.; Kolev, N.; Balev, D.; Dragoev, S.; Ignatova, M. Performance, Carcass Composition, and Meat Quality during Frozen Storage in Male Layer-Type Chickens. *Agriculture* **2024**, *14*, 185. https://doi.org/10.3390/agriculture14020185

Academic Editors: Petru Alexandru Vlaicu, Arabela Elena Untea and Mihaela Saracila

Received: 23 December 2023
Revised: 22 January 2024
Accepted: 24 January 2024
Published: 26 January 2024

## 1. Introduction

Meat is an important component in the human diet. It provides high value protein, iron, vitamin B12, zinc, selenium, and phosphorus. Removing it from the diet poses a risk of severe nutritional deficiencies, imbalanced diet, and impaired human health [1]. Poultry meat makes no exception, and it is even considered healthier compared to red meats, having a lower content of saturated fatty acids but still being a very good source of protein. The inclusion of poultry meat in the diet, especially when substituting red meat, has been associated with a reduced risk of developing cardiovascular diseases [2], obesity, and type 2 diabetes mellitus [3,4]. Furthermore, poultry consumption might decrease the risk of certain cancers [5,6]. Chicken meat is the most consumed among all the poultry meats [7]. Fast-growing broilers have been the main source of chicken meat; however, in recent decades, consumers have become more and more aware of the benefits of eating meat from slow-growing chickens. In comparison to fast-growing broilers, the latter have a more favorable nutritional profile in terms of protein and lipid content [8] and fatty acid composition [9]. An alternative source of this meat could be male layer-type chickens. In fact, their rearing for meat has been considered as an option in addition to in ovo sex determination and the development of dual-purpose breeds to avoid the common practice of culling them immediately after hatching [10]. Based on previous studies, showing that the meat of male layer-type chickens is not inferior to fast- and slow-growing chickens, we focused on their age at slaughter as a factor to produce high-quality meat and meat products, as well as achieve more economically efficient production. Since the raising of male layer-type chickens is a longer and less profitable process, to avoid accumulating higher costs, we applied fragmentation of the stocking density at a certain age to the birds, that being the age of first slaughter. This method has been modified in the Institute of Animal Science—Kostinbrod, and its advantages are described in more detail elsewhere [11]. Our previous studies have shown that the most appropriate age for applying density fragmentation is 5 weeks. However, since the carcasses of the slaughtered birds at this age are quite small (approx. 200 g) [11], for the purposes of this study, we decided to decrease the stocking density of the birds at 6 instead of 5 weeks of age and to assess the economic efficiency of the process.

After harvesting the meat, it is of crucial importance to store it properly so that its quality and nutritional value are maximally preserved. Freezing is one of the most important preservation methods that allows for a longer shelf life of meat and meat products to be generated, including poultry. However, during frozen storage, meat components can be involved in many reactions, such as oxidation. Oxidative reactions in meat can significantly contribute towards quality deterioration, negatively affecting color, pigments, flavor, texture, and nutritional value. Poultry meat is particularly susceptible to oxidation due to its high contents of polyunsaturated fatty acids [12]. Although extensive research has been carried out on the changes in the quality of poultry meat during low-temperature long-term storage, such studies on male layer-type chickens remain rather scarce. So far, studies on the meat of male layer-type chickens have focused on comparisons with fast-growing genotypes and local or dual-purpose breeds [13–16], as well as on the effect of production systems [17]. This study was conducted to present the growth performance of male layer-type chickens, raised to 6 and 9 weeks of age, economical evaluation of this rearing, as well as the changes in the quality characteristics of the meat during frozen storage.

## 2. Materials and Methods

### 2.1. Experimental Birds and Housing

This trial was carried out in the experimental poultry farm of the Institute of Animal Science—Kostinbrod, Bulgaria, involving a total of 800 one-day-old Lohmann Brown Classic male chickens, supplied by Bulagro AD. The birds were distributed into 5 pens, each containing 160 chickens, and were reared conventionally until 9 weeks of age at an initial stocking density of 9 chickens/m$^2$. When the chickens reached 6 weeks of age, fragmentation of the stocking density was applied to reach 3 birds/m$^2$. The decrease in the stocking density of the chickens was achieved through the preliminary weighing of each bird and differentiation by live weight. By applying the fragmentation of the density, approximately $\frac{1}{4}$ of the chickens set for trial remained and were reared until 9 weeks of age. For each of the groups, the limit weight was individually determined, and all the birds of a lower weight ($\leq$540 g) were slaughtered. Male layer-type chickens were reared in deep litter and a controlled microclimate. The lighting regime was 3 h light and 3 h dark, repeated during the 24 h cycle. Feed and water for the birds were provided ad libitum. Their diet consisted of standard broiler feed containing maize, wheat, sunflower meal, and soybean meal (crude protein 20%, crude fat 5.99%, crude fiber 4.40%, Ca 0.94%, P 0.80%, Lys 1.30%, Met 0.53%, metabolizable energy 3000 kcal/kg) and was purchased from a certified feed supplier. Water was provided through gravity drinkers. During the trial period, the live weight (LW) and feed intake (FI) of the birds were controlled weekly. These indicators were further used to calculate the body weight gain for the controlled periods (BWG) and the feed conversion ratio (FCR) of the male layer-type chickens.

## 2.2. Slaughtering, Carcass Analysis, Sampling, and Storage

The chickens were slaughtered at the age of 6 and 9 weeks in a certified poultry abattoir in Pazadrzik, Southern Bulgaria. After stunning, decapitating, and bleeding, the carcasses were plucked and eviscerated. Their feet and edible viscera (heart, liver, gizzard) were removed in order to obtain the ready-to-cook carcass. The carcasses were then placed in a refrigerator at 4 °C for 24 h. On the next day, ten 9-week-old chickens were subjected to carcass analysis. Each of the carcasses were weighed and cut to neck, breast, thighs, wings, and back. Sampling for meat quality analysis and storage was performed in the breast and thighs of chickens at both ages ($n = 10$). Briefly, the breast and thighs of each carcass were collected, and their skin and bones were removed. A part of the samples was used for the analysis of the meat quality and oxidative stability on the same day before freezing (day 0), while the others were frozen and stored at $-18$ °C for subsequent analysis after 60 and 120 days.

## 2.3. Economic Evaluation

Methodology of Calculation of the Costs:

The total costs for obtaining kg live weight and kg product by the chickens were calculated according to the following equations:

$$\text{LWC} = \frac{\left(\sum_{i=1}^{n} FC_i + \sum_{j=1}^{m} VC1_j\right)}{TLW} \tag{1}$$

$$\text{TC} = \frac{\left(\sum_{i=1}^{n} FC_i + \sum_{j=1}^{m} VC1_j + \sum_{k=1}^{l} VC2_k\right)}{TPW}, \tag{2}$$

where *TLW* is the total live weight of all the chickens (kg), *TPW* is the total weight of the ready product obtained (ready-to-cook carcass) (kg); LWC denotes the costs for rearing the chickens before slaughter (EUR/kg)—as all the costs are further presented in the same units—and TC denotes the total costs to obtain kg of ready product.

$FC_i$ are fixed costs, comprising the rent (depreciation) of the building used for rearing the chickens and preparation of the premises, including covering and preliminary heating the one-day-old chickens.

$VC1_j$ are a group of variable costs that are used to calculate the cost price of the live weight of the chickens and comprise labor costs; price of a one-day-old chicken; cost for vaccines and decontamination of the premises; energy costs; and feed costs.

$VC2_k$ are a second group of variable costs used to calculate the price cost of the ready product: costs for slaughter; transportation; and packaging and freezing.

## 2.4. Analysis of Meat Quality

The analysis of the meat quality parameters was carried out in the Department of Technology of Meat and Fish Products at the University of Food Technology, Plovdiv, Bulgaria.

### 2.4.1. Measurement of Meat pH and Color

pH and color were measured on the fresh meat (day 0) and on day 60 and day 120 of the frozen storage. The pH measurements were performed using a portable pH meter

(HI99163, Hanna Instruments, Inc., Smithfield, RI, USA) equipped with a specialized probe (FCO99) with a conical tip and stainless-steel blade. Calibration prior to use at pH 4.0 and 7.0 was performed. The color of the breast and thigh meat surface was described by the CIELab color space. The lightness (L*), redness (a*), and yellowness (b*) were measured using a chroma meter CR-410 (Konica Minolta Holding, Inc., Ewing, NJ, USA) at the following settings: aperture = 8 mm, standard observer 2°, and illuminant D65. The instrument was calibrated using a standard white plate (Y = 94.3, x = 0.3134 and y = 0.3197). Measurements of the pH and color were performed at three locations in the meat samples, and the results were averaged.

### 2.4.2. Free Amino Groups

The content of the free amino groups was determined in muscle proteins after extraction, as described Khan et al. [18], with slight modifications. Meat samples (2.5 g) were homogenized with 48.5 mL PBS buffer (49 mM $Na_2HPO_4.7H_2O$, 4.5 mM $NaH_2PO_4.H_2O$, KCl to obtain I = 0.55). The homogenate was kept in a refrigerator for 12 h and then centrifuged at $1000 \times g$ for 15 min. The free amino groups were measured according to Vassilev et al. [19] as follows: 2 mL of supernatant was transferred in a test tube and mixed with 1 mL solution of ninhydrin reagent (0.5% ninhydrin; $10\%Na_2HPO_2 \cdot 12H_2O$; $6\%KH_2PO_4$; 0.3% fructose). The mixture was heated in boiling water for 16 min and then cooled at room temperature for 20 min and diluted with 5 mL water–ethanol (3:2) solution of $KIO_3$ (2%). The absorbance was read at 570 nm against the blank sample using a Camspec, M 550 two-beam UV-VIS spectrophotometer (Spectronic Camspec Ltd., Leeds, UK). The concentration of free amino groups was calculated using a standard curve made by leucine and presented as mg leucine/g meat.

### 2.4.3. Lipid Oxidation

The lipids from the breast and thigh meat were extracted according to the method of Bligh and Dyer [20].

The peroxide value was determined as described by Shantha and Decker [21]. The extracted lipid (0.1 g) was mixed in a glass tube with 50 µL iron (II) solution and 50 µL $NH_4$ SCN (300 mg/mL) and $CHCl_3:CH_3OH$ (3:5, *v/v*) to form a final volume of 10 mL. The samples were incubated for 5 min at room temperature, and the absorbance was determined by a spectrophotometer at 507 nm against a blank containing all the reagents except the sample. The results were presented as $meqO_2$/kg lipid.

Thiobarbituric acid reactive substances (TBARSs) were determined according to the method described by Botsoglou et al. [22] with slight modifications. A total of 10 g of meat was homogenized with 50 mL NaCl (0.9%) and left for 5 min. Further, 50 mL of trichloroacetic acid (10%) was added, and the samples were filtered through grade 391 filter paper. The filtered samples (4 mL) were then mixed with 1mL 2-thiobabituric acid (1%) and were incubated at 70 °C for 30 min. After cooling to room temperature, the absorbance of the samples was determined at 532 nm against a blank containing distilled water instead of the sample. The TBARS concentrations were calculated using 1, 1, 3, 3 tetraethoxypropane as standard. The results were expressed as mg MDA/kg meat or TBARS units.

### 2.5. Statistical Evaluation

Statistical evaluation of the data was performed using JMP v.7 software package [23]. A one-way ANOVA was applied to assess the effect of the storage duration on the examined quality traits and oxidation in the breast and thigh meat of the chickens slaughtered at 6 and 9 weeks of age. In the case of significance, the difference between means was further assessed through post hoc comparisons (Tukey HSD, $p < 0.05$). Data were presented as means and standard deviations.

## 3. Results

### 3.1. Growth Performance and Carcass Composition

The initial live weight of the chickens was 36.43 g. At 6 weeks of age, when density fragmentation was applied, the average live weight of the male layer-type chickens was 601.81 g, and until the end of the experiment (9 weeks of age), their weight increased to 1115.93 g (Figure 1).

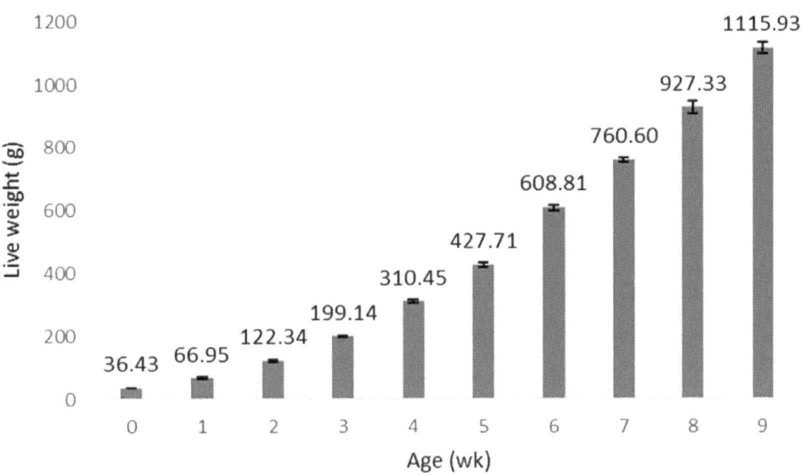


**Figure 1.** Live weight of the male layer-type chickens during the experiment.

As seen from Table 1, the weight gain of the male layer-type chickens increased until the 6th week.

**Table 1.** Dynamics of growth performance traits of the male layer-type chickens during the trial period.

| Trait | BWG, g | FI, g/Bird | FI Cumulative g/Bird | FCR Feed/Gain |
|---|---|---|---|---|
| 1 week | 30.52 (3.16) | 65.00 (1.20) | 65.00 (1.20) | 2.15 (0.25) |
| 2 week | 55.40 (1.73) | 74.32 (2.62) | 139.32 (3.06) | 1.34 (0.06) |
| 3 week | 76.79 (1.12) | 183.07(2.47) | 322.38 (4.47) | 2.38 (0.04) |
| 4 week | 111.32 (3.98) | 250.04 (4.21) | 572.43 (8.14) | 2.25 (0.10) |
| 5 week | 117.25 (7.74) | 349.40 (2.68) | 921.83 (8.46) | 2.99 (0.18) |
| 6 week | 181.10 (11.04) | 369.00 (0.99) | 1290.83 (7.53) | 2.04 (0.13) |
| 7 week | 151.79 (12.67) | 464.79 (16.55) | 1755.61 (15.71) | 3.07 (0.15) |
| 8 week | 166.73 (20.75) | 597.96 (21.28) | 2353.57 (35.63) | 3.63 (0.42) |
| 9 week | 188.60 (22.11) | 618.92 (30.37) | 2972.49 (63.31) | 3.31 (0.31) |

On the 7th and 8th week of the trial, the values of this trait decreased; however, body weight constantly increased. As presented by the data, despite the lower weight gain measured at the age of 7 and 8 weeks, the feed intake remained higher and was associated with higher values of FCR. Maximum weight gain was reached at the end of the experimental period (9 weeks). The chickens gained 572.38 g to reach the maximum weight for the first slaughter at 6 weeks with FCR 2.25, while the FCR for the whole period was 2.75.

The weight at slaughter for 6-week-old chickens was 527.42 g, and for the birds at 9 weeks it was 1110.13 g. The carcass yield for the younger chicks was 57.58%, while for the 9-week-old birds the value of this parameter was 65.18%. The carcass analysis of the male layer-type chickens showed that thighs had the highest proportion of all parts, followed by breast and back (Figure 2).

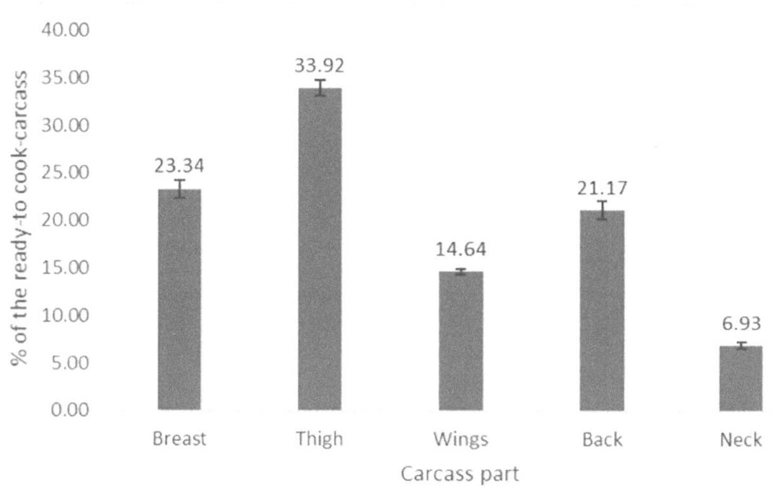


**Figure 2.** Proportion of the carcass parts in male layer-type chickens (9 weeks old).

*3.2. Economic Evaluation of the Rearing of Male Layer-Type Chickens*

Comparisons of the costs to produce 1 kg of live weight between the chickens raised to 6 and 9 weeks of age are shown in Table 2. The differences between both ages are small (0.07 EUR/kg) in favor of the chickens raised until 9 weeks old.

**Table 2.** Costs for producing 1 kg live weight of male layer-type chickens.

| Group of Costs | Parameters | Costs (EUR/kg Live Weight) | | |
| --- | --- | --- | --- | --- |
| | | 6 Weeks Old | 9 Weeks Old | Difference |
| FC (Fixed costs) | Preparing the premises | 0.43 | 0.22 | 0.21 |
| | Rent | 0.76 | 1.07 | −0.31 |
| VC1 (Variable costs 1) | Price of one day chicken | 0.29 | 0.14 | 0.16 |
| | Energy | 0.76 | 0.70 | 0.06 |
| | Labor | 0.96 | 1.26 | −0.30 |
| | Feed | 1.49 | 1.41 | 0.07 |
| | Medication | 0.35 | 0.16 | 0.18 |
| TC (Total costs) | | 5.03 | 4.96 | 0.07 |
| Live weight (kg) | | 304.698 | 237.440 | |

Source: own calculation.

Table 3 presents the comparison of the costs made to produce 1 kg ready product between the two age groups. The differences are again in favor of the older birds; however, here they are considerable (1.84 EUR/kg).

**Table 3.** Costs for producing 1 kg ready product of male layer-type chickens.

| Group of Costs | Parameters | Cost (EUR/kg Product Weight) | | |
| --- | --- | --- | --- | --- |
| | | 6 Weeks Old | 9 Weeks Old | Difference |
| FC (Fixed costs) | Preparation of premises | 0.74 | 0.33 | 0.41 |
| | Rent | 1.32 | 1.64 | −0.32 |
| VC1 (Variable costs 1) | Price of one-day-old chick | 0.51 | 0.21 | 0.30 |
| | Energy | 1.32 | 1.07 | 0.25 |
| | Labor | 1.66 | 1.93 | −0.27 |
| | Feed | 2.58 | 2.16 | 0.42 |
| | Medication | 0.60 | 0.25 | 0.35 |
| VC2 (Variable costs 2) | Transport | 0.73 | 0.83 | −0.10 |
| | Slaughter | 1.34 | 0.56 | 0.78 |
| | Package | 0.17 | 0.14 | 0.03 |
| TC (Total costs) | | 10.97 | 9.13 | 1.84 |
| Product weight (kg) | | 175.439 | 154.770 | |

Source: own calculation.

### 3.3. Meat Quality

### 3.3.1. pH and Color

The values of the pH in breast and thigh (Table 4) meat were significantly affected ($p < 0.0001$) by the storage period in the chickens slaughtered at 6 and 9 weeks of age. The pH of the breast and thigh meat samples of the 6-week-old birds stored for 60 days increased, but the values decreased in the meat stored for 120 days. The meat of the 9-week-old birds, however, had a significant increase in pH at the end of storage (120 days).

**Table 4.** pH and color of meat in male layer-type chickens.

| Trait | Age (Weeks) | Storage (Days) | | | Significance ($p$) |
| --- | --- | --- | --- | --- | --- |
| Breast | | 0 | 60 | 120 | |
| pH | 6 | 6.06 (0.004) [b] | 6.13 (0.02) [a] | 5.87 (0.06) [c] | <0.0001 |
| | 9 | 5.87 (0.04) [b] | 5.94 (0.07) [b] | 6.24 (0.03) [a] | <0.0001 |
| L* | 6 | 58.51 (1.25) [b] | 61.86(1.81) [a] | 58.53 (1.99) [b] | 0.0001 |
| | 9 | 56.51 (0.28) [b] | 59.76 (0.65) [a] | 54.96 (1.20) [c] | <0.0001 |
| a* | 6 | 13.60 (0.75) [a] | 13.19 (0.99) [a] | 9.54 (1.29) [b] | <0.0001 |
| | 9 | 14.44 (0.17) [a] | 7.66 (0.89) [b] | 13.70 (1.36) [a] | <0.0001 |
| b* | 6 | 8.46 (0.41) [b] | 10.90 (0.68) [a] | 6.19 (0.65) [c] | <0.0001 |
| | 9 | 7.62 (0.19) [a] | 6.76 (0.43) [b] | 6.49 (0.85) [b] | 0.0003 |
| Thigh | | | | | |
| pH | 6 | 6.55 (0.007) [b] | 6.67 (0.03) [a] | 6.54 (0.04) [b] | <0.0001 |
| | 9 | 6.42 (0.03) [b] | 6.41 (0.02) [b] | 6.59 (0.008) [a] | <0.0001 |
| L* | 6 | 52.06 (0.90) [a] | 52.97 (1.60) [a] | 47.91 (2.32) [b] | <0.0001 |
| | 9 | 47.95 (1.42) [b] | 53.10 (2.88) [a] | 49.06 (1.52) [b] | <0.0001 |
| a* | 6 | 16.52 (0.23) | 16.01 (1.26) | 16.26 (1.25) | 0.5501 |
| | 9 | 18.49 (0.81) [a] | 13.55 (1.49) [c] | 16.96 (1.36) [b] | <0.0001 |
| b* | 6 | 5.23 (0.67) [c] | 8.65 (0.84) [a] | 6.89 (0.88) [b] | <0.0001 |
| | 9 | 4.86 (0.75) [b] | 4.22 (0.95) [b] | 7.76 (0.83) [a] | <0.0001 |

Means connected with different letters within an age group differ significantly ($p < 0.05$).

The storage time had a significant effect on most of the color parameters of the meat ($p < 0.0001$); however, different trends for the change in their values were observed both according to the age of the birds and the type of meat. Lightness (L*) significantly increased in the breast meat of 6- and 9-week-old chickens stored for 60 days and then decreased until 120 days of storage. Such a trend was registered in the thigh meat of 9-week-old chickens as well. Similar to breast, the lightness of the thigh meat showed the lowest values at the end of the storage period. The dynamics of the changes in the redness (a*) and yellowness (b*) of the meat showed patterns in the breast and thighs of the chickens that differed in the different age groups. The redness of the breast of the 6-week-old chickens underwent negligible changes for 60 days of storage; however, this then significantly decreased for 120 days. Likewise, the yellowness of the breast was lowest in the samples stored for 120 days; however, the values of this parameter peaked in the middle of the storage period compared to those that were measured initially and at the end of storage. In the birds slaughtered at 9 weeks old, both redness and yellowness decreased over the course of storage as the redness showed substantially decreased values in the samples stored for 60 days.

The redness of the thigh meat was only affected by the storage period in 9-week-old birds, and its values decreased gradually in the samples stored for 120, with minimal values on day 60. On the other hand, b* was affected by the storage of the thigh meat of both 6- and 9-week-old birds. The former displayed more yellow meat when stored for 60 days and decreased values of this trait for 120 days of storage. The thigh meat of the 9-week-old birds had higher b* values at the end of storage.

### 3.3.2. Proteolysis in Meat

The content of the free amino groups showed similar trends of change during storage in breast and thigh meat depending on the age at slaughter of the chickens and was affected by frozen storage (Figure 3). There was a significant increase in the proteolysis in the breast and thighs of the younger birds stored for 60 days and a gradual decrease afterwards until day 120. In the meat samples stored for 60 days, the content of the free amino groups showed maximum values. In the meat of the older birds, we observed a constant increase in this parameter over the course of the frozen storage as the highest content was registered in the meat samples stored for 120 days.

### 3.3.3. Lipid Oxidation

Lipid oxidation in terms of PV showed that the content of peroxides was affected significantly by the storage period and showed similar trends of changes in the birds slaughtered at 6 and 9 weeks. The highest content of lipid peroxides was measured in the breast and thigh meat stored for 60 days. In the samples stored for 120 days, the levels of the peroxides significantly decreased (Figure 4).

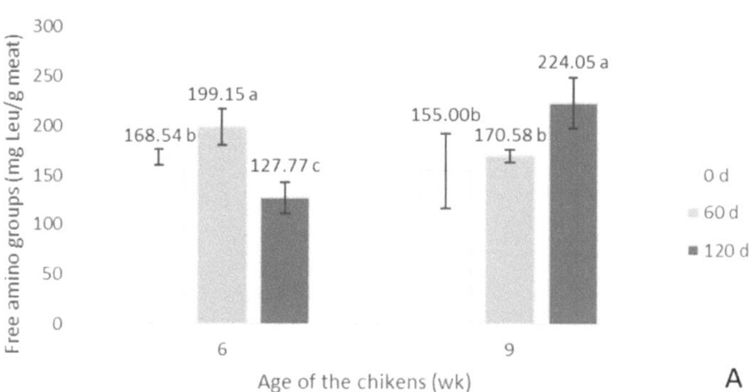


**Figure 3.** *Cont.*

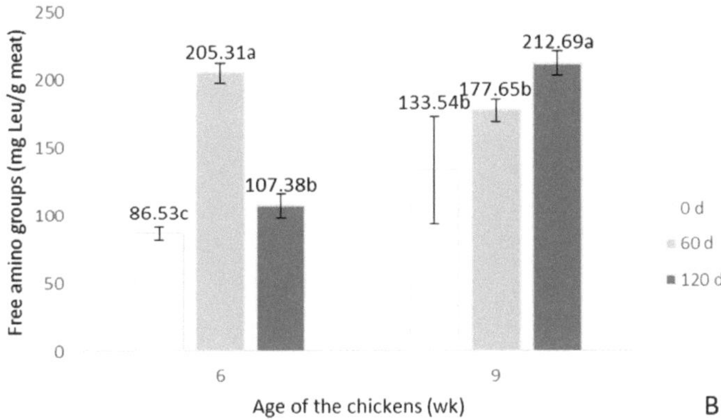



**Figure 3.** Content of free amino groups in the breast (**A**) and thigh (**B**) meat of male layer-type chickens during frozen storage. Means connected with different letters within an age group differ significantly ($p < 0.05$).

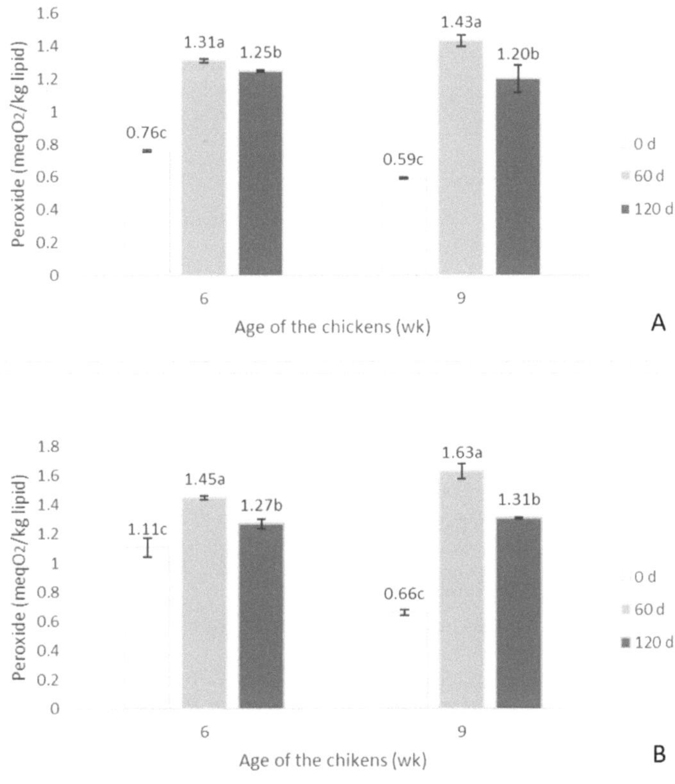



**Figure 4.** Content of peroxides in the breast (**A**) and thigh (**B**) meat of male layer-type chickens during frozen storage. Means connected with different letters within an age group differ significantly ($p < 0.05$).

The TBARS values showed a constant significant increase ($p < 0.0001$) over the course of storage in the breast and thigh (Figure 5) regardless of the age of the birds. The maximum TBARS contents were measured in the meat stored for 120 days and varied within 1.99–2.46 for the breast and 2.31–2.33 for the thigh meat.

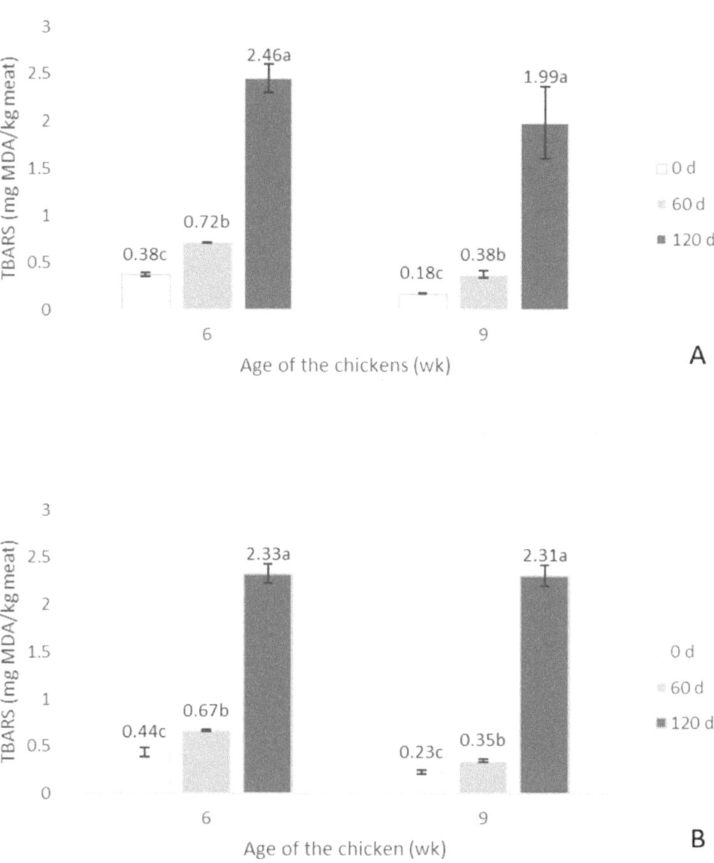


**Figure 5.** Content of TBARSs in the breast (**A**) and thigh (**B**) meat of male layer-type chickens during frozen storage. Means connected with different letters within an age group differ significantly ($p < 0.05$).

## 4. Discussion

Raising male layer-type chickens for meat is one option to avoid their culling right after hatching. Yet, despite studies reporting the positive characteristics of their meat [13,24], this option is still considered challenging by many, mainly due to economic reasons. There are certain factors, however, that can affect the rearing of this type of bird to achieve lower costs. Stocking density and the suitable age for slaughter are important for economically efficient production. In a previous study [11], we focused on the growth performance of male layer-type chickens reared at an initial stocking density of 22 birds/m$^2$ until 5 weeks of age, when a part of the birds was slaughtered. For the present trial, we opted for a much lower initial density and higher age of first slaughter to obtain a higher weight for the ready-to-cook carcass, particularly for small chickens. After applying fragmentation of the stocking density after 6 weeks of age to reduce it to 3 birds/m$^2$, the average final live weight of the birds was 1115.93 g. Though the difference in the final body weight is negligible, the body weight of the birds reared at a lower density before and after fragmentation remained higher than that previously reported for the birds reared at a higher stocking density [11]. The live weight of the male layer-type chickens that we measured in this study was considerably lower than that of Habig et al. [25] in Lohmann Brown males for the whole trial period. Similarly, other studies with male layer-type chickens reported live weights higher than ours until the 4th [26] and 5th week [27]. Additionally, Putra et al. [27] reported a lower growth rate of the birds after the 6th week, which we can explain with the fragmentation of the stocking density that we applied to select birds with a higher live weight for further rearing. The final live weight of the male layer-type chickens at 9 weeks of age was comparable to that of 3-week-old fast-growing broilers [14]. For the whole period until 9 weeks of age, the FCR was 2.75. This value was lower than that reported by Mueller et al. [15] for Lohmann Brown chickens reared for 67 and 84 days, but it is closer to that determined for dual-purpose breeds and slow-growing broilers. It can be considered acceptable when compared to the FCR for rearing male layer-type chickens reported in previous studies [28,29].

The carcass composition of the male layer-type chickens in this study showed that thighs had the highest proportion of all parts, followed by the breast and back. The content of the thighs is approximately 10% higher than that of the breast. This is in agreement with our previous research on layer cockerels [11], slow-growing chickens [30], and dual-purpose crosses [31]. A higher percentage of thighs in male layer-type chickens has also been reported in other studies regardless of age [16,32]. In contrast, in fast-growing broilers, both conventionally and organically reared, the yield of breast was reported to be 10% higher than that of thighs (37% vs. 27%) [33]. The body composition of the layer strains makes them less attractive to consumers compared to fast-growing broilers at the same slaughter weight. On the other hand, when slaughtered at an early age, male layer-type chickens might be used as "coquelets", as shown by Koenig et al. [26] in a study with two different layer genotypes (medium–heavy and light). It was found that this was economically feasible; however, the carcass composition differed between the genotypes in favor of the heavier strain.

The economic analysis of the production of male layer-type chickens revealed that the total costs for producing 1 kg live weight were lower for birds reared until 9 weeks of age. A detailed breakdown of the individual parameters revealed that rearing the chickens to older age was associated with higher labor and rental costs. The latter was a direct consequence of the reduced stocking density, whereas the increase in labor costs was due to the negligible difference in the activities for rearing to 6 weeks (for total weight of 304.698 kg) and 9 weeks of age (for total weight 237.44 kg). Higher costs were compensated by the rest of the parameters, the preparation of the premises, the price of a one-day-old chick, and the costs for medication, having the highest share. The first two costs were made once at the beginning of the experiment, and as the weight of the chickens increased, it was obvious that their proportion decreased. On the other hand, medication costs included decontamination of the premises (constant parameter) and vaccine costs, which were used mainly until the chickens were 6 weeks old.

Again, when evaluating the efficiency of the production of 1 kg ready product, we found that the total costs were lower for the 9-week-old birds. This was mainly due to two factors. First was the lower cost for slaughter. At the same price for slaughtering a chicken (as it is in our case), it is clear that it is economically more efficient to slaughter chickens of higher weight. The second factor is the higher dressing percentage of 9-week-old chickens (65.18%) compared to that of 6-week-old chickens (57.58%). The dressing percentage obtained by us is slightly higher than the one presented in the study of Mueller et al. [15], where the value is 57.28% of Lohmann Brown Classic chickens slaughtered at the age of 67 days. It could be suggested that lower costs can also be obtained by slaughtering at an older age, but not more than 12 weeks, because it has been found that after this age, the growth rates for this breed decrease significantly, and the economic costs are unjustified.

The pH of meat is an important factor in its quality. It reflects the rate of the post-mortem glycolysis, and variations in the pH values affect the ability of meat to retain water, as well as its color. The highest pH values in chickens slaughtered at 6 weeks were observed in meat stored for 60 days, followed by a decrease, while in 9-week-old birds, the values of pH peaked in the meat stored for 120 days. The trends of the changes in the pH values of the meat in this study were consistent with the changes in the free amino groups over the course of the frozen storage. The increase in the pH values during storage corresponded with the increased content of the free amino groups, and point towards the association of pH, with proteolysis occurring in meat [19]. So far, the results reporting the effect of the frozen storage on meat pH remain contradictory. Our results, showing an increase in pH, coincide with those for goose meat when subjected to frozen storage [34]. Other studies, however, report no change [35,36] or decrease [37,38] in pH during frozen storage. Moreover, the different patterns of the change in pH in the age groups regardless of the type of meat suggest an interaction between storage and age. Since we refer to 6- and 9-week-old chickens as a source of two different products, we have not examined the effect of age, and in this study, both age groups are separated in regard to meat quality characteristics. However, in our previous study, when comparing the meat quality traits of layer cockerels slaughtered at 5 and 9 weeks of age, we found significant differences in the pH of breast and thigh meat between the two ages [24].

During the course of frozen storage, we observed a significant increase in L* values in the breast of 6- and 9-week-old birds stored for 60 days, as well as in the thigh of 9-week-old male chickens. The thighs of the chickens slaughtered at 6 weeks old also displayed this trend. The increased values of the L* in the breast and thighs of the 9-week-old chickens measured after 60 days of frozen storage corresponded to the lower degree of redness. On the other hand, the dynamics of the changes in b* showed similar patterns in birds at 6 weeks of age in the breast and thigh, with increased values in the samples stored for 60 days. The lower values of a* and increased values of b* suggest denaturation of the myoglobin, particularly during 60 days of frozen storage. The effect of the frozen storage on the color parameters has been reported to be different among studies. Constant increased values of L* and b* during 18 months of frozen storage were reported for pork [39]; however, for a*, the authors observed an increase from 3 to 12 months and then a decrease until 18 months of frozen storage. In chicken meat, Lee at al. [40] observed a gradual decrease in L* until 6 months of storage and an increase afterwards until the 8th month. The authors reported a similar trend for a* and a constant decrease in b*. Our results coincide with those of Agustynska-Prejsnar et al. [41], that demonstrated a decrease in L* in chicken meat during long periods of frozen storage compared to shorter ones. On the other hand, Ali et al. [42] reported an increase in L* in chicken meat for 6 months of frozen storage, which is in line with our findings. The authors also observed increased b* values for this period, that we only observed in the meat of 6-week-old chickens, again suggesting an interaction between storage with age.

The changes in the content of the free amino groups reflect the accumulation of the end products of proteolysis. The initial and final content of free amino groups in this study were higher in the breast than in the thigh meat. Vassilev et al. [19] also reported a higher content of free amino groups in breast meat compared to leg when refrigerated for 7 days. These observations indicate that the trends of proteolysis are dependent on the different muscle fiber characteristics [43]. It could be seen that the concentration of the free amino groups showed different trends during the frozen storage process in the meat of the chickens slaughtered at 6 and 9 weeks of age. The highest levels of amino groups were measured after 60 days of storage in the breasts and thighs of younger chicks, while in the meat of 9-week-old chickens, the value of this trait reached maximum at day 120 of storage. It could be suggested that proteolytic processes have different intensities depending on the age of the birds due to the different activity of the enzyme systems [44].

In this study, the peroxide value and TBARSs were selected to represent the oxidative processes that occur in the chicken meat during storage. The peroxide value is a measure of the formation of peroxides that are primary products of lipid oxidative processes and determine their extent at the initial stages. We observed considerable augmentation of the content of the peroxides for 60 days of frozen storage, regardless of the type of the meat or the age of the birds, followed by a reduction in the peroxide values. The decrease in peroxide formation during frozen storage is not unusual. In line with our results, a reduction in PV was reported after frozen storage in breast and leg meat, respectively, after 2 and 3 months [45]. Another study on chicken meat also reported a dramatic decrease in peroxide formation at 6 months of frozen storage regardless of the presence of antioxidants [46]. As stated by Yi et al. [47], the ability to form peroxides is relevant to meat quality since it precedes the formation of an off-flavor as well as the crosslinking of proteins. Furthermore, it was showed [48] that even in small concentrations, lipid peroxides can exert toxic effects on the cell. Lipid peroxides are unstable and susceptible to decomposition [49] and the formation of other products, such as alcohols, ketones, and aldehydes. This could explain the decrease in their content during frozen storage that we observed in this study, corresponding to the dramatic increase in the TBARSs that we measured in the meat stored for 120 days. We observed a constant increase in the TBARS values over the course of storage. This is in line with the results obtained in goose meat stored for 365 days [50], poultry meat frozen for 3 months [51], as well as in the meat of other species—lamb [52], horse [53], and pork [39]. The major TBA reactive substance is malondialdehyde (MDA)—an important secondary product of the oxidation of polyunsaturated fatty acids. It is associated with rancid odors in meat even at low concentrations. According to some studies [54,55], MDA, in an amount of 2.5 mg/kg meat, has been set as a limit below which no rancidity was detected. The TBARSs content presented in this study did not exceed the mentioned threshold; hence, frozen storage was not associated with negative changes in the flavor of the meat of male layer-type chickens.

## 5. Conclusions

This study aimed to demonstrate the potential of male layer-type chickens to produce meat as an option to avoid their culling after hatching and to further investigate the influence of frozen storage on the quality of their meat derived at two ages of slaughter. Over the course of frozen storage, no considerable deviations in the quality parameters indicating deterioration of the meat were observed. Yet, it had significant effect on pH, color, proteolysis, and lipid oxidative processes in the breast and thigh meat, and in some of these parameters the changes were different according to the age of slaughter. Further, our results showed that the application of some rearing practices such as fragmentation of stocking density, as well as finding a suitable age for slaughter, are very important for obtaining good economic results. We found that rearing male layer-type chickens until an older age (9 weeks) had lower costs and higher economic efficiency.

**Author Contributions:** Conceptualization, T.P., E.P., K.D. and D.V.-V.; methodology, T.P., E.P., D.V.-V., D.B. and N.K.; formal analysis, T.P., K.D. and N.K.; investigation, T.P., E.P., K.D., M.I., D.V.-V., D.B., N.K. and S.D.; resources, T.P., E.P., S.D. and M.I.; data curation, T.P., E.P., K.D. and N.K.; writing—original draft preparation, T.P. and K.D.; writing—review and editing, S.D. and M.I.; supervision, S.D. and M.I.; project administration, T.P.; funding acquisition, T.P. All authors have read and agreed to the published version of the manuscript.

**Funding:** This research was funded by the Bulgarian National Science Fund, Ministry of Education and Science in Bulgaria (Project INOVAMESPRO, Contract No. KP-06-N56/10, 12 November 2021).

**Institutional Review Board Statement:** The experimental protocol used in this study was designed in compliance with the guidelines of the European and Bulgarian legislation regarding the protection of animals used for experimental and other scientific purposes (Directive 2010/63; EC, 2010–put into law in Bulgaria with Regulation 20/2012). The protocol was based on the permit for use of animals in experiments No. 277 of the Bulgarian Food Safety Agency (Statement No. 193 of the Bulgarian Animal Ethics Committee, prot.No. 18/02.07.2020).

**Data Availability Statement:** The data presented in this study are available upon request from the corresponding author.

**Conflicts of Interest:** The authors declare no conflicts of interest.

## References


1. Pereira, P.M.C.C.; Vicente, A.F.R.B. Meat nutritional composition and nutritive role in the human diet. *Meat Sci.* **2013**, *93*, 586–592. [CrossRef]
2. Papp, R.E.; Hasenegger, V.; Ekmekcioglu, C.; Schwingshackl, L. Association of poultry consumption with cardiovascular diseases and all-cause mortality: A systematic review and dose response meta-analysis of prospective cohort studies. *Crit. Rev. Food Sci. Nutr.* **2023**, *63*, 2366–2387. [CrossRef]
3. Bonpoor, J.; Petermann-Rocha, F.; Parra-Soto, S.; Pell, J.P.; Gray, S.R.; Celis-Morales, C.; Ho, F.K. Types of diet, obesity, and incident type 2 diabetes: Findings from the UK Biobank prospective cohort study. *Diabetes Obes. Metab.* **2022**, *24*, 1351–1359. [CrossRef] [PubMed]
4. Ibsen, D.B.; Warberg, C.K.; Würtz, A.M.L.; Overvad, K.; Dahm, C.C. Substitution of red meat with poultry or fish and risk of type 2 diabetes: A Danish cohort study. *Eur. J. Nutr.* **2019**, *58*, 2705–2712. [CrossRef]
5. Daniel, C.R.; Cross, A.J.; Graubard, B.I.; Hollenbeck, A.R.; Park, Y.; Sinha, R. Prospective investigation of poultry and fish intake in relation to cancer risk. *Cancer Prev. Res.* **2011**, *4*, 1903–1911. [CrossRef]
6. Lo, J.J.; Park, Y.-M.M.; Sinha, R.; Sandler, D.P. Association between meat consumption and risk of breast cancer: Findings from the Sister Study. *Int. J. Cancer* **2020**, *146*, 2156–2165. [CrossRef] [PubMed]
7. FAOSTAT. Available online: https://www.fao.org/faostat/en/#data/QCL (accessed on 20 November 2023).
8. Valenta, J.; Chodová, D.; Tůmová, E.; Ketta, M. Carcass characteristics and breast meat quality in fast-, medium- and slow-growing chickens. *Czech J. Anim. Sci.* **2022**, *67*, 286–294. [CrossRef]
9. Sirri, F.; Castellini, C.; Bianchi, M.; Petracci, M.; Meluzzi, A.; Franchini, A. Effect of fast-, medium- and slow-growing strains on meat quality of chickens reared under the organic farming method. *Animal* **2011**, *5*, 312–319. [CrossRef]
10. Popova, T.; Petkov, E.; Ignatova, M.; Vlahova-Vangelova, D.; Balev, D.; Dragoev, S.; Kolev, N. Male layer-type chickens—An alternative source for high quality poultry meat: A review on the carcass composition, sensory characteristics and nutritional profile. *Braz. J. Poult. Sci.* **2022**, *24*, 1–10. [CrossRef]
11. Popova, T.; Petkov, E.; Ignatova, M.; Dragoev, S.; Vlahova-Vangelova, D.; Balev, D.; Kolev, N. Growth Performance, carcass composition and tenderness of meat in male layer-type chickens slaughtered at different age. *C. R. Acad. Bulg. Sci.* **2023**, *76*, 156–164.
12. Angelovičová, M.; Angelovič, M.; Čapla, J.; Zajác, P.; Folvarčíková, P.; Čurlej, J. The effect of oregano essential oil on chicken meat lipid oxidation and peroxidation. *Potravin. S. J. Food Sci.* **2021**, *15*, 1056–1068. [CrossRef]
13. Lichovníková, M.; Jandásek, J.; Jůzl, M.; Draèková, E. The meat quality of layer males from free range in comparison with fast growing chickens. *Czech J. Anim. Sci.* **2009**, *54*, 490–497. [CrossRef]
14. Mueller, S.; Kreuzer, M.; Siegrist, M.; Mannale, K.; Messikommer, R.E.; Gangnat, I.D.M. Carcass and meat quality of dual-purpose chickens (Lohmann Dual, Belgian Malines, Schweizerhuhn) in comparison to broiler and layer chicken types. *Poult. Sci.* **2018**, *97*, 3325–3336. [CrossRef]
15. Mueller, S.; Taddei, L.; Albiker, D.; Kreuzer, M.; Siegrist, M.; Messikommer, R.E.; Gangnat, I.D.M. Growth, carcass, and meat quality of 2 dual-purpose chickens and a layer hybrid grown for 67 or 84 D compared with slow-growing broilers. *J. Appl. Poult. Res.* **2020**, *29*, 185–196. [CrossRef]
16. Choo, Y.K.; Oh, S.T.; Lee, K.W.; Kang, C.W.; Kim, H.W.; Kim, C.J.; Kim, E.J.; Kim, H.S.; An, B.K. The growth performance, carcass characteristics, and meat quality of egg-type male growing chicken and White-Mini Broiler in comparison with commercial broiler (Ross 308). *Korean J. Food Sci. Anim. Resour.* **2014**, *34*, 622–629. [CrossRef]
17. Evaris, E.F.; Sarmiento-Franco, L.; Sandoval-Castro, C.A. Meat and bone quality of slow-growing male chickens raised with outdoor access in tropical climate. *J. Food Compos. Anal.* **2021**, *98*, 103802. [CrossRef]
18. Khan, A. Extraction and fractionation of proteins in fresh chicken muscle. *J. Food Sci.* **1962**, *27*, 430–434. [CrossRef]
19. Vassilev, K.; Ivanov, G.; Balev, D.; Dobrev, G. Protein changes of chicken light and dark muscles during chilled storage. *J. EcoAgriTourism* **2012**, *8*, 263–268.
20. Bligh, E.G.; Dyer, W.J. A rapid method of total lipid extraction and purification. *Can. J. Biochem. Physiol.* **1959**, *37*, 911–917. [CrossRef]
21. Shantha, N.C.; Decker, E.A. Rapid, sensitive, iron-based spectrophotometric methods for determination of peroxide values of food lipids. *J. AOAC Int.* **1994**, *7*, 421–424. [CrossRef]
22. Botsoglou, N.A.; Fletouris, D.J.; Papageorgiou, G.E.; Vassilopoulos, V.N.; Mantis, A.J.; Trakatellis, A.G. Rapid, sensitive, and specific thiobarbituric acid method for measuring lipid peroxidation in animal tissue, food, and feedstuff samples. *J. Agric. Food Chem.* **1994**, *42*, 1931–1937. [CrossRef]
23. *JMP*, version 7; SAS Institute Inc.: Cary, NC, USA, 2007.

Popova, T.; Petkov, E.; Ignatova, M.; Vlahova-Vangelova, D.; Balev, D.; Dragoev, S.; Kolev, N.; Dimov, K. Meat quality of male layer-type chickens slaughtered at different ages. *Agriculture* **2023**, *13*, 624. [CrossRef]

Habig, C.; Bayerbach, M.; Kember, N. Comparative analyses of layer males, dual purpose males and mixed sex broilers kept for fattening purposes regarding their floor space covering, weight-gain and several animal health traits. *Arc. Geflügelkd.* **2016**, *80*, 1–10.

Koenig, M.; Hahn, G.; Damme, K.; Schmutz, M. Utilization of laying-type cockerels as coquelets: Influence of genotype and diet characteristics on growth performance and carcass composition. *Arc. Geflügelkd.* **2012**, *76*, 197–202.

Putra, W.P.B.; Riaz, R.; Gunawan, A.A.; Orman, A. Comparison of growth curve in male layer chickens. *J. Res. Vet. Med.* **2021**, *40*, 49–53. [CrossRef]

Evaris, F.E.; Sarmiento, F.L.; Sandoval, C.C.; Segura, C.J.; Caamal, M.J.A. Male Layer chicken's response to dietary *Moringa oleifera* meal in a tropical climate. *Animals* **2022**, *12*, 1843. [CrossRef] [PubMed]

De Silva, P.; Wickramasinghe, Y.; Kalubowila, D. Growth performance and carcass quality of layer type cockerels and broiler chicken. *Iran. J. Appl. Anim. Sci.* **2016**, *6*, 429–433.

Petkov, E.; Popova, T.; Ignatova, M. Carcass and meat composition in f1 crosses of two lines of slow-growing chickens reared in conventional or alternative system with access to pasture. *Int. J. Innov. Approaches Agric. Res.* **2018**, *2*, 359–374. [CrossRef]

Petkov, E.; Popova, T.; Ignatova, M.; Sharkova, V.; Dimov, K. Development of dual-purpose cross for meat and egg production I. Growth performance and carcass composition of the crossbred chickens in comparison to the parent lines. *Arch. Zootech.* **2022**, *25*, 119–129. [CrossRef]

Murawska, D.; Bochno, R. Comparison of the slaughter quality of Layer-type cockerels and broiler chicken. *J. Poult. Sci.* **2007**, *44*, 105–110. [CrossRef]

Sarica, M.; Yamak, U.S.; Boz, M.A.; Erensoy, K.; Cilavdaroglu, E.; Noubandiguim, M. Performance of fast, medium and slow growing broilers in indoor and free-range production systems. *S. Afr. J. Anim. Sci.* **2019**, *49*, 1127–1138. [CrossRef]

Wereńska, M.; Okruszek, A. Impact of frozen storage on some functional properties and sensory evaluation of goose meat. *Poult. Sci.* **2023**, *102*, 102894. [CrossRef] [PubMed]

Śmiecińska, K.; Hnatyk, N.; Daszkiewicz, T.; Kubiak, D.; Matusevičius, P. The effect of frozen storage on the quality of vacuum-packaged Turkey meat. *Vet. Zootech.* **2015**, *71*, 61–66.

Kluth, I.K.; Teuteberg, V.; Ploetz, M.; Krischek, C. Effects of freezing temperatures and storage times on the quality and safety of raw turkey meat and sausage products. *Poult. Sci.* **2021**, *100*, 101305. [CrossRef]

Wei, R.; Wang, P.; Han, M.; Chen, T.; Xu, X.; Zhou, G. Effect of freezing on electrical properties and quality of thawed chicken breast meat. *Asian-Australas J. Anim. Sci.* **2017**, *30*, 569–575. [CrossRef]

Saewa, S.; Khidhir, Z.; Al Bayati, M. The impact of storage duration and conditions on the formation of biogenic amines and microbial content in poultry meat. *Iraqi J. Vet. Sci.* **2021**, *35*, 183–188. [CrossRef]

Medić, H.; Kušec, I.D.; Pleadin, J.; Kozačinski, L.; Njari, B.; Hengl, B.; Kušec, G. The impact of frozen storage duration on physical, chemical and microbiological properties of pork. *Meat Sci.* **2018**, *140*, 119–127. [CrossRef] [PubMed]

Lee, Y.S.; Saha, A.; Xiong, R.; Owens, C.M.; Meullenet, J.F. Changes in broiler breast fillet tenderness water-holding capacity and colour attributes during long-term frozen storage. *J. Food Sci.* **2008**, *73*, 162–168. [CrossRef]

Augustyńska-Prejsnar, A.; Hanus, P.; Sokołowicz, Z.; Kačániová, M. Assessment of technological characteristics and microbiological quality of marinated turkey meat with the use of dairy products and lemon juice. *Anim. Biosci.* **2021**, *34*, 2003–2011. [CrossRef]

Ali, S.; Rajput, N.; Li, C.; Zhang, W.; Zhou, G. Effect of Freeze-thaw cycles on lipid oxidation and myowater in broiler chickens. *Braz. J. Poult. Sci.* **2016**, *18*, 35–40. [CrossRef]

Cheng, H.; Song, S.; Park, T.S.; Kim, G.D. Proteolysis and changes in meat quality of chicken pectoralis major and iliotibialis muscles in relation to muscle fiber type distribution. *Poult. Sci.* **2022**, *108*, 102185. [CrossRef]

Teye, G.A.; Okutu, I. Effect of ageing under tropical conditions on the eating qualities of beef. *Afr. J. Food Agric. Nutr. Dev.* **2010**, *9*, 1901–1913. [CrossRef]

Soyer, A.; Özalp, B.; Dalmış, Ü.; Bilgin, V. Effects of freezing temperature and duration of frozen storage on lipid and protein oxidation in chicken meat. *Food Chem.* **2010**, *120*, 1025–1030. [CrossRef]

Lai, M.M.C.; Zhang, H.A.; Kitts, D.D. Ginseng prong added to broiler diets reduces lipid peroxidation in refrigerated and frozen stored poultry meats. *Molecules* **2021**, *26*, 4033. [CrossRef] [PubMed]

Yi, G.; Haug, A.; Nyquist, N.F.; Egelandsdal, B. Hydroperoxide formation in different lean meats. *Food Chem.* **2013**, *141*, 2656–2665. [CrossRef] [PubMed]

Angeli, J.P.F.; Garcia, C.C.M.; Sena, F.; Freitas, F.P.; Miyamoto, S.; Medeiros, M.H.G.; Di Mascio, P. Lipid hydroperoxide-induced and hemoglobin-enhanced oxidative damage to colon cancer cells. *Free Radic. Biol. Med.* **2011**, *51*, 503–515. [CrossRef] [PubMed]

Talbot, G. The stability and shelf life of fats and oils. In *The Stability and Shelf Life of Food*; Subramaniam, P., Ed.; Elsevier: Amsterdam, The Netherlands, 2016; pp. 461–503.

Wereńska, M.; Okruszek, A.; Haraf, G.; Wołoszyn, J.; Goluch, Z. Impact of frozen storage on oxidation changes of some components in goose meat. *Poult Sci.* **2022**, *101*, 101517. [CrossRef] [PubMed]

51. Konieczka, P.; Czauderna, M.; Rozbicka-Wieczorek, A.; Smulikowska, S. The effect of dietary fat, vitamin E and selenium concentrations on the fatty acid profile and oxidative stability of frozen stored broiler meat. *J. Anim. Feed Sci.* **2015**, *24*, 244. [CrossRef]
52. Pinheiro, R.S.B.; Francisco, C.L.; Lino, D.M.; Borba, H. Meat quality of Santa Inês lamb chilled-then-frozen storage up to 12 months. *Meat Sci.* **2019**, *148*, 72–78. [CrossRef]
53. Seong, P.N.; Seo, H.W.; Kim, J.H.; Kang, G.H.; Cho, S.H.; Chae, H.S.; Park, B.Y.; Van Ba, H. Assessment of frozen storage duration effect on quality characteristics of various horse muscles. *Asian-Australas J. Anim. Sci.* **2017**, *30*, 1756–1763. [CrossRef]
54. Campo, M.M.; Nute, G.R.; Hughes, S.I.; Enser, M.; Wood, J.D.; Richardson, R.I. Flavour perception of oxidation in beef. *Meat* **2006**, *72*, 303–311. [CrossRef] [PubMed]
55. Zhang, Y.; Holman, B.W.B.; Ponnampalam, E.N.; Kerr, M.G.; Bailes, K.L.; Kilgannon, A.K.; Collins, D.; Hopkins, D.L. Understanding beef flavour and overall liking traits using two different methods for determination of thiobarbituric acid reactive substances (TBARS). *Meat Sci.* **2019**, *149*, 114–119. [CrossRef] [PubMed]

MDPI AG
Grosspeteranlage 5
4052 Basel
Switzerland
Tel.: +41 61 683 77 34

*Agriculture* Editorial Office
E-mail: agriculture@mdpi.com
www.mdpi.com/journal/agriculture